AF382871

Time-Variant Systems and Interpolation

Edited by

I. Gohberg

Springer Basel AG

Editors' address:

I. Gohberg
Raymond and Beverly Sackler Faculty of Exact Sciences
School of Mathematical Sciences
Tel Aviv University
69978 Tel Aviv, Israel

Deutsche Bibliothek Cataloging-in-Publication Data

Time-variant systems and interpolation / ed. by I. Gohberg. –
Basel ; Boston ; Berlin : Birkhäuser, 1992
  (Operator Theory ; Vol. 56)
    ISBN 978-3-0348-9701-3   ISBN 978-3-0348-8615-4 (eBook)
    DOI 10.1007/978-3-0348-8615-4
NE: Gochberg, Izrail' [Hrsg.]; GT

© 1992 Springer Basel AG
Originally published by Birkhäuser Verlag Basel in 1992
Softcover reprint of the hardcover 1st edition 1992

ISBN 978-3-0348-9701-3

v

Table of Contents

EDITORIAL INTRODUCTION

This volume consists of six papers dealing with the theory of linear time-varying systems and time-varying analogues of interpolation problems. All papers are dedicated to generalizations to the time-variant setting of results and theorems from operator theory, complex analysis and system theory, well-known for the time-invariant case. Often this is connected with a complicated transition from functions to infinite dimensional operators, from shifts to weighted shifts and from Toeplitz to non-Toeplitz operators (in the discrete or continuous form). The present volume contains a cross-section of recent progress in this area.

The first paper, "Nevanlinna-Pick interpolation for time-varying input-output maps: The discrete case" of J.A. Ball, I. Gohberg and M.A. Kaashoek, generalizes for time-varying input-output maps the results for the Nevanlinna-Pick interpolation problem for strictly contractive rational matrix functions. This paper is based on a system theoretic point of view. The time-variant version of the homogeneous interpolation problem developed in the same paper, plays an important role.

The second paper, also of J.A. Ball, I. Gohberg and M.A. Kaashoek, is entitled "Nevanlinna-Pick interpolation for time-varying input-output maps: The continuous time case". The previous paper contains a time-varying analogue of the Nevanlinna-Pick interpolation for the disk. This paper contains the time-varying analogue for the half plane, and hence the latter results may be viewed as appropriate continuous analogues of the results of the first paper. Here, as well as in the previous paper, all solutions are described via a linear fractional formula.

In the third paper, "Dichotomy of systems and invertibility of linear ordinary differential operators" of A. Ben-Artzi and I. Gohberg, are considered linear ordinary differential operators of first order with bounded matrix coefficients on the half line and on the full line. Conditions are found when these operators are invertible or Fredholm on the half line. The main theorems are stated in terms of dichotomy. In the case of invertibility, the main operator is a direct sum of two generators of semigroups, one is supported on the negative half line and the other on the positive half line.

The fourth paper, "Inertia theorems for block weighted shifts and applications" of A. Ben-Artzi and I. Gohberg, contains time-variant versions of the well-known inertia theorem from linear algebra. These theorems are connected with linear time dependent dynamical systems and are stated in terms of dichotomy and Fredholm characteristics of weighted block shifts.

The fifth paper, "Interpolation for upper triangular operators" of P. deWilde and H. Dym, treats for the time-varying case the tangential problems of Nevanlinna-Pick and Caraththéodory-Fejér, as well as more complicated ones for operator-valued functions. Here both the cantractive and the strictly contractive cases are considered. The description of all solutions in a linear fractional form is given. The general case of varying coordinate spaces is analysed. The main method is based on an appropriate generalization of the theory of reproducing kernel spaces.

The sixth paper, "Minimality and realization of discrete time-varying systems" of I. Gohberg, M.A. Kaashoek and L. Lerer, analyses time-varying finite dimensional linear systems with time-varying state space. A theory which is an analogue of the classical minimality and realization theory for time independent systems, is developed. Special attention is paid to periodical systems.

I. Gohberg

Operator Theory:
Advances and Applications, Vol. 56
© 1992 Birkhäuser Verlag Basel

NEVANLINNA-PICK INTERPOLATION FOR TIME-VARYING INPUT-OUTPUT MAPS: THE DISCRETE CASE

J.A. Ball*, I. Gohberg and M.A. Kaashoek

This paper presents the conditions of solvability and describes all solutions of the matrix version of the Nevanlinna-Pick interpolation problem for time-varying input-output maps. The system theoretical point of view is employed systematically. The technique of solution generalizes the method for finding rational solutions of the time-invariant version of the problem which is based on reduction to a homogeneous interpolation problem.

0. INTRODUCTION

The simplest interpolation problem of Nevanlinna-Pick type reads as follows. Given N different points $z_1, \ldots, z_N$ in the open unit disc $\mathbf{D}$ of the complex plane and arbitrary complex numbers $w_1, \ldots, w_N$, determine a function f, analytic in $\mathbf{D}$, such that

(i) $f(z_j) = w_j, \; j = 1, \ldots, N,$

(ii) $\sup_{|z|<1} |f(z)| < 1.$

This problem can be restated as a problem involving double infinite Toeplitz matrices acting on $\ell^2(\mathbf{Z})$, namely find a lower triangular Toeplitz matrix T,

$$(0.1) \qquad T = \begin{pmatrix} \ddots & \ddots & \ddots & \ddots & \ddots \\ \ddots & f_0 & 0 & 0 & \ddots \\ \ddots & f_1 & \boxed{f_0} & 0 & \ddots \\ \ddots & f_2 & f_1 & f_0 & \ddots \\ \ddots & \ddots & \ddots & \ddots & \ddots \end{pmatrix},$$

such that

---

* The first author thanks the Netherlands organization for scientific research (NWO) for supporting his research.

(i)' $\mathrm{diag}((I - Z_j)^{-1}T) = W_j$, $j = 1, \ldots, N$,

(ii)' $\|T\| < 1$.

Here $Z_j = z_j S^*$, where S is the forward shift on $\ell^2(\mathbf{Z})$, and $W_j = w_j I$, $j = i, \ldots, N$. In this operator form the problem has a natural non-Toeplitz version, in which the operators T, Z_j, and W_j appearing in (i)' and (ii)' are replaced by

$$(0.2) \qquad T = \begin{pmatrix} \ddots & \ddots & \ddots & \ddots & \ddots \\ \ddots & f_{-1,-1} & 0 & 0 & \ddots \\ \ddots & f_{0,-1} & \boxed{f_{0,0}} & 0 & \ddots \\ \ddots & f_{1,-1} & f_{1,0} & f_{1,1} & \ddots \\ \ddots & \ddots & \ddots & \ddots & \ddots \end{pmatrix},$$

$Z_j = \Delta_j S^*$, where Δ_j is the diagonal matrix

$$(0.3) \qquad \Delta_j = \mathrm{diag}\,(\ldots, z_{-1}^{(j)}, z_0^{(j)}, z_1^{(j)}, \ldots),$$

and W_j is the diagonal matrix $W_j = \mathrm{diag}\,(\ldots, w_{-1}^{(j)}, w_0^{(j)}, w_1^{(j)}, \ldots)$, $j = 1, \ldots, N$. More precisely, the non-Toeplitz version of the problem reads as follows. One has given diagonal matrices Δ_j, $j = 1, \ldots, N$, as in (0.3) such that

$$(0.4) \qquad \lim_{k \to \infty} (\sup_{i \in \mathbf{Z}} |z_i^{(j)} z_{i+1}^{(j)} \cdots z_{i+k-1}^{(j)}|)^{\frac{1}{k}} < 1, \quad j = 1, \ldots, N,$$

and

$$(0.5) \qquad \left(1 + \sum_{\nu=1}^{\infty} z_k^{(i)} \cdots z_{k+\nu-1}^{(i)} \bar{z}_{k+\nu-1}^{(j)} \cdots \bar{z}_k^{(j)}\right)_{i,j=1}^{N} \geq \epsilon I_N, \quad k \in \mathbf{Z},$$

where ϵ is a positive number independent of k and I_N is the $N \times N$ identity matrix. In the classical case condition (0.4) means that the interpolation points are inside the unit disc and the positive definiteness in (0.5) replaces the condition that the points are different. Now the problem is to find a lower triangular double infinite matrix T as in (0.2) such that T acts as a bounded linear operator on $\ell^2(\mathbf{Z})$, $\|T\| < 1$, and for $j = 1, \ldots, N$ the following interpolation requirements are fulfilled:

$$(0.6) \qquad f_{k,k} + \sum_{\nu=1}^{\infty} z_k^{(j)} \cdots z_{k+\nu-1}^{(j)} f_{k+\nu,k} = w_k^{(j)}, \quad k \in \mathbf{Z}.$$

It turns out that this generalized Nevanlinna-Pick problem is solvable whenever for some $\epsilon > 0$ we have

$$(0.7)$$
$$\left(1 - w_k^{(i)} \bar{w}_k^{(j)} + \sum_{\nu=1}^{\infty} z_k^{(i)} \cdots z_{k+\nu-1}^{(i)} (1 - w_{k+\nu}^{(i)} \bar{w}_{k+\nu}^{(j)}) \bar{z}_{k+\nu-1}^{(j)} \cdots \bar{z}_k^{(j)} \right)_{i,j=1}^{N} \geq \epsilon I_N, \quad k \in \mathbf{Z}.$$

The left hand side of (0.6) is the generalized point evaluation which appears (for the scalar case) in [1] in the role of W-transform and has been developed further in [2] for triangular operators with matrix or operator coefficients, in connection with time-variant lossless inverse scattering problems. The non-Toeplitz Nevanlinna-Pick problem mentioned above with the entries f_{ij} being matrices and no tangential restrictions has been treated in [8]. The solution of the most general version of the problem, i.e., with operator entries f_{ij} and tangential interpolation conditions appears in [9] (in the present volume). In these papers the main tools come from an appropriate generalization of the reproducing kernel space method. The starting point for the present research was H. Dym's lecture at Oberwolfach in October 1989 in which he discussed Nevanlinna-Pick interpolation in the general context of upper triangular operators.

In the Toeplitz-version of the problem there is a special interest in solutions f that are rational. The latter means that the operator T in (0.1) is the input-output map of a causal time-invariant discrete system. In that case the relation

$$(0.8) \qquad T \begin{pmatrix} \vdots \\ u_{-1} \\ u_0 \\ u_1 \\ \vdots \end{pmatrix} = \begin{pmatrix} \vdots \\ y_{-1} \\ y_0 \\ y_1 \\ \vdots \end{pmatrix}$$

is given in the following way:

$$\begin{cases} x_{n+1} = A x_n + B u_n, \quad n = 0, \pm 1, \pm 2, \ldots, \\ y_n = C x_n + D u_n, \end{cases}$$

where A, B, C and D are matrices of appropriate sizes. In the non-Toeplitz version the requirement that the solution is rational is replaced by the condition that the operator T is the input-output map of a time-variant system, i.e., the action of T in (0.8) is given by

$$\begin{cases} x_{n+1} = A_n x_n + B_n u_n, \ n = 0, \pm 1, \pm 2, \ldots, \\ y_n = C_n x_n + D_n u_n, \end{cases}$$

where now the input, output and state matrices may vary in time.

In the description of the solutions of the non-Toeplitz version of the tangential Nevanlinna-Pick interpolation problem we are mainly interested in input-output maps of time-variant finite dimensional systems. This restriction allowed us to use for the technique of solution a modification of the method for finding rational solutions for the time-invariant version of the problem as developed in [6] (see also the book [7]). In this way we derive an algorithm to compute the solutions explicitly. Also the conditions of solvability are obtained. The results are illustrated by computing in detail the solutions for one example, namely when $N = 1$ and $\Delta_1 = (\cdots, 0, 0, w, 0, 0, \cdots)$. Throughout the paper our aim is to keep the exposition and techniques elementary at the level of linear algebra and standard matrix computations. This means that some proofs would be shorter if one uses known results concerning the geometry of Krein spaces (see [3], [10]).

The main theorem of this paper with an application to sensitivity minimization of time-variant systems was presented at MTNS-91 (see [4]).

The classical Nevanlinne-Pick interpolation problem has a half plane version. In the system theoretical approach to the latter problem the role of discrete time systems is taken over by continuous time systems (see [7]). This led us to a time-variant version of the half plane interpolation problem which we treat in another paper [5] (also in this volume).

A few words about notation. An identity operator is denoted by I; from the context it should be clear on which space it acts. For an invertible matrix or Hilbert space operator A we let A^{-*} denote the adjoint of A^{-1}.

Acknowledgement: The authors are grateful to J. Kos for his careful reading of the manuscript and his useful remarks.

1. PRELIMINARIES

1.1 Input-output maps of causal time-varying linear systems. Consider a discrete time-varying linear system

$$(1.1) \qquad \Sigma \begin{cases} x_{k+1} & = A_k x_k + B_k u_k, \\ y_k & = C_k x_k + D_k u_k. \end{cases}$$

The input sequence (u_k) is assumed to take values in the input space $\mathcal{U} = \mathbf{C}^r$, the state vector sequence (x_k) takes values in the state space $X_k = \mathbf{C}^{n_k}$ whose dimension we allow to depend on the time k and the output sequece (y_k) takes values in the output space $\mathcal{Y} = \mathbf{C}^m$. Thus A_k, B_k, C_k, D_k are matrices of respective sizes $n_{k+1} \times n_k$, $n_{k+1} \times r$, $m \times n_k$ and $m \times r$. Sometimes it is convenient to view $\mathcal{U}$, X_k and $\mathcal{Y}$ in a more coordinate-free form as simply finite dimensional linear spaces; then A_k, B_k, C_k, D_k are linear transformations acting between finite dimensional spaces and become matrices only after some particular choice of basis.

We shall assume that the system Σ in (1.1) is at rest until some time instant κ; thus we take $x_k = 0$ for $k \leq \kappa$ and $u_k = 0$ for $k < \kappa$. Then the equations in (1.1) give $y_k = 0$ for $k < \kappa$. By ℓ_+^r we denote the space of all (input) sequences $(u_k)_{-\infty}^{\infty}$ with values in $\mathbf{C}^r$ such that $u_k = 0$ for all $k < \kappa$ for some integer κ depending on the sequence. Such sequences are said to have *finite negative support*. To handle the sequence of state vectors, denote by $\ell_+^{(n_k)}$ the space of all sequences $(x_k)_{-\infty}^{\infty}$ such that $x_k \in \mathbf{C}^{n_k}$ for $k = \dots, -1, 0, 1, \dots$ and $x_k = 0$ for all $k < \kappa$ for some integer κ (depending on the sequence). Given a $\vec{u} - (u_k)_{-\infty}^{\infty} \in \ell_+^r$ with $u_k = 0$ for $k < \kappa$, if the system (1.1) is initialized with $x_\kappa = 0$, the equations (1.1) generate a state vector sequence $\vec{x} = (x_k)_{-\infty}^{\infty} \in \ell_+^{n_k}$ and an output sequence $\vec{y} = (y_k)_{-\infty}^{\infty} \in \ell_+^m$ such that $y_k = 0$ and $x_k = 0$ for all $k \in \kappa$ depending on the sequence (i.e., the bound κ on the negative support is the same for all three sequences $\vec{u}$, $\vec{x}$, $\vec{y}$).

Let $S : \ell_+^{(n_k)} \to \ell_+^{S(n_k)}$ be the (bilateral) forward shift operator on state vector sequences $\vec{x}$ defined by $(S\vec{x})_k = x_{k-1}$, where $\vec{x} = (x_k)_{-\infty}^{\infty}$, and let $S^{-1} : \ell_+^{(n_k)} \to \ell_+^{S^{-1}(n_k)}$ be the (bilateral) backward shift. Note S^{-1} is the inverse of the forward shift from $\ell_+^{S^{-1}(n_k)}$ to $\ell_+^{(n_k)}$ which we also denote by S. Then the system (1.1) can be expressed as a system of equations on the sequences $\vec{u} \in \ell_+^r$, $\vec{x} \in \ell_+^{(n_k)}$ and $\vec{y} \in \ell_+^m$ in the form:

$$(1.2) \qquad \begin{cases} S^{-1}\vec{x} & = \mathcal{A}\vec{x} + \mathcal{B}\vec{u}, \\ \vec{y} & = \mathcal{C}\vec{x} + \mathcal{D}\vec{u}. \end{cases}$$

Here $\mathcal{A}$, $\mathcal{B}$, $\mathcal{C}$, $\mathcal{D}$ are block diagonal operators,

$$\mathcal{A} : \ell_+^{(n_k)} \to \ell_+^{S^{-1}(n_k)}, \quad \mathcal{B} : \ell_+^r \to \ell_+^{S^{-1}(n_k)}, \quad \mathcal{C} : \ell_+^{(n_k)} \to \ell_+^m, \quad \mathcal{D} : \ell_+^r \to \ell_+^m,$$

given by

$$\mathcal{A} = \mathrm{diag}\,(A_k)_{-\infty}^\infty, \quad \mathcal{B} = \mathrm{diag}\,(B_k)_{-\infty}^\infty, \quad \mathcal{C} = \mathrm{diag}\,(C_k)_{-\infty}^\infty, \quad \mathcal{D} = \mathrm{diag}\,(D_k)_{-\infty}^\infty.$$

Note that action of the operator $I - S\mathcal{A} : \ell_+^{(n_k)} \to \ell_+^{(n_k)}$ may be described by block matrix multiplication in the following way:

$$(I - S\mathcal{A})^{-1}\vec{x} = \vec{y} \iff \begin{pmatrix} \ddots & & \ddots & & & \\ & A_{-2} & I & & & \\ & & A_{-1} & \boxed{I} & & \\ & & & A_0 & I & \\ & & & & \ddots & \ddots \end{pmatrix} \begin{pmatrix} \vdots \\ x_{-1} \\ \boxed{x_0} \\ x_1 \\ \vdots \end{pmatrix} = \begin{pmatrix} \vdots \\ y_{-1} \\ \boxed{y_0} \\ y_1 \\ \vdots \end{pmatrix}.$$

Here the unspecified entries are zero. Thus as a block matrix $I - S\mathcal{A}$ is block lower triangular with block main diagonal entries equal to I. It follows that $I - S\mathcal{A}$ is invertible in the algebra of linear operators acting on $\ell_+^{(n_k)}$ with inverse given by

$$(1.3) \qquad\qquad (I - S\mathcal{A})^{-1}\vec{x} = \sum_{j=0}^\infty (S\mathcal{A})^j \vec{x}, \qquad \vec{x} \in \ell_+^{(n_k)}.$$

Since $x_k = 0$ for all $k < \kappa$ for some integer κ, the sequence $(S\mathcal{A})^j \vec{x}$ is equal to the zero sequence for j sufficiently large. Therefore the infinite series in (1.3) degenerates to a finite sum and hence is well-defined. Thus one can solve the first equation in (1.2) for $\vec{x}$ to get

$$\vec{x} = (I - S\mathcal{A})^{-1} S\mathcal{B}\vec{u}$$

and plug this value into the second equation to get

$$\vec{y} = [\mathcal{D} + \mathcal{C}(I - S\mathcal{A})^{-1} S\mathcal{B}]\vec{u}.$$

This gives us a closed form expression for the input-output map $T_\Sigma : \vec{u} \mapsto \vec{y}$ from ℓ_+^r into ℓ_+^m generated by the linear time-varying system Σ given by (1.1).

We note that T_Σ is *causal* in the sense that

$$(1.4) \qquad Q_k T_\Sigma = Q_k T_\Sigma Q_k, \qquad k = 0, \pm 1, \pm 2, \ldots,$$

where $\vec{v} = Q_k \vec{w}$ is defined by

$$v_j = \begin{cases} w_j, & \text{if } j \leq k, \\ 0, & \text{if } j > k. \end{cases}$$

Here $\vec{v} = (v_k)_{-\infty}^\infty$ and $\vec{w} = (w_k)_{-\infty}^\infty$. In more physical language, causality means that the output up to time k is independent of inputs after time k.

A linear transformation $T : \ell_+^r \to \ell_+^m$, which is of the form $T = T_\Sigma$ for some linear finite dimensional time-varying system Σ as in (1.1), will be called a *causal, rational* (linear) transformation. The terminology is inspired by the time-invariant case where a block Toeplitz matrix $T = (a_{i-j})_{i,j=-\infty}^\infty$ is causal and rational in the above sense if and only if T is lower triangular (i.e., $a_k = 0$ for $k < 0$) and $r(z) = \sum_{k=0}^\infty a_k z^k$ is a rational matrix function. The realization theorems in [12], Section 5, give the necessary and sufficient conditions in order that a linear transformation $T : \ell_+^r \to \ell_+^m$ is causal and rational, and they also provide an algorithm to construct a realization of T (i.e, a time-varying system Σ as in (1.1) such $T_\Sigma = T$) when these conditions are fulfilled.

1.2 Anticausal time-varying systems. When working in the algebra of linear operators mapping ℓ_+^r into ℓ_+^m, the matrix associated with any such operator is necessarily of the form $S^{-k}T$, where T is lower triangular and S is the forward shift. Therefore to generate more examples we consider also input-output maps for anticausal systems.

By an anticausal discrete time-varying system we mean a system of the form

$$(1.5) \qquad \sigma \begin{cases} x_k = A_k x_{k+1} + B_k u_k, \\ y_k = C_k x_{k+1} + D_k u_k, \end{cases}$$

where the state evolves in backwards time. Here at each point in time x_k is an element of the state space $X_k = \mathbb{C}^{n_k}$, the vector u_k is an element of the input space $\mathcal{U} = \mathbb{C}^r$ and y_k is an element of the output space $\mathcal{Y} = \mathbb{C}^m$. Furthermore, $A_k : X_{k+1} \to X_k$, $B_k : \mathcal{U} \to X_k$, $C_k : X_{k+1} \to \mathcal{Y}$ and $D_k : \mathcal{U} \to \mathcal{Y}$ are linear transformations.

If the system σ in (1.5) is initialized at some time instant $k = \kappa$ by $x_{k+1} = 0$ and then inputs $u_k, u_{k-1}, \ldots$ are fed in in decreasing time, then the equations (1.5) generate a well-defined sequence $(\ldots 0, 0, x_\kappa, x_{\kappa-1}, \ldots)$ of state vectors and a well-defined sequence $(\ldots 0, 0, y_\kappa, y_{\kappa-1}, \ldots)$ of output vectors. The associated input-output map $T = T_\sigma$ is naturally defined as a map from ℓ^r_- into ℓ^m_- where ℓ^ν_- is the set of sequences $\vec{w} = (w_k)^\infty_{-\infty}$ with values in $\mathbb{C}^\nu$ such that $w_k = 0$ for all $k > \kappa$ for some κ, where κ depends on the sequence $\vec{w}$. Any such input-output map T_σ is anticausal in the sense that

$$(1.6) \qquad\qquad\qquad P_k T_\sigma = P_k T_\sigma P_k$$

for all integers k where $P_k = I - Q_{k-1}$ is the projection operator defined on biinfinite sequences $\vec{w} = (w_j)^\infty_{-\infty}$ by $P_k(\vec{w}) = \vec{v}$ with $\vec{v} = (v_j)^\infty_{-\infty}$ given by

$$v_j = \begin{cases} w_j & \text{for } j \geq k, \\ 0 & \text{for } j < k. \end{cases}$$

The theory developed in Section 1.1 for causal linear time-varying systems and causal maps has an analogue for anticausal linear time-varying systems and anticausal maps. In particular σ is an anticausal linear time-varying system as in (1.5), then the associated input-output map $T_\sigma : \ell^r_- \to \ell^m_-$ can be expressed as the upper-triangular matrix

$$(1.7) \qquad T_\sigma = \mathcal{D} + \mathcal{C}S^{-1}(I - \mathcal{A}S^{-1})^{-1}\mathcal{B} = \mathcal{D} + \sum_{j=0}^\infty \mathcal{C}S^{-1}(\mathcal{A}S^{-1})^j \mathcal{B}$$

where $\mathcal{A}$, $\mathcal{B}$, $\mathcal{C}$, $\mathcal{D}$ are the block diagonal matrices

$$(1.8) \quad \mathcal{A} = \text{diag } (A_k)^\infty_{-\infty}, \quad \mathcal{B} = \text{diag } (B_k)^\infty_{-\infty}, \quad \mathcal{C} = \text{diag } (C_k)^\infty_{-\infty}, \quad \mathcal{D} = \text{diag } (D_k)^\infty_{-\infty},$$

acting between the spaces

$$(1.9) \qquad \mathcal{A} : \ell^{S^{-1}(n_k)}_- \to \ell^{(n_k)}_-, \quad \mathcal{B} : \ell^r_- \to \ell^{(n_k)}_-, \quad \mathcal{C} : \ell^{S^{-1}(n_k)}_- \to \ell^m_-, \quad \mathcal{D} : \ell^r_- \to \ell^m_-,$$

and, as before, S is the forward bilateral shift.

1.3 Input-output maps defined on ℓ_2. In this section we consider a system Σ as in (1.1) assuming additionally that

$$(1.10) \qquad \sup_k \|A_k\| < \infty, \quad \sup_k \|B_k\| < \infty, \quad \sup_k \|C_k\| < \infty, \quad \sup_k \|D_k\| < \infty.$$

Here $\|M\|$ denotes the spectral norm of the matrix M, i.e., $\|M\|$ is the largest singular value of M. The first inequality in (1.10) implies that

$$SA = \begin{pmatrix} \ddots & & \ddots & & & \\ & A_{-2} & & 0 & & \\ & & A_{-1} & & \boxed{0} & \\ & & & A_0 & & 0 \\ & & & & \ddots & & \ddots \end{pmatrix}$$

defines a bounded linear operator on $\ell_2^{(n_k)}$, the Hilbert space of all doubly infinite norm-square summable sequences with entry x_k at time k in $\mathbf{C}^{n_k}$. We shall also assume that $\rho(SA)$, the spectral radius of SA, is strictly less than one or, more explicitly, that

$$(1.11) \qquad \limsup_{\nu \to \infty} (\sup_j \|A_{j+\nu-1} \ldots A_j\|)^{\frac{1}{\nu}} < 1.$$

Then $I - SA$ is invertible as an operator on $\ell_2^{(n_k)}$ with inverse given by

$$(1.12) \qquad (I - SA)^{-1} = \sum_{\nu=0}^{\infty} (SA)^{\nu}.$$

Because of condition (1.11), the series in the right hand side converges in the operator norm for operators on $\ell_2^{(n_k)}$.

From (1.12) and the boundedness conditions on B_k, C_k and D_k in (1.10) it follows that the input-output map T_Σ, initially defined only on sequences in ℓ_2^r having finite negative support, extends uniquely by continuity to a bounded linear operator from ℓ_2^r into ℓ_2^m which we also denote by T_Σ. Note that any such T_Σ defined on ℓ_2^r also has the causality property (1.4), and hence is given by a lower triangular matrix. The expansion

$$(1.13) \qquad T_\Sigma = \mathcal{D} + \sum_{\nu=0}^{\infty} \mathcal{C}(SA)^{\nu} S\mathcal{B}$$

gives an expansion of T_Σ in terms of its diagonals (the main diagonal $\mathcal{D}$ together with all diagonals below the main diagonal). Exactly which linear discrete time-varying systems Σ have input-output maps T_Σ which acts as bounded operators from ℓ_2^r into ℓ_2^m we leave as a topic for future work.

To get bounded upper-triangular input-output maps from ℓ_2^r into ℓ_2^m we have to consider the anticausal systems from Section 1.2. Let σ be the anticausal discrete time-varying system (1.5), and assume that the boundedness conditions in (1.10) are fulfilled. Condition (1.11) is now replaced by

$$(1.14) \qquad \limsup_{\nu \to \infty} (\sup_i \| A_i\, A_{i+1} \ldots A_{i+\nu-1} \|)^{\frac{1}{\nu}} < 1.$$

which is equivalent to the requirements that $\rho(\mathcal{A}S^{-1})$, the spectral radius of $\mathcal{A}S^{-1}$, is strictly less than one. Under the conditions (1.10) and (1.14) the associated input-output map T_σ given by (1.7) defines a bounded operator from ℓ_2^r into ℓ_2^m which has an upper triangular matrix representation. We shall say that a linear bounded operator from ℓ_2^r into ℓ_2^m is a *time-varying rational operator* if it can be written as the sum of a bounded input-output map of a causal linear time-varying system and a bounded input-output map of an anticausal linear time-varying system.

1.4 A time-varying calculus for lower triangular matrices. Denote by $\ell_2(\mathbf{Z})$ the space of doubly infinite square summable complex-valued sequences $(x_k)_{k=-\infty}^{\infty}$ (indexed by the set of all integers $\mathbf{Z}$). For m a positive integer, $\ell_2^m(\mathbf{Z})$ denotes the set of block sequences $(x_k)_{k=-\infty}^{\infty}$ where each entry x_k is in the space $\mathbf{C}^m$ of complex m tuples (viewed as column vectors). The space of all bounded linear operators on ℓ_2 we denote by $\mathcal{X}$; note that each element F in $\mathcal{X}$ has a biinfinite matrix representation

$$F = (F_{ij})_{i,j=-\infty}^{\infty} = \begin{pmatrix} & \vdots & \vdots & \vdots & \\ \ldots & F_{-1,-1} & F_{-1,0} & F_{-1,1} & \ldots \\ \ldots & F_{0,-1} & \boxed{F_{0,0}} & F_{0,1} & \ldots \\ \ldots & F_{1,-1} & F_{1,0} & F_{1,1} & \ldots \\ & \vdots & \vdots & \vdots & \end{pmatrix}$$

such that $\vec{y} = F\vec{x}$ is given by

$$y_i = \sum_{j=-0}^{\infty} F_{ij} x_j$$

if $\vec{x} = (x_j)_{j=-\infty}^{\infty} \in \ell_2(\mathbf{Z})$ and $\vec{y} = (y_j)_{j=-\infty}^{\infty} \in \ell_2(\mathbf{Z})$. For m and r positive integers, $\mathcal{X}^{m \times r}$ denotes the bounded linear operators from ℓ_2^r into ℓ_2^m. These similarly can be identified with doubly infinite block matrices $F = (F_{ij})_{i,j=-\infty}^{\infty}$ where each block has size $m \times r$. We call $(F_{ij})_{i,j=-\infty}^{\infty}$ the (standard) block matrix representation of F or just the matrix of F. Denote by $\mathcal{L}$ the subclass of $\mathcal{X}$ consisting of all operators F in $\mathcal{X}$ which leave invariant each of the subspaces

$$\ell_2([k,\infty)) := \{(x_j)_{j=-\infty}^{\infty} \in \ell_2 : x_j = 0 \text{ if } j < k\}$$

for all $k \in \mathbf{Z}$. This class $\mathcal{L}$ coincides with the set of F's in $\mathcal{X}$ having a lower triangular matrix representation, i.e., $F = (F_{ij})_{i,j=-\infty}^{\infty}$ with $F_{ij} = 0$ if $i < j$.

Analogously, we define transposed versions of these spaces, $\mathcal{U}$, $\mathcal{U}^{m \times r}$ which consist of upper triangular matrices, for example, $\mathcal{U}^{m \times r}$ consits of all operators F in $\mathcal{X}^{m \times r}$ which leave invariant the subspace

$$\ell_2^r((-\infty, k]) := \{(x_j)_{-\infty}^{\infty} \in \ell_2^r : x_j = 0 \text{ for } j > k\}$$

for all $k \in \mathbf{Z}$. The intersection $\mathcal{L}^{m \times r} \cap \mathcal{U}^{m \times r}$ consists of diagonal matrices; this class we denote by $\mathcal{D}^{m \times r}$.

A key operator for the setup which we now describe is the bilateral forward shift operator

$$S = \begin{pmatrix} \ddots & \ddots & & & \\ & I & 0 & & \\ & & I & \boxed{0} & \\ & & & I & 0 \\ & & & & \ddots & \ddots \end{pmatrix}$$

consisting of the identity matrix (of a size determined by the context) on the diagonal below the main diagonal. We use the same symbol S to denote the shift operator in $\mathcal{L}^{m \times m}$ for any positive integer m. If $G = (G_{ij})_{i,j=-\infty}^{\infty}$ is in $\mathcal{X}^{m \times r}$, then $SG \in \mathcal{X}^{m \times r}$ and for the (i,j)-th entry in the block matrix representation of SG we have

$$(SG)_{ij} = G_{i-1,j}, \qquad i,j = 0, \pm 1, \pm 2, \ldots$$

Thus the block matrix representation of SG is equal to the one of G with each row shifted one down. Note that

$$S^{-1} = S^* = \begin{pmatrix} \ddots & & \ddots & & & \\ & 0 & I & & & \\ & & \boxed{0} & I & & \\ & & & 0 & I & \\ & & & & \ddots & \ddots \end{pmatrix},$$

and thus the block matrix representation of $S^{-1}G$ is the one of G with each row shifted one up.

For $G \in \mathcal{X}^{m \times r}$ and k an integer we let $G_{[k]}$ be the block diagonal operator in $\mathcal{D}^{m \times r}$ of which the main diagonal entries are given by $(G_{[k]})_{jj} = G_{j+k,j}$ for each $j \in \mathbf{Z}$. We have (cf., [2], Lemma 2.7)

$$\|G_{[k]}\| = \sup_{j} \|G_{j+k,j}\| \leq \|G\|.$$

Note that

$$(S^k G_{[k]})_{ij} = \begin{cases} G_{ij}, & i - j = k, \\ 0, & \text{otherwise.} \end{cases}$$

It follows that G admits the series expansion:

$$(1.15) \qquad G = \sum_{k=-\infty}^{\infty} S^k G_{[k]},$$

where the convergence is entrywise. The right hand side of (1.15) amounts to breaking G up along diagonals parallel to the main diagonal. The series expansion in (1.15) corresponds to the Fourier series expansion of a bounded measurable function on the unit circle $\mathbf{T}$. Indeed, if g is such a function with Fourier series

$$g(z) = \sum_{j=-\infty}^{\infty} g_j z^j,$$

then for the Laurent operator G on $\ell_2(\mathbf{Z})$ with symbol g we have

$$G = (g_{i-j})_{i,j=-\infty}^{\infty} = \sum_{k=-\infty}^{\infty} S^k(g^k I)$$

where $g_k I$ is the diagonal matrix with constant value g_k along the main diagonal.

Now we introduce the generalized point evaluation map for operators in $\mathcal{L}^{m \times r}$ studied by Alpay-Dewilde-Dym [2]. Let $F \in \mathcal{L}^{m \times r}$ and $\Delta \in \mathcal{D}^{m \times m}$. We suppose that ΔS^{-1} (whose block matrix representation has all nonzero entries on the diagonal immediately above the main diagonal) has spectral radius $\rho(\Delta S^{-1})$ strictly less than 1 as an operator on $\ell_2^m(\mathbf{Z})$, i.e.,

$$(1.16) \qquad \rho(\Delta S^{-1}) := \limsup_{j \to \infty} \|(\Delta S^{-1})^j\|^{1/j} < 1.$$

Then $S - \Delta$ is an invertible operator on $\ell_2^m(\mathbf{Z})$, and its inverse on $\ell_m^2(\mathbf{Z})$ is given by

$$(S - \Delta)^{-1} = S^{-1}(I - \Delta S^{-1})^{-1} = \sum_{j=0}^{\infty} S^{-1}(\Delta S^{-1})^{-j}$$

with convergence in the operator norm. We define $\widehat{F}(\Delta)$ to be the block diagonal operator in $\mathcal{D}^{m \times r}$ with main diagonal equal to the main diagonal of $S(S - \Delta)^{-1}F$. The map $\Delta \mapsto \widehat{F}(\Delta)$ is the natural generalization of the point evaluation map for analytic functions on the unit disc. The analogy appears if we consider the diagonal expansion

$$(1.17) \qquad\qquad F = \sum_{j=0}^{\infty} S^j F_{[j]}$$

of $F \in \mathcal{L}^{m \times r}$. In terms of (1.17) we have

$$(1.18) \qquad\qquad \widehat{F}(\Delta) = \sum_{j=0}^{\infty} (\Delta S^{-1})^j S^j F_{[j]},$$

where the series concerges in the operator norm. If $F_{[j]} = f_j I$ and $\Delta = \lambda I$ with $|\lambda| < 1$ are all constant diagonals, then the shift operators in the right hand side of (1.18) cancel and (1.18) collapses to

$$(1.19) \qquad\qquad \widehat{F}(\lambda I) = \sum_{j=0}^{\infty} \lambda^j f_j I = f(\lambda)I.$$

From another point of view, for any $G \in \mathcal{X}^{m \times r}$ define the *total residue* $\mathcal{R}(G)$ of G to be the coefficient of S^{-1} in its diagonal expansion (1.15), i.e., put $\mathcal{R}(G) = G_{[-1]}$. Then $\widehat{F}(\Delta)$ can alternatively be defined as

$$(1.20) \qquad\qquad \widehat{F}(\Delta) = \mathcal{R}\{(S - \Delta)^{-1}F\},$$

because $(SG)_{[k]} = G_{[k-1]}$ for any $G \in \mathcal{X}^{m \times r}$ and $k \in \mathbf{Z}$. In the time invariant case where $F = (f_{i-j})_{i,j=-\infty}^{\infty}$ and $\Delta = \lambda I$ whith $|\lambda| < 1$, it is readily checked that

$$\mathcal{R}\{(S - \Delta)^{-1}F\} = \mathcal{R}\{(z - \lambda)^{-1}f(\lambda)\}I = f(\lambda)I,$$

where $f(\lambda) = \sum_{j=0}^{\infty} \lambda^j f_j$ and $\mathcal{R}$ in the second term denotes the sum of the residues inside the unit disk in the usual complex variables sense. We shall see more striking parallels with this more general noncommutative time-varying calculus and the standard calculus for analytic functions as we proceed.

We now present some basic properties of the time varying point evaluation $F \mapsto \widehat{F}(\Delta)$ which we shall need in the sequel. These results appear in [2]; we include proofs for the sake of completeness.

PROPOSITION 1.1. *Let $F \in \mathcal{L}^{m \times r}$ and $\Delta \in \mathcal{D}^{m \times m}$ be given, where $\rho(\Delta S^{-1}) < 1$. Then $\widehat{F}(\Delta)$ is the unique element of $\mathcal{D}^{m \times r}$ such that $(S - \Delta)^{-1}(F - \widehat{F}(\Delta)) \in \mathcal{L}^{m \times r}$.*

PROOF. We first settle the uniqueness issue. If D_1 and D_2 were two diagonal operators in $\mathcal{D}^{m \times r}$ such that $(S - \Delta)^{-1}(F - D_j) \in \mathcal{L}^{m \times r}$ for $j = 1, 2$, then by linearity $(S - \Delta)^{-1}E \in \mathcal{L}^{m \times r}$ where $E = D_1 - D_2 \in \mathcal{D}^{m \times r}$. However $(S - \Delta)^{-1} = \sum_{j=0}^{\infty} S^{-1}(\Delta S^{-1})^j$ is strictly upper triangular, and hence remains so when multiplied by a diagonal E. This forces $(S - \Delta)^{-1}E = 0$, and hence $E = D_1 - D_2 = 0$. This establishes uniqueness.

In general, if $E \in \mathcal{D}^{m \times r}$ is diagonal, then (1.18) shows that $\widehat{E}(\Delta) = E$. By linearity of the map $H \mapsto \widehat{H}(\Delta)$, we see that whenever $F \in \mathcal{L}^{m \times r}$, then $G = F - \widehat{F}(\Delta) \in \mathcal{L}^{m \times r}$ has $\widehat{G}(\Delta) = 0$. The lower triangularity is thus settled if we show that $(S - \Delta)^{-1}G \in \mathcal{L}^{m \times r}$ whenever $G \in \mathcal{L}^{m \times r}$ and $\widehat{G}(\Delta) = 0$.

To do this we must show that the ℓ-th diagonal above the main diagonal of $(S - \Delta)^{-1}G$ is zero for $\ell = 1, 2, \dots$. Write

$$W := (S - \Delta)^{-1}G = \sum_{k=-\infty}^{\infty} S^k W_{[k]}.$$

For $(S - \Delta)^{-1}$ and G we have the diagonal expansions

$$(S - \Delta)^{-1} = \sum_{j=0}^{\infty} S^{-(j+1)}(S^j(\Delta S^{-1})^j), \quad G = \sum_{\nu=0}^{\infty} S^{\nu} G_{[\nu]}.$$

It follows that for $k = -1, -2, \dots$

$$W_{[k]} = \sum_{\nu=0}^{\infty} S^{-k-1}(\Delta S^{-1})^{\nu-k-1} S^{\nu} G_{[\nu]}$$

$$= S^{-k-1}(\Delta S^{-1})^{-k-1} \Big(\sum_{\nu=0}^{\infty} (\Delta S^{-1})^{\nu} S^{\nu} G_{[\nu]} \Big)$$

$$= S^{-k-1}(\Delta S^{-1})^{-k-1} \widehat{G}(\Delta),$$

because of (1.18). Since $\widehat{G}(\Delta) = 0$, we see that $W_{[k]} = 0$ for $k = -1, -2, \ldots$, and hence $(S - \Delta)^{-1}G$ is lower triangular. $\qquad\square$

PROPOSITION 1.2. *If* $F \in \mathcal{L}^{m \times r}$, $G \in \mathcal{L}^{r \times p}$ *and* $\Delta \in \mathcal{D}^{m \times m}$ *with* $\rho(\Delta S^{-1}) < 1$, *then*

$$(FG)^{\wedge}(\Delta) = \{\widehat{F}(\Delta)G\}^{\wedge}(\Delta).$$

PROOF. Since $(S - \Delta)^{-1}(FG - \widehat{FG}(\Delta)) \in \mathcal{L}^{m \times p}$, we see from Proposition 1.1 that it suffices to show that

$$(S - \Delta)^{-1}(FG - \widehat{F}(\Delta)G) \in \mathcal{L}^{r \times p}.$$

We calculate

$$(S - \Delta)^{-1}(FG - \widehat{F}(\Delta)G) = [(S - \Delta)^{-1}(F - \widehat{F}(\Delta))]G = HG \in \mathcal{L}^{m \times p},$$

since $H = (S - \Delta)^{-1}(F - \widehat{F}(\Delta)) \in \mathcal{L}^{m \times r}$ by Proposition 1.1 and $G \in \mathcal{L}^{r \times p}$ by assumption. $\square$

It is convenient also to consider certain classes of Hilbert-Schmidt matrices. By $\mathcal{X}_2^{m \times r}$ we denote the class of all doubly infinite block matrices

$$F = (F_{ij})_{i,j=-\infty}^{\infty} = \begin{pmatrix} & \vdots & \vdots & \vdots & \\ \cdots & F_{-1,-1} & F_{-1,0} & F_{-1,1} & \cdots \\ \cdots & F_{0,-1} & \boxed{F_{0,0}} & F_{0,1} & \cdots \\ \cdots & F_{1,-1} & F_{1,0} & F_{1,1} & \cdots \\ & \vdots & \vdots & \vdots & \end{pmatrix}$$

where each block has size $m \times r$ such that

$$(1.21) \qquad\qquad\qquad \sum_{i,j} \|F_{ij}\|_2^2 < \infty.$$

where $\|F_{ij}\|_2^2$ is the sum of the squares of the moduli of the matrix entries of F_{ij} (i.e., the square of the Hilbert-Schmidt norm of F_{ij}). Since $\|F_{ij}\| \leq \|F_{ij}\|_2$ for all i and j, such an F defines a bounded linear operator from ℓ_2^r into ℓ_2^m which we also denote by F. In fact, F is a Hilbert-Schmidt operator and the quantity in (1.21) is the square of the Hilbert-Schmidt norm of F. In particular

$$\mathcal{X}_2^{m \times r} = \{F \in \mathcal{X}^{m \times r} \mid F \text{ is Hilbert} - \text{Schmidt}\}.$$

By $\mathcal{L}_2^{m\times r}$ we denote the subclasses of F in $\mathcal{X}_2^{m\times r}$ which are lower triangular ($F_{ij} = 0$ for $i < j$). Similarly, $\mathcal{U}_2^{m\times r}$ consists of the upper triangular matrices in $\mathcal{X}_2^{m\times r}$. Our primary focus will be on block column lower triangular matrices of this type ($\mathcal{L}_2^{r\times 1}$ or $\mathcal{L}_2^{m\times 1}$). Note that $\mathcal{L}_2^{r\times 1}$ is a Hilbert space in the natural inner product

$$\langle f, g \rangle_{\mathcal{L}_2^{r\times 1}} = \mathrm{tr}\ (g^* f), \qquad f, g \in \mathcal{L}_2^{r\times 1}.$$

If $X \in \mathcal{D}^{1\times r}$, $\Delta \in \mathcal{D}^{1\times 1}$ such that $\rho(\Delta S^{-1}) < 1$ and $D \in \mathcal{D}_2^{1\times 1}$, then $(Xf)^\wedge(\Delta) \in \mathcal{D}_2^{1\times 1}$ (c.f., Lemma 7.4 in [2]) and

$$f \mapsto\ \mathrm{tr}\ (D^*(Xf)^\wedge(\Delta)) = \langle (Xf)^\wedge(\Delta), D \rangle_{\mathcal{D}_2^{1\times 1}}$$

is a bounded linear functional on the Hilbert space $\mathcal{L}_2^{r\times 1}$, hence by the Riesz representation theorem there must be an element $k = k(X, \Delta, D)$ in $\mathcal{L}_2^{r\times 1}$ for which

$$\langle (Xf)^\wedge(\Delta), D \rangle_{\mathcal{D}_2^{1\times 1}} = \langle f, k \rangle_{\mathcal{L}_2^{r\times 1}}.$$

The following proposition identifies this element $k(X, \Delta, D)$; in the time-invariant case, the element $k(X, \Delta, D)$ is associated with the kernel function $k(z, w) = (1 - z\overline{w})^{-1}$ for the Hardy space H^2 . A form of this result also appears in [2].

PROPOSITION 1.3. *Let* $X \in \mathcal{D}^{1\times r}$, $\Delta \in \mathcal{D}^{1\times 1}$ *with* $\rho(\Delta S^{-1}) < 1$ *and* $D \in \mathcal{D}_2^{1\times 1}$ *be given. Then* $X^*(I - S\Delta^*)^{-1}D \in \mathcal{L}_2^{r\times 1}$ *and satisfies the identity*

$$\langle f, X^*(I - S\Delta^*)^{-1}D \rangle_{\mathcal{L}_2^{r\times 1}} = \langle (Xf)^\wedge(\Delta), D \rangle_{\mathcal{D}_2^{1\times 1}}.$$

for all $f \in \mathcal{L}_2^{r\times 1}$.

PROOF. Indeed,

$$\begin{aligned}
\langle f, X^*(I - S\Delta^*)^{-1}D \rangle_{\mathcal{L}_2^{r\times 1}} &= \mathrm{tr}\ D^*(I - \Delta S^{-1})^{-1}Xf \\
&= \sum_i D_{ii}^*[(I - \Delta S^{-1})^{-1}Xf]_{ii} = \sum_i D_{ii}^*[(S - \Delta)^{-1}Xf]_{i+1,i} \\
&= \sum_i D_{ii}^*[(Xf)^\wedge(\Delta)]_{ii} = \mathrm{tr}\ (D^*(Xf)^\wedge(\Delta)) \\
&= \langle (Xf)^\wedge(\Delta), D \rangle_{\mathcal{D}_2^{1\times 1}}. \qquad \square
\end{aligned}$$

If $F \in \mathcal{L}^{m \times r}$, then the operator $L_f : f \mapsto Ff$ of multiplication by F on the left is a bounded operator on $\mathcal{L}_2^{r \times 1}$. The adjoint L_F^*,

$$(L_F)^* : g \mapsto P_{\mathcal{L}_2^{r \times 1}}(F^* g),$$

requires the orthogonal projection $P_{\mathcal{L}_2^{r \times 1}}$ from $\mathcal{X}_2^{r \times 1}$ onto $\mathcal{L}_2^{r \times 1}$ following left multiplication by F^*. It is useful to know that $(L_F)^*$ can be computed explicitly on the "kernel function" elements $X^*(I - S\Delta^*)^{-1}D$.

PROPOSITION 1.4. *Let* $X \in \mathcal{D}^{1 \times m}$, $\Delta \in \mathcal{D}^{1 \times 1}$ *with* $\rho(\Delta S^{-1}) < 1$ *and* $D \in \mathcal{D}_2^{1 \times 1}$ *be given. Suppose* $F \in \mathcal{L}^{m \times r}$ *and* $Y = (XF)^\wedge(\Delta) \in \mathcal{D}^{1 \times r}$. *Denote by* L_F *the operator on* $\mathcal{L}_2^{r \times 1}$ *given by multiplication on the left by* F. *Then*

$$(L_F)^* X^*(I - S\Delta^*)^{-1}D = Y^*(I - S\Delta^*)^{-1}D$$

PROOF. Take f in $\mathcal{L}_2^{r \times 1}$. We compute

$$\langle f, (L_F)^* X^*(I - S\Delta^*)^{-1}D \rangle_{\mathcal{L}_2^{r \times 1}} = \langle Ff, X^*(I - S\Delta^*)^{-1}D \rangle_{\mathcal{L}_2^{m \times 1}}$$

$$= \langle (XFf)^\wedge(\Delta), D \rangle_{\mathcal{D}_2^{1 \times 1}},$$

where we used Proposition 1.3. But then we can apply Proposition 1.2 to show that

$$\langle f, (L_F)^* X^*(I - S\Delta^*)^{-1}D \rangle_{\mathcal{L}_2^{r \times 1}} = \langle \{(XF)^\wedge(\Delta)f\}^\wedge(\Delta), D \rangle_{\mathcal{D}_2^{1 \times 1}}$$

$$= \langle (Yf)^\wedge(\Delta), D \rangle_{\mathcal{D}_2^{1 \times 1}},$$

$$= \langle f, Y^*(I - S\Delta^*)^{-1}D \rangle_{\mathcal{L}_2^{r \times 1}},$$

where we used again Proposition 1.2. $\square$

2. J-UNITARY OPERATORS ON ℓ_2

2.1 J-unitary and J-contractive operators. Suppose $\mathcal{H}_+$ and $\mathcal{H}_-$ are two separable Hilbert spaces and J is the self-adjoint and unitary operator on the orthogonal direct sum space $\mathcal{H} = \mathcal{H}_+ \oplus \mathcal{H}_-$ defined by $J = I_{\mathcal{H}_+} - I_{\mathcal{H}_-}$. Here $I_\mathcal{K}$ denotes the identity operator on the space $\mathcal{K}$. A bounded linear operator Θ on $\mathcal{H}$ is said to be J-*isometric* if $\Theta^* J \Theta = J$, or equivalently, Θ preserves the indefinite inner product induced by J:

$$\langle J\Theta h, \Theta h \rangle = \langle Jh, h \rangle, \qquad h \in \mathcal{H}.$$

We say that Θ is J-*unitary* if both Θ and Θ^* are J-isometric; equivalently, Θ is J-isometric and Im $\Theta = \mathcal{H}$. We say that Θ is J-*contractive* (or Θ is a J-*contraction*) if $\Theta^* J \Theta \leq J$, or equivalently

$$\langle J\Theta h, \Theta h \rangle \leq \langle Jh, h \rangle, \qquad h \in \mathcal{H}.$$

If both Θ and Θ^* are J-contractive, then Θ is called J-*bicontractive*. The Hilbert space $\mathcal{H}$ with the Hilbert space inner product

$$\left\langle \begin{pmatrix} h_+ \\ h_- \end{pmatrix}, \begin{pmatrix} k_+ \\ k_- \end{pmatrix} \right\rangle_{\mathcal{H}} = \langle h_+, k_+ \rangle_{\mathcal{H}_+} + \langle h_-, k_- \rangle_{\mathcal{H}_-}$$

replaced by the indefinite inner product

$$\left[\begin{pmatrix} h_+ \\ h_- \end{pmatrix}, \begin{pmatrix} k_+ \\ k_- \end{pmatrix} \right]_{\mathcal{H}} = \langle h_+, k_+ \rangle_{\mathcal{H}_+} - \langle h_-, k_- \rangle_{\mathcal{H}_-} = \langle J \begin{pmatrix} h_+ \\ h_- \end{pmatrix}, \begin{pmatrix} k_+ \\ k_- \end{pmatrix} \rangle_{\mathcal{H}}$$

is known as a Krein space, and J-isometric, J-unitary, J-contractive and J-bicontractive operators correspond to isometric, unitary, contractive and bicontractive operators in the Krein space sense. A systematic study of the operator theory and geometry associated with such operators is given in [3]; a useful summary can be found in [10]. Here we set down only a few basic properties, well known among specialists but not so well known in general, which we shall need. In the next subsection, we specialize to the setting where $\mathcal{H}_+ = \ell_2^m$ and $\mathcal{H}_- = \ell_2^r$. Although these results can be gleaned from the more general results of [3] and [10], we include simple, direct proofs to keep the exposition self-contained.

We begin with a result concerning J-bicontractions.

THEOREM 2.1. *Suppose* $\Theta = \begin{pmatrix} \Theta_{11} & \Theta_{12} \\ \Theta_{21} & \Theta_{22} \end{pmatrix}$ *is a J-contraction. Then Θ is a J-bicontraction if and only if Θ_{22} is invertible. In this case, $\|\Theta_{22}^{-1}\Theta_{21}\| < 1$.*

PROOF. Suppose first that Θ is a J-bicontraction. Then the relations

$$\Theta^* J \Theta \leq J, \qquad \Theta J \Theta^* \leq J$$

yield

$$(2.1) \qquad \begin{pmatrix} \Theta_{11}^* \Theta_{11} - \Theta_{21}^* \Theta_{21} & \Theta_{11}^* \Theta_{12} - \Theta_{21}^* \Theta_{22} \\ \Theta_{12}^* \Theta_{11} - \Theta_{22}^* \Theta_{21} & \Theta_{12}^* \Theta_{12} - \Theta_{22}^* \Theta_{22} \end{pmatrix} \leq \begin{pmatrix} I & 0 \\ 0 & -I \end{pmatrix}$$

and

$$(2.2) \qquad \begin{pmatrix} \Theta_{11}\Theta_{11}^* - \Theta_{12}\Theta_{12}^* & \Theta_{11}\Theta_{21}^* - \Theta_{12}\Theta_{22}^* \\ \Theta_{21}\Theta_{11}^* - \Theta_{22}\Theta_{12}^* & \Theta_{21}\Theta_{21}^* - \Theta_{22}\Theta_{22}^* \end{pmatrix} \leq \begin{pmatrix} I & 0 \\ 0 & -I \end{pmatrix}$$

From the $(2,2)$-entries in (2.1) and (2.2) we get

$$(2.3) \qquad \Theta_{22}^*\Theta_{22} \geq I + \Theta_{12}^*\Theta_{12},$$

$$(2.4) \qquad \Theta_{22}\Theta_{22}^* \geq I + \Theta_{21}\Theta_{21}^*.$$

From (2.4) we see that Θ_{22} is onto while (2.3) gives that Θ_{22} is one-to-one. Hence Θ_{22} is invertible on $\mathcal{H}_-$ as asserted. Moreover (2.4) yields

$$\Theta_{22}^{-1}\Theta_{21}\Theta_{21}^*\Theta_{22}^{-*} \leq I - \Theta_{22}^{-1}\Theta_{22}^{-*}.$$

This implies $\|\Theta_{22}^{-1}\Theta_{21}\| < 1$.

Next, suppose that Θ is a J-contraction and Θ_{22} is invertible. Then we can solve the system of equations

$$(2.5) \qquad \Theta_{11}u + \Theta_{12}y = z, \qquad \Theta_{21}u + \Theta_{22}y = w$$

for (z,y) in terms of (w,u). The result is

$$(2.6) \qquad U_{11}w + U_{12}u = z, \qquad U_{21}w + U_{22}u = y,$$

where

$$(2.7) \qquad U = \begin{pmatrix} U_{11} & U_{12} \\ U_{21} & U_{22} \end{pmatrix} = \begin{pmatrix} \Theta_{12}\Theta_{22}^{-1} & \Theta_{11} - \Theta_{12}\Theta_{22}^{-1}\Theta_{21} \\ \Theta_{22}^{-1} & -\Theta_{22}^{-1}\Theta_{21} \end{pmatrix}.$$

Moreover, the J-contractive property of Θ implies

$$\|z\|^2 - \|w\|^2 \leq \|u\|^2 - \|y\|^2$$

whenever u, y, z, w in $\mathcal{H}_+, \mathcal{H}_-, \mathcal{H}_+, \mathcal{H}_-$ respectively satisfy (2.5). We conclude that

$$\|z\|^2 + \|y\|^2 \leq \|w\|^2 + \|u\|^2$$

whenever u, y, z, w satisfy (2.6). This gives us that U defined by (2.7) is an ordinary Hilbert space contraction. For the case $J = I$ there is no distiction between contractions and J-contractions; hence also

$$\widehat{U} = \begin{pmatrix} 0 & -I \\ I & 0 \end{pmatrix} U^* \begin{pmatrix} 0 & -I \\ I & 0 \end{pmatrix} = \begin{pmatrix} \Theta_{21}^* \Theta_{22}^{-*} & \Theta_{11}^* - \Theta_{21}^* \Theta_{22}^{-*} \Theta_{12}^* \\ \Theta_{22}^{-*} & -\Theta_{22}^{-*} \Theta_{12}^* \end{pmatrix}$$

is contractive. Thus

$$(2.8) \qquad \qquad \|z_1\|^2 + \|y_1\|^2 \leq \|w_1\|^2 + \|u_1\|^2$$

whenever

$$(2.9) \qquad \widehat{U}_{11} w_1 + \widehat{U}_{12} u_1 = z_1, \qquad \widehat{U}_{21} w_1 + \widehat{U}_{22} u_1 = y_1.$$

But since $\widehat{U}_{21} = \Theta_{22}^{-*}$ is invertible, we can solve (2.9) for (z_1, w_1) in terms of (u_1, y_1); indeed, note that $\widehat{U}$ has the same form as U but with $\Theta^* = \begin{pmatrix} \Theta_{11}^* & \Theta_{21}^* \\ \Theta_{12}^* & \Theta_{22}^* \end{pmatrix}$ in place of Θ. The result is

$$(2.10) \qquad \Theta_{11}^* u_1 + \Theta_{21}^* y_1 = z_1, \qquad \Theta_{12}^* u_1 + \Theta_{22}^* y_1 = w_1.$$

Since (2.10) and (2.9) are equivalent systems of equations, from (2.8) we see that

$$\|z_1\|^2 - \|w_1\|^2 \leq \|u_1\|^2 - \|y_1\|^2$$

whenever (u_1, y_1, z_1, w_1) satisfy (2.10), i.e., Θ^* is J-contractive. This verifies all assertions in Theorem 2.1. $\qquad \square$

2.2 J-unitary and J-inner operators on ℓ_2^{m+r}. In this subsection we specialize the results of the previous section to the case where $\mathcal{H}_+ = \ell_2^m$ and $\mathcal{H}_- = \ell_2^r$. The signature operator J then is given by $J = I_{\ell_2^m} \oplus -I_{\ell_2^r}$ on $\ell_2^m \oplus \ell_2^r \cong \ell_2^{m+r}$. We abuse notation and denote also by J the signature operator $J = I_m \oplus -I_r$ on the finite dimensional space $\mathbf{C}^{m+r}$; this should cause no confusion as the meaning will be clear from the context. By a J-unitary map on ℓ_2^{m+r} we therefore mean a bounded linear operator $\Theta : \ell_2^{m+r} \to \ell_2^{m+r}$ such that $\Theta^* J \Theta = J$ and $\Theta J \Theta^* = J$. If we write Θ in block form as $\Theta = \begin{pmatrix} \Theta_{11} & \Theta_{12} \\ \Theta_{21} & \Theta_{22} \end{pmatrix}$, Theorem 2.1 guarantees that Θ_{22} is invertible on ℓ_2^r and $\Theta_{22}^{-1}\Theta_{21} : \ell_2^m \to \ell_2^r$ has $\|\Theta_{22}^{-1}\Theta_{21}\| < 1$ if Θ is J-unitary.

In this section we wish to derive some basic properties of J-unitary maps Θ which have an additional property with respect to the time structure of ℓ_2^{m+r}. If Θ is a J-unitary map on ℓ_2^{m+r} we say that Θ is J-inner if Θ also satisfies

$$(2.11) \qquad Q_k\Theta^*JQ_k\Theta Q_k \leq JQ_k$$

for all integer k. Recall that Q_k is the projection operator defined by

$$(Q_k\vec{u})_j = \begin{cases} u_j & \text{if } j \leq k, \\ 0 & \text{if } j > k. \end{cases}$$

If Θ is also lower triangular (i.e., in the terminology of subsection 1.4 we have $\theta \in \mathcal{L}^{(m+r)\times(m+r)}$), then $Q_k\Theta = Q_k\Theta Q_k$ and (2.11) can be simplified to

$$(2.12) \qquad \Theta^*JQ_k\Theta \leq JQ_k.$$

The following gives an equivalent formulation of the J-inner property in terms of the projections $P_k = I - Q_{k-1}$ rather than Q_k.

PROPOSITION 2.2. *Suppose Θ is a lower triangular J-unitary map on ℓ_2^{m+r}. Then Θ is J-inner if and only if*

$$\Theta JP_k\Theta^* \leq JP_k, \qquad k = \ldots, -1, 0, 1, \ldots.$$

PROOF. Suppose that Θ is lower triangular J-unitary. After reindexing in (2.12) we see that Θ is J-inner if and only if

$$(2.13) \qquad \Theta^*JQ_{k-1}\Theta \leq JQ_{k-1}$$

for $k = \ldots, -1, 0, 1, \ldots$. Substitute $Q_{k-1} = I - P_k$ and use that $\Theta^*J\Theta = J$ to get that (2.13) is equivalent to

$$(2.14). \qquad \Theta^*JP_k\Theta \geq JP_k$$

Multiply on the left by ΘJ and on the right by $J\Theta^*$ and we use that $\Theta J\Theta^* = J$ to get $JP_k \geq \Theta J P_k \Theta^*$ as required. $\qquad\square$

The following is a useful characterization of the J-inner property for lower triangular J-unitary maps.

THEOREM 2.3. *Suppose* $\Theta = \begin{pmatrix} \Theta_{11} & \Theta_{12} \\ \Theta_{21} & \Theta_{22} \end{pmatrix}$ *is a lower triangular J-unitary map on ℓ_2^{m+r}; in particular, Θ_{22} is invertible on ℓ_2^r. Then Θ is J-inner if and only if Θ_{22}^{-1} is lower triangular.*

PROOF. Note that Θ_{22} is invertible by Theorem 2.1. Suppose that Θ is lower triangular and J-unitary. We consider $[\Theta]_k = \Theta|\ell_2^{m+r}([k,\infty))$ as a mapping on $\ell_2^{m+r}([k,\infty))$. Since Θ is lower triangular and J-unitary, $[\Theta]_k$ is $(J|\operatorname{Im} P_k)$-isometric

$$([\Theta]_k)^* J P_k [\Theta]_k = \Theta^* J\Theta|\operatorname{Im} P_k = J|\operatorname{Im} P_k.$$

In particular, $[\Theta]_k$ is $(J|\operatorname{Im} P_k)$-contractive. The content of Proposition 2.2 is that a lower triangular J-unitary Θ is J-inner if and only if $(([\Theta])_k)^*$ is $(J|\operatorname{Im} P_k)$-contractive, i.e., if and only if $[\Theta]_k$ is a $(J|\operatorname{Im} P_k)$-bicontraction for every $k = \ldots, -1, 0, 1, \ldots$. On the other hand, by Theorem 2.1 such a Θ has the property that $[\Theta]_k$ is a $(J|\operatorname{Im} P_k)$-bicontraction if and only if $[\Theta_{22}]_k = \Theta_{22}|\ell_2^r([k,\infty))$ is invertible for every k. This last condition is equivalent to Θ_{22}^{-1} mapping $\ell_2^r([k,\infty))$ into itself, i.e., to Θ_{22}^{-1} being lower triangular. $\qquad\square$

2.3 Realization of J-unitary and J-inner maps. In this section we define a class of systems which yield lower-triangular J-unitary or J-inner operators on ℓ_2^{m+r} as its input-output maps. By a (causal) *stable J-unitary* time-varying system we mean a linear time-varying system

$$(2.15) \qquad\qquad \Sigma \begin{cases} x_{k+1} & = A_k x_k + B_k u_k \\ y_k & = C_k x_k + D_k u_k \end{cases}$$

with state space $X_k = X = \mathbb{C}^{n_0}$ independent of k which satisfies the sufficient conditions (1.10) and (1.11) to generate a bounded lower triangular input-output map on ℓ_2^{m+r} and for which there exist invertible Hermitian linear transformations H_k on $X_k = X = \mathbb{C}^{n_0}$ such that

$$(2.16) \qquad\qquad \sup_{k \in \mathbb{Z}} \{ \|H_k\|, \|H_k^{-1}\| \} < \infty,$$

$$(2.17) \qquad \begin{pmatrix} A_k^* & C_k^* \\ B_k^* & D_k^* \end{pmatrix} \begin{pmatrix} H_{k+1} & 0 \\ 0 & J \end{pmatrix} \begin{pmatrix} A_k & B_k \\ C_k & D_k \end{pmatrix} = \begin{pmatrix} H_k & 0 \\ 0 & J \end{pmatrix}, k \in \mathbf{Z}.$$

We say that the stable J-unitary system Σ is J-inner if in addition the Hermitian matrix H_k is positive definite for all k. The following result will be a basic tool in our solution of the time-varying version of the tangential Nevanlinna-Pick interpolation problem.

THEOREM 2.4. *The input-output map* $\Theta = T_\Sigma$ *of a stable J-unitary system Σ is a J-unitary map on ℓ_2^{m+r}. Furthermore, Σ is J-inner as a system if and only if $\Theta = T_\Sigma$ is J-inner as an operator on ℓ_2^{m+r}.*

PROOF. Suppose Σ as in (2.15) is a stable J-unitary system, and suppose $\vec{u} = (\dots, 0, 0, u_\kappa, u_{\kappa-1}, \dots)$ is an input sequence in $\ell_+^{m+r} \cap \ell_2^{m+r}$. Then the J-unitary property of Σ implies the equality

$$\rho_{H_{k+1}}(x_{k+1}) + \rho_J(y_k) = \rho_{H_k}(x_k) + \rho_J(u_k),$$

or equivalently

$$(2.18) \qquad \rho_{H_{k+1}}(x_{k+1}) - \rho_{H_k}(x_k) = \rho_J(u_k) - \rho_J(y_k)$$

at each point k in time. Here, in general, $\rho_W(w) = \langle Ww, w \rangle$ is the Hermitian form induced by the Hermitian matrix W acting on the vector w. Summing from $k = \kappa$ to $k = j$ in (2.18) and using $x_\kappa = 0$ gives

$$(2.19) \qquad \rho_{H_{j+1}}(x_{j+1}) = \sum_{k=-\infty}^{j} \rho_J(u_k) - \sum_{k=-\infty}^{j} \rho_J(y_k).$$

Since the system Σ is stable by assumption, the sequence $\vec{x} = (x_k)_{-\infty}^{\infty} = (I - SA)^{-1} SB\vec{u}$ is in $\ell_2^{n_0}$; hence, in particular, $\lim_{k \to \infty} x_k = 0$. By assumption (2.16) it follows that $\lim_{k \to \infty} \rho_{H_k}(x_k) = 0$ as well. From (2.19) we conclude that

$$(2.20) \qquad \sum_{k=-\infty}^{\infty} \rho_J(y_k) = \sum_{k=-\infty}^{\infty} \rho_J(u_k)$$

whenever $\vec{y} = \Theta\vec{u}$ and $\vec{u} \in \ell_+^{m+r} \cap \ell_2^{m+r}$. By an approximation argument it follows that (2.20) continues to hold for all $u \in \ell_2^{m+r}$. We conclude that Θ is J-isometric.

To show that Θ is J-unitary, we must show that Θ^* is also a J-isometry. If Θ is the input-output map of the causal stable system (2.15), it is straightforward to see that Θ^* is the input-output map of the anticausal antistable system

$$(2.21) \qquad \Sigma^* \begin{cases} x_k &= A_k^* x_{k+1} + C_k^* u_k, \\ y_k &= B_k^* x_{k+1} + D_k^* u_k. \end{cases}$$

From (2.16) we see that

$$\begin{pmatrix} H_{k+1} & 0 \\ 0 & J \end{pmatrix} \begin{pmatrix} A_k & B_k \\ C_k & D_k \end{pmatrix} \begin{pmatrix} H_k^{-1} & 0 \\ 0 & J \end{pmatrix}$$

is a right inverse for $\begin{pmatrix} A_k^* & C_k^* \\ B_k^* & D_k^* \end{pmatrix}$. Since all these matrices are square, a right inverse is also a left inverse. Hence,

$$(2.22) \qquad \begin{pmatrix} A_k & B_k \\ C_k & D_k \end{pmatrix} \begin{pmatrix} H_k^{-1} & 0 \\ 0 & J \end{pmatrix} \begin{pmatrix} A_k^* & C_k^* \\ B_k^* & D_k^* \end{pmatrix} = \begin{pmatrix} H_{k+1}^{-1} & 0 \\ 0 & J \end{pmatrix}$$

for all k. Since Σ satisfies (1.11), Σ^* satisfies (1.14). Now one can proceed to show that Θ^* is J-isometric in the same way that Θ was shown to be J-isometric above, with the minor modification that the direction of time should be reversed.

Now suppose in addition that Σ is J-inner as a system, i.e., $H_k > 0$ for all k. Then (2.19) gives

$$(2.23) \qquad \sum_{k=-\infty}^{j} \rho_J(y_k) \le \sum_{k=-\infty}^{j} \rho_J(u_k).$$

for all j whenever $\vec{y} = \Theta \vec{u}$. Rewriting (2.23) in operater form gives

$$\Theta^* Q_j J \Theta \le Q_j J, \qquad j \in \mathbb{Z},$$

and hence Θ is J-inner.

To establish the reverse implication we first show that a stable J-unitary system is completely reachable, that is (see [12], Section 2), for any time $\ell + 1$ and $x \in \mathbb{C}^{n_0}$ there is a sequence of inputs

$$(2.24) \qquad \vec{u} = (\ldots, 0, 0, u_\kappa, u_{\kappa+1}, \ldots, u_{\ell-1}, u_\ell)$$

so that with $x_\kappa = 0$ the resulting state $x_{\ell+1}$ at time $\ell+1$ is equal to x. To prove this, note that (2.22) implies that

$$H_{k+1}^{-1} - A_k H_k^{-1} A_k^* = B_k J B_k^*, \qquad k \in \mathbf{Z}.$$

It follows that for $k < \ell$

$$H_{\ell+1}^{-1} - A_\ell \cdots A_k H_k^{-1} A_k^* \cdots A_\ell^* = B_\ell J B_\ell^* + \sum_{\nu=k}^{\ell-1} A_\ell \cdots A_{\nu+1} B_\nu J B_\nu^* A_{\nu+1}^* \cdots A_\ell^*.$$

Since (1.11) holds, there exist $0 < \beta < 1$ and an integer $\ell_0 < \ell$ such that

$$\|A_\ell \cdots A_{\nu+1}\| \le \beta^{\ell-\nu}, \qquad \nu \le \ell_0.$$

Recall that the sequences $(\|B_k\|)_{-\infty}^{\infty}$ and $(\|H_k^{-1}\|)_{-\infty}^{\infty}$ are bounded. So we may conclude that

$$(2.25) \qquad H_{\ell+1}^{-1} = B_\ell J B_\ell^* + \sum_{\nu=-\infty}^{\ell-1} A_\ell \cdots A_{\nu+1} B_\nu J B_\nu^* A_{\nu+1}^* \cdots A_\ell^*,$$

where the convergence is in the operator norm for operators on $X_{\ell+1} = \mathbf{C}^{n_0}$. The left hand side of (2.25) is invertible, and hence we may use the fact that the set of invertible operators is open to conclude that for some integer $\kappa < \ell$ the operator

$$B_\ell J B_\ell^* + \sum_{\nu=\kappa}^{\ell-1} A_\ell \cdots A_{\nu+1} B_\nu J B_\nu^* A_{\nu+1}^* \cdots A_\ell^*$$

is also invertible. So, given $x \in \mathbf{C}^{n_0}$, there exist vectors $z_\kappa, \ldots, z_\ell$ in $\mathbf{C}^{n_0}$ such that

$$x = B_\ell J B_\ell^* z_\ell + \sum_{\nu=\kappa}^{\ell-1} A_\ell \cdots A_{\nu+1} B_\nu J B_\nu^* A_{\nu+1}^* \cdots A_\ell^* z_\nu.$$

Now, put $u_j = J B_j^* A_{j+1}^* \cdots A_j^* z_j$ for $j = \kappa, \ldots, \ell-1$ and $u_\ell = J B_\ell^* z_l$. Then

$$x = B_\ell u_\ell + \sum_{\nu=\kappa}^{\ell-1} A_\ell \cdots A_{\nu+1} B_\nu u_\nu,$$

which implies that for this choice of $u_\kappa, \ldots, u_\ell$ the input sequence $(\ldots, 0, 0, u_\kappa, \ldots, u_\ell)$ has the desired property.

Now, assume that Θ is J-inner as an operator on ℓ_2^{m+r}. Fix an integer ℓ and let x be an arbitrary vector in $X_{\ell+1} = \mathbb{C}^{n_0}$. By the result of the previous paragraph, we can find an input sequence $\vec{u}$ as in (2.24) so that with $x_\kappa = 0$ the resulting state $x_{\ell+1}$ at time $\ell + 1$ is equal to x. We consider $\vec{u}$ as an element of ℓ_2^{m+r} by setting $u_k = 0$ for $k > \ell$. Put $\vec{y} = \Theta\vec{u}$. Since Θ is J-inner, we have

$$\sum_{k=-\infty}^{\ell} \rho_J(y_k) \le \sum_{k=-\infty}^{\ell} \rho_J(u_k).$$

But from the identity (2.19) this in turn yields

$$(2.26) \qquad \rho_{H_{\ell+1}}(x) = \rho_{H_{\ell+1}}(x_{\ell+1}) \ge 0.$$

By assumption, $H_{\ell+1}$ is invertible. Since x is an arbitrary element of the state space $\mathbb{C}^{n_0}$, formula (2.26) implies that $H_{\ell+1}$ is positive definite for all ℓ. $\square$

Note that J-unitary maps on ℓ_2^{m+r} arising as the input-output map of a stable J-unitary system are necessarily lower triangular. We can produce upper triangular J-unitary maps on ℓ_2^{m+r} by considering input-output maps of systems evolving in backwards time; we have already seen that Θ^* is such a map whenever Θ is the input-output map of a causal, stable, J-unitary system. In general, let

$$(2.27) \qquad \sigma \begin{cases} x_k = A_k x_{k+1} + B_k u_k \\ y_k = C_k x_{k+1} + D_k u_k \end{cases}$$

be a linear time-varying system evolving in backwards time. We shall say that σ is an (anticausal) *anti-stable J-unitary* system if σ satisfies the sufficient conditions (1.10) and (1.14) to induce an input-output map T_σ which is bounded on ℓ_2^{m+r} and, in addition, there exist a sequence $(H_k)_{k=-\infty}^{\infty}$ of invertible Hermitian linear transformations (or matrices) on the state space $\mathbb{C}^{n_0}$ such that

$$(2.28) \qquad \sup_{k \in \mathbb{Z}}\{\|H_k\|, \|H_k^{-1}\|\} < \infty,$$

$$(2.29) \qquad \begin{pmatrix} A_k^* & C_k^* \\ B_k^* & D_k^* \end{pmatrix} \begin{pmatrix} H_k & 0 \\ 0 & J \end{pmatrix} \begin{pmatrix} A_k & B_k \\ C_k & D_k \end{pmatrix} = \begin{pmatrix} H_{k+1} & 0 \\ 0 & J \end{pmatrix}$$

for all $k \in \mathbf{Z}$. If, in addition, H_k is positive definite for all k, we shall say that σ is anti-J-inner. At the input-output level, we call a J-unitary map Θ on ℓ_2^{m+r} anti-J-inner if

$$\Theta^* J P_k \Theta \le J P_k, \qquad k \in \mathbf{Z},$$

where $P_k = I - Q_{k-1}$. The following result is the analogue of Theorem 2.4 for systems evolving in backwards time; as the proof is also completely analogous, it is omitted.

THEOREM 2.5. *The input-output map $\Theta = T_\sigma$ of an antistable J-unitary system σ is a J-unitary map on ℓ_2^{m+r}. Furthermore, σ is anti-J-inner as a system if and only if Θ is anti-J-inner as an operator.*

Proposition 2.2 shows that there is a simple connection between lower triangular J-inner maps and upper triangular anti-J-inner maps (independent of any realizations as input-output maps); we state the result explicitly in the next proposition.

PROPOSITION 2.6. *A lower triangular J-unitary map Θ is J-inner if and only if Θ^* is an upper triangular anti-J-inner map.*

PROOF. Apply Theorem 2.3. $\qquad \square$

3. TIME-VARYING NEVANLINNA-PICK INTERPOLATION

In this section we consider the time-varying (tangential) Nevanlinna-Pick interpolation problem. We are given $2N$ row diagonal matrices $X_j \in \mathcal{D}^{1 \times m}$ and $Y_j \in \mathcal{D}^{1 \times r}$ $(j = 1, \dots, N)$ and N scalar diagonal matrices $\Delta_j \in \mathcal{D}^{1 \times 1}$ for which the spectral radius of $\Delta_j S^{-1}$ is strictly less than one $(j = 1, \dots, N)$. Put

$$X = \begin{pmatrix} X_1 \\ \vdots \\ X_N \end{pmatrix}, \qquad \Delta = \begin{pmatrix} \Delta_1 & & \\ & \ddots & \\ & & \Delta_N \end{pmatrix},$$

and consider the operator

$$\Xi = \begin{pmatrix} X & \Delta S^{-1} X & (\Delta S^{-1})^2 X & \dots \end{pmatrix} : \oplus_0^\infty \ell_2^m \to \ell_2^N,$$

where $\oplus_0^\infty \ell_2^m$ stands for the Hilbert space of square summable sequences $(x_0, x_1, x_2, \dots)$ with entries in ℓ_2^m. Since $\rho(\Delta S^{-1})$ is strictly less than one, Ξ is a well-defined bounded operator. In what follows we also require that the pair (X, Δ) is *exactly controllable* in the sense that

$$(3.1) \qquad\qquad \Xi\Xi^* > 0$$

as an operator on ℓ_2^N. Note that $\Xi\Xi^*$ may be written as an $N \times N$ operator matrix of which the entries are bounded linear operators on ℓ_2. In fact, in terms of the original data we have

$$\Xi\Xi^* = \left(\{X_i X_j^*(I - S\Delta_j^*)^{-1}\}^\wedge(\Delta_i) \right)_{i,j=1}^N.$$

The *time-varying (tangential) Nevanlinna-Pick interpolation* (TVNPI) *problem is: Find necessary and sufficient conditions for the existence of a lower triangular matrix* $F \in \mathcal{L}^{m \times r}$ *such that*

$$(3.2) \qquad\qquad \|F\| < 1$$

and

$$(3.3) \qquad\qquad (X_j F)^\wedge(\Delta_j) = Y_j, \quad \text{for } j = 1, \ldots, N.$$

When these conditions are satisfied describe all such F.

The norm in (3.2) is the induced operator norm of F as an operator from ℓ_2^r into ℓ_2^m. We shall refer to $\{X_i, Y_i, \Delta_i : i = 1, \ldots, N\}$ as an *admissible TVNPI data set* provided (3.1) is satisfied.

The following theorem settles the existence problem

THEOREM 3.1. *Let* $\{X_i, Y_i, \Delta_i : i = 1, \ldots, N\}$ *be an admissible TVNPI data set. Then the associated TVNPI problem has a solution if and only if the Hermitian matrix*

$$\Lambda(\{X_i, Y_i, \Delta_i\}) = \left(\{(X_i X_j^* - Y_i Y_j^*)(I - S\Delta_j^*)^{-1}\}^\wedge(\Delta_i) \right)_{i,j=1}^N$$

is positive definite on ℓ_2^N.

PROOF OF NECESSITY. Suppose that $F \in \mathcal{L}^{m \times r}$ is a solution of the TVNPI problem. Then by Proposition 1.4 and the interpolation conditions (3.3) we see that

$$(3.4) \qquad\qquad L_F^*(X_j^*(I - S\Delta_j^*)^{-1} D_j) = Y_j^*(I - S\Delta_j^*)^{-1} D_j$$

for $j = 1, \ldots, N$, where L_F is the operator from $\mathcal{L}_2^{r \times 1}$ into $\mathcal{L}_2^{m \times 1}$ given by $L_F(f) = Ff$ and $D_1, \ldots, D_N$ belong to $\mathcal{D}_2^{1 \times 1}$. It is straightforward to check that the induced operator norm $\|L_F\|$ of L_F as an operator from $\mathcal{L}_2^{r \times 1}$ to $\mathcal{L}_2^{m \times 1}$ is the same as the induced operator norm $\|F\|$ of F as an operator from ℓ_2^r to ℓ_2^m. By condition (3.2) it follows that $\|L_F^*\| = \|L_F\| = \|F\| < 1$. Hence, if $D_1, \ldots, D_N$ are any diagonal matrices in $\mathcal{D}_2^{1 \times 1}$, then

$$\| \sum_{j=1}^{N} X_j^*(I - S\Delta_j^*)^{-1} D_j \|_2^2 - \|(L_F)^* \sum_{j=1}^{N} X_j^*(I - S\Delta_j^*)^{-1} D_j \|_2^2 \geq \delta^2 \| \sum_{j=1}^{N} X_j^*(I - S\Delta_j^*)^{-1} D_j \|_2^2$$

for some $\delta > 0$. Writing the squares of norms in terms of inner products, expanding and using (3.4) gives

$$
\begin{aligned}
(3.5) \quad \sum_{i=1}^{N}\sum_{j=1}^{N} & \langle X_j^*(I - S\Delta_j^*)^{-1} D_j, \quad X_i^*(I - S\Delta_i^*)^{-1} D_i \rangle_{\mathcal{L}_2^{m \times 1}} \\
& - \sum_{i=1}^{N}\sum_{j=1}^{N} \langle Y_j^*(I - S\Delta_j^*)^{-1} D_j, \quad Y_i^*(I - S\Delta_i^*)^{-1} D_i \rangle_{\mathcal{L}_2^{r \times 1}} \\
& = \sum_{i=1}^{N}\sum_{j=1}^{N} \langle \{(X_i X_j^* - Y_i Y_j^*)(I - S\Delta_j^*)^{-1}\}^{\wedge}(\Delta_i) D_j, D_i \rangle \\
& \geq \delta^2 \sum_{i=1}^{N}\sum_{j=1}^{N} \langle \{X_i X_j^*(I - S\Delta_j^*)^{-1}\}^{\wedge}(\Delta_i) D_j, D_i \rangle_{\mathcal{L}_2^{1 \times 1}}
\end{aligned}
$$

Since $D_1, \ldots, D_N$ are arbitrary diagonal matrices in $\mathcal{D}_2^{1 \times 1}$, and we are assuming that $\{(X_j, \Delta_j) : j = 1, \ldots, N\}$ satisfies the exact controllability assumption (3.1), it follows from (3.5) that $\Lambda = \Lambda(\{X_i, Y_i, \Delta_i : i = 1\})$ is positive definite as asserted. $\qquad\square$

The proof of sufficiency will be postponed until the next section, where moreover a linear fractional description for the set of all solutions will be constructed.

Let us remark here that condition (3.1) is automatically fulfilled if the generalized Pick matrix $\Lambda(\{X_i, Y_i, \Delta_i : i = 1, \ldots, N\})$ is positive definite. This follows from the equality

$$\Xi\Xi^* - \Lambda(\{X_i, Y_i, \Delta_i : i = 1, \ldots, N\}) = \left(\sum_{\nu=0}^{\infty} (\Delta_i S^{-1})^{\nu} Y_i Y_j^* (S\Delta_j^*)^{\nu} \right)_{i,j=1}^{N}.$$

and the fact that the latter operator is positive semi-definite.

In the rest of this section we set down some general principles concerning the connections between linear fractional maps and solutions of the TVNPI problem.

Following the approach of [7], we first show how the interpolation conditions (3.3) can be reduced to a homogeneous interpolation problem for a matrix $\Theta \in \mathcal{L}^{(m+r)\times(m+r)}$ which can be used to parametrize the set of all solutions. Indeed, if F satisfies (3.3), then $\begin{pmatrix} F \\ I \end{pmatrix} \in \mathcal{L}^{(m+r)\times(m+r)}$ satisfies the following set of homogeneous interpolation conditions:

$$(3.6) \qquad \left\{ \begin{pmatrix} X_j & -Y_j \end{pmatrix} \begin{pmatrix} F \\ I \end{pmatrix} \right\}^{\wedge}(\Delta_j) = 0, \qquad j = 1,\dots,N.$$

Moreover, by the property for the time-varying calculus given by Proposition 1.2, (3.6) implies that

$$(3.7) \qquad \left\{ \begin{pmatrix} X_j & -Y_j \end{pmatrix} \begin{pmatrix} F \\ I \end{pmatrix} H \right\}^{\wedge}(\Delta_j) = 0, \qquad j = 1,\dots,N.$$

for all $H \in \mathcal{L}^{(m+r)\times 1}$. If we construct a lower triangular matrix Θ in $\mathcal{L}^{(m+r)\times(m+r)}$ such that

$$(3.8) \quad \Theta \mathcal{L}^{(m+r)\times 1} = \{ H \in \mathcal{L}^{(m+r)\times 1} : \left\{ \begin{pmatrix} X_j & -Y_j \end{pmatrix} H \right\}^{\wedge}(\Delta_j) = 0 \quad \text{for } j = 1, 2, \dots, N \},$$

then we see that

$$(3.9) \qquad\qquad \begin{pmatrix} F \\ I \end{pmatrix} = \Theta \begin{pmatrix} G_1 \\ G_2 \end{pmatrix}$$

for some $\begin{pmatrix} G_1 \\ G_2 \end{pmatrix} \in \mathcal{L}^{(m+r)\times r}$. This is the first step to obtaining a parametrization of the set of all solutions $F \in \mathcal{L}^{m \times r}$ of (3.3). More precisely we have the following result.

THEOREM 3.2. *Let* $\{X_j, Y_j, \Delta_j : j = 1, \dots, N\}$ *be an admissible TVNPI data set, and suppose that* $\Theta \in \mathcal{L}^{(m+r)\times(m+r)}$ *satisfies* (3.8). *Then an operator* F *in* $\mathcal{L}^{m \times r}$ *satisfies the interpolation conditions* (3.3) *if and only if* F *has a representation of the form*

$$F = (\Theta_{11}G_1 + \Theta_{12}G_2)(\Theta_{21}G_1 + \Theta_{22}G_2)^{-1}$$

for some pair of lower triangular matrices $G_1 \in \mathcal{L}^{m \times r}$ *and* $G_2 \in \mathcal{L}^{r \times r}$ *such that* $\Theta_{21}G_1 + \Theta_{22}G_2 \in \mathcal{L}^{r \times r}$ *is invertible with inverse again in* $\mathcal{L}^{r \times r}$.

PROOF. Suppose that $F \in \mathcal{L}^{m \times r}$ satisfies (3.3). Write $\begin{pmatrix} F \\ I \end{pmatrix}$ in the form $(f_1, f_2, \ldots, f_r)$ where each $f_i \in \mathcal{L}^{(m+r) \times 1}$. Then (3.3) implies $\{(X_j \quad -Y_j) f_i\}^{\wedge}(\Delta_j) = 0$ for $j = 1, \ldots, N$ and for each $i = 1, \ldots, r$. Hence by property (3.8) of Θ, each $f_i \in \Theta \mathcal{L}^{(m+r) \times 1}$ for each $i = 1, \ldots, r$. We conclude that $\begin{pmatrix} F \\ I \end{pmatrix}$ itself has a factorization (3.9) with $\begin{pmatrix} G_1 \\ G_2 \end{pmatrix} \in \mathcal{L}^{(m+r) \times r}$. In particular, $G_1 \in \mathcal{L}^{m \times r}$, $G_2 \in \mathcal{L}^{r \times r}$ and $\Theta_{21} G_1 + \Theta_{22} G_2 = I$ has a lower triangular inverse. We conclude that F has a representation as asserted.

Conversely, suppose that $G_1 \in \mathcal{L}^{m \times r}$, $G_2 \in \mathcal{L}^{r \times r}$ and $\Theta_{21} G_1 + \Theta_{22} G_2$ is invertible with inverse in $\mathcal{L}^{m \times r}$, and

$$F = (\Theta_{11} G_1 + \Theta_{12} G_2)(\Theta_{21} G_1 + \Theta_{22} G_2)^{-1}.$$

Then F is the product of lower triangular matrices, so itself is in $\mathcal{L}^{m \times r}$. Moreover

$$\begin{pmatrix} F \\ I \end{pmatrix} = \Theta \begin{pmatrix} G_1 \\ G_2 \end{pmatrix} (\Theta_{21} G_1 + \Theta_{22} G_2)^{-1}.$$

Since $\begin{pmatrix} G_1 \\ G_2 \end{pmatrix} (\Theta_{21} G_1 + \Theta_{22} G_2)^{-1} \in \mathcal{L}^{(m+r) \times r}$, each column is in $\mathcal{L}^{(m+r) \times 1}$. By the defining characteristic (3.8) of Θ, we see that each column f_ℓ of $\begin{pmatrix} F \\ I \end{pmatrix}$ satisfies

$$\{(X_j \quad -Y_j) f_\ell\}^{\wedge}(\Delta_j) = 0, \qquad j = 1, \ldots, N.$$

From this we see that $\begin{pmatrix} F \\ I \end{pmatrix}$ satisfies the homogeneous form (3.6) of the interpolation conditions, and hence F satisfies (3.3). $\qquad \square$

The next step is to adapt Theorem 3.2 to handle the norm constraint (3.2).

THEOREM 3.3. *Let* $\{X_j, Y_j, \Delta_j : j = 1, \ldots, N\}$ *be an admissible TVNPI data set. Suppose that the lower triangular matrix*

$$\Theta = \begin{pmatrix} \Theta_{11} & \Theta_{12} \\ \Theta_{21} & \Theta_{22} \end{pmatrix} \in \mathcal{L}^{(m+r) \times (m+r)},$$

in addition to (3.8), *is J-inner. Then there exist solutions* $F \in \mathcal{L}^{m \times r}$ *of the TVNPI problem. Moreover any solution F is given by*

$$F = (\Theta_{11} G + \Theta_{12})(\Theta_{21} G + \Theta_{22})^{-1}$$

where G is any lower triangular matrix in $\mathcal{L}^{m \times r}$ with $\|G\| < 1$.

PROOF. Suppose first that $F \in \mathcal{L}^{m \times r}$ is a solution of the TVNPI problem. In particular F satisfies (3.3). Since Θ satisfies (3.8), it follows that $\begin{pmatrix} F \\ I \end{pmatrix}$ has a factorization (3.9) with $\begin{pmatrix} G_1 \\ G_2 \end{pmatrix} \in \mathcal{L}^{m \times r}$.

We argue next that G_2 is invertible with G_2^{-1} also lower triangular. Indeed, since F satisfies (3.2) and Θ is a J-isometry we have

$$0 > F^*F - I = (F^* \quad I) J \begin{pmatrix} F \\ I \end{pmatrix}$$
$$= (G_1^* \quad G_2^*) \Theta^* J \Theta \begin{pmatrix} G_1 \\ G_2 \end{pmatrix}$$
$$= (G_1^* \quad G_2^*) J \begin{pmatrix} G_1 \\ G_2 \end{pmatrix}$$
$$= G_1^* G_1 - G_2^* G_2.$$

So we have proved that

$$(3.10) \qquad\qquad G_2^* G_2 - G_1^* G_1 > 0.$$

If $x \in \ell_2^r$ and $G_2 x = 0$, then (3.10) forces $G_1 x = 0$. But then we can use (3.9) to show that

$$x = (0 \quad I) \begin{pmatrix} F \\ I \end{pmatrix} x = (0 \quad I) \Theta \begin{pmatrix} G_1 \\ G_2 \end{pmatrix} x = 0,$$

and we conclude that G_2 has a trivial kernel. What's more, since Θ is J-unitary, Θ has a bounded inverse Θ^{-1} on ℓ_2^{m+r}. Hence

$$(G_1^* \quad G_2^*) \begin{pmatrix} G_1 \\ G_2 \end{pmatrix} = (F^* \quad I) \Theta^{-*} \Theta^{-1} \begin{pmatrix} F \\ I \end{pmatrix}$$
$$\geq m(F^*F + I) \geq mI$$

for some $m > 0$. But from (3.10) we have

$$2G_2^* G_2 > (G_1^* \quad G_2^*) \begin{pmatrix} G_1 \\ G_2 \end{pmatrix}.$$

We conclude that $G_2^* G_2 > \frac{1}{2} mI$, and so G_2 has closed range. To show that G_2 is invertible with inverse lower triangular, it remains only to show that $G_2 \ell_2^r([k, \infty))$ is a dense subset of $\ell_2^r([k,\infty))$ for every integer k. Therefore, suppose $x \in \ell_2^r([k,\infty))$ is orthogonal to $G_2\ell_2^r([k,\infty))$ for some k. Set

$$\begin{pmatrix} y_1 \\ y_2 \end{pmatrix} = \Theta \begin{pmatrix} 0 \\ x \end{pmatrix} \in \ell_2^{m+r}([k,\infty)).$$

Then for all $w \in \ell_2^r([k,\infty))$,

$$\begin{aligned} \langle J \begin{pmatrix} F \\ I \end{pmatrix} w, \begin{pmatrix} y_1 \\ y_2 \end{pmatrix} \rangle &= \langle J\Theta \begin{pmatrix} G_1 \\ G_2 \end{pmatrix} w, \Theta \begin{pmatrix} 0 \\ x \end{pmatrix} \rangle \\ &= \langle J \begin{pmatrix} G_1 \\ G_2 \end{pmatrix} w, \begin{pmatrix} 0 \\ x \end{pmatrix} \rangle \\ &= -\langle G_2 w, x \rangle = 0. \end{aligned}$$

In particular, this holds with $w = y_2$. Hence

$$\begin{aligned} \|y_1 - Fy_2\|^2 &= \langle J\{ \begin{pmatrix} y_1 \\ y_2 \end{pmatrix} - \begin{pmatrix} F \\ I \end{pmatrix} y_2 \}, \begin{pmatrix} y_1 \\ y_2 \end{pmatrix} - \begin{pmatrix} F \\ I \end{pmatrix} y_2 \rangle \\ &= \langle J \begin{pmatrix} y_1 \\ y_2 \end{pmatrix}, \begin{pmatrix} y_1 \\ y_2 \end{pmatrix} \rangle + \langle J \begin{pmatrix} F \\ I \end{pmatrix} y_2, \begin{pmatrix} F \\ I \end{pmatrix} y_2 \rangle \\ &\leq \langle J\Theta \begin{pmatrix} 0 \\ x \end{pmatrix}, \Theta \begin{pmatrix} 0 \\ x \end{pmatrix} \rangle \\ &= \langle J \begin{pmatrix} 0 \\ x \end{pmatrix}, \begin{pmatrix} 0 \\ x \end{pmatrix} \rangle = -\|x\|^2 \leq 0, \end{aligned}$$

where the inequality stems from the norm constraint (3.2). We conclude that $y_1 = Fy_2$. Thus

$$\Theta \begin{pmatrix} 0 \\ x \end{pmatrix} = \begin{pmatrix} y_1 \\ y_2 \end{pmatrix} = \begin{pmatrix} F \\ I \end{pmatrix} y_2 = \Theta \begin{pmatrix} G_1 \\ G_2 \end{pmatrix} y_2,$$

from which we get $x = G_2 y_2$. As x was chosen to be orthogonal to $G_2\ell_2^r([k,\infty))$, we conclude that $x = 0$ as needed. We now have shown that $G_2^{-1} \in \mathcal{L}^{r \times r}$.

Since $G_2^{-1} \in \mathcal{L}^{r \times r}$, we see that $G := G_1 G_2^{-1} \in \mathcal{L}^{m \times r}$. Moreover (3.10) implies that $\|G\| < 1$. Finally from the identity (3.9) we read off

$$F = \Theta_{11} G_1 + \Theta_{12} G_2 = (\Theta_{11} G + \Theta_{12}) G_2$$

and

$$I = \Theta_{21} G_1 + \Theta_{22} G_2 = (\Theta_{21} G + \Theta_{22}) G_2.$$

Hence F has a representation $F = (\Theta_{11} G + \Theta_{12})(\Theta_{21} G + \Theta_{22})^{-1}$ of the required form.

Conversely, suppose that Θ is a J-inner matrix satisfying (3.8), and let G be any lower triangular matrix in $\mathcal{L}^{m \times r}$ with $\|G\| < 1$. Define $\begin{pmatrix} F_1 \\ F_2 \end{pmatrix} \in \mathcal{L}^{(m+r) \times n}$ by

$$(3.11) \qquad \begin{pmatrix} F_1 \\ F_2 \end{pmatrix} = \Theta \begin{pmatrix} G \\ I \end{pmatrix}.$$

From Theorem 2.3 we know that $\Theta_{22}^{-1} \in \mathcal{L}^{r \times r}$, and according to Theorem 2.1 we have $\|\Theta_{22}^{-1}\Theta_{21}\| < 1$. Since $\Theta_{22}^{-1}\Theta_{21}G \in \mathcal{L}^{m \times r}$ with $\|\Theta_{22}^{-1}\Theta_{21}G\| < 1$, by the Neumann series expansion we see that $\Theta_{22}^{-1}\Theta_{21}G + I$ has a lower triangular inverse. Hence

$$F_2 = \Theta_{21}G + \Theta_{22} = \Theta_{22}(\Theta_{22}^{-1}\Theta_{21}G + I)$$

has a lower triangular inverse $F_2^{-1} \in \mathcal{L}^{r \times r}$. Hence $F := F_1 F_2^{-1} \in \mathcal{L}^{m \times r}$ and (3.11) give

$$\begin{pmatrix} F \\ I \end{pmatrix} = \Theta \begin{pmatrix} G \\ I \end{pmatrix} F_2^{-1}.$$

Since Θ satisfies (3.8), we see from Theorem 3.2 that F satisfies the interpolation conditions (3.3). Moreover, since Θ is a J-isometry,

$$\begin{aligned}
F^*F - I = (F^* \quad I)J\begin{pmatrix} F \\ I \end{pmatrix} &= F_2^{-*}(G^* \quad I)\Theta^*J\Theta\begin{pmatrix} G \\ I \end{pmatrix}F_2^{-1} \\
&= F_2^{-*}(G^* \quad I)J\begin{pmatrix} G \\ I \end{pmatrix}F_2^{-1} = F_2^{-*}(G^* \quad I)J\begin{pmatrix} G \\ I \end{pmatrix}F_2^{-1} \\
&= F_2^{-*}(G^*G - I)F_2^{-1} < 0,
\end{aligned}$$

and so $\|F\| < 1$. $\qquad \square$

4. SOLUTION OF THE TIME-VARYING TANGENTIAL NEVANLINNA-PICK INTERPOLATION PROBLEM

In this section we deal with the construction of a J-unitary Θ meeting the hypotheses of Theorem 3.3 and go on to complete the proof of Theorem 3.1, including a parametrization of the set of all solutions.

Throughout this section $\{X_j, Y_j, \Delta_j : j = 1, \ldots, N\}$ is an admissible TVNPI data set as defined in Section 3. Introduce the following operators

$$(4.1) \qquad \mathcal{A} = \text{diag}\,(A_k)_{k=-\infty}^{\infty} \in \mathcal{D}^{N \times N}, \quad A_k = \begin{pmatrix} \Delta_1^{(k)} & & \\ & \ddots & \\ & & \Delta_N^{(k)} \end{pmatrix},$$

$$
(4.2) \qquad \mathcal{B} = \mathrm{diag}\,(B_k)_{k=-\infty}^{\infty} \in \mathcal{D}^{N\times(m+r)}, \quad B_k = \begin{pmatrix} X_1^{(k)} & -Y_1^{(k)} \\ \vdots & \vdots \\ X_N^{(k)} & -Y_N^{(k)} \end{pmatrix}.
$$

Here $X_j^{(k)}, Y_j^{(k)}$ and $\Delta_j^{(k)}$ denote the k-th diagonal entries of X_j, Y_j and Δ_j, respectively.

The next theorem provides a useful sufficient condition for the existence of a $\Theta \in \mathcal{L}^{(m+r)\times(m+r)}$ satisfying at least (3.8).

THEOREM 4.1. *Let* $\{X_j, Y_j, \Delta_j : j = 1,\dots,N\}$ *be an admissible TVNPI data set, and suppose* $\Theta \in \mathcal{L}^{(m+r)\times(m+r)}$ *has inverse* $\Theta^{-1} \in \mathcal{U}^{(m+r)\times(m+r)}$ *which admits a representation of the form*

$$
(4.3) \qquad \Theta^{-1} = \mathcal{D} + \mathcal{C}(S - \mathcal{A})^{-1}\mathcal{B} = \mathcal{D} + \sum_{\nu=0}^{\infty} \mathcal{C}S^{-1}(\mathcal{A}S^{-1})^{\nu}\mathcal{B},
$$

where $\mathcal{A}$ *and* $\mathcal{B}$ *are as in (4.1) and (4.2) respectively,* $\mathcal{C} \in \mathcal{D}^{(m+r)\times N}$ *is such that*

$$
(4.4) \qquad \cap_{j=0}^{\infty} \mathrm{Ker}\, \mathcal{C}S^{-1}(\mathcal{A}S^{-1})^{j} = (0),
$$

and $\mathcal{D}$ *is any element of* $\mathcal{D}^{(m+r)\times(m+r)}$. *Then* Θ *satisfies (3.8).*

PROOF. Let $H \in \mathcal{L}^{(m+r)\times 1}$ be given. Then $H \in \Theta\mathcal{L}^{(m+r)\times 1}$ if and only if $\Theta^{-1}H \in \mathcal{L}^{(m+r)\times 1}$. Since H is lower triangular, H has a series expansion

$$
H = \sum_{j=0}^{\infty} S^j H_{[j]}
$$

where $H_{[j]} \in \mathcal{D}^{(m+r)\times 1}$. From the expansion (4.3) for Θ^{-1}, we see that the j-th diagonal of $\Theta^{-1}H$ above the main diagonal is given by

$$
S^{-j}(\Theta^{-1}H)_{[-j]} = \sum_{\ell=0}^{\infty} \mathcal{C}S^{-1}(\mathcal{A}S^{-1})^{\ell+j-1}\mathcal{B}S^{\ell}H_{[\ell]}
$$
$$
= \mathcal{C}S^{-1}(\mathcal{A}S^{-1})^{j-1}\left(\sum_{\ell=0}^{\infty}(\mathcal{A}S^{-1})^{\ell}\mathcal{B}S^{\ell}H_{[\ell]}\right).
$$

Thus $\Theta^{-1}H \in \mathcal{L}^{(m+r)\times 1}$ if and only if

$$
(4.5) \qquad \mathcal{C}S^{-1}(\mathcal{A}S^{-1})^{j-1} \cdot \sum_{\ell=0}^{\infty}(\mathcal{A}S^{-1})^{\ell}\mathcal{B}S^{\ell}H_{[\ell]} = 0, \quad j = 1, 2, \dots
$$

From (4.4) we see that (4.5) in turn is equivalent to

$$(4.6) \qquad \sum_{\ell=0}^{\infty} (\mathcal{A}S^{-1})^{\ell} \mathcal{B} S^{\ell} H_{[\ell]} = 0.$$

Recalling now the definitions of $\mathcal{A}$ and $\mathcal{B}$ in (4.1) and (4.2), we see that (4.6) is the same as

$$\sum_{\ell=0}^{\infty} (\Delta_j S^{-1})^{\ell} (\, X_j \quad -Y_j \,) S^{\ell} H_{[\ell]} = 0, \qquad j = 1, \ldots, N.$$

In other words

$$\{(\, X_j \quad -Y_j \,) H\}^{\wedge}(\Delta_j) = 0, \qquad j = 1, \ldots, N,$$

as required. $\qquad \square$

Next we obtain a sufficient condition for the existence of a J-inner Θ satisfying condition (3.8); happily this sufficient condition coincides with the necessary condition for existence of solutions of the TVNPI problem already established (necessity in Theorem 3.1). In what follows we write $\mathcal{H}$ for the operator $\Lambda(\{X_i, Y_i, \Delta_i\})$ introduced in Theorem 3.1. Thus

$$(4.7) \qquad \mathcal{H} = \left(\{X_i X_j^* - Y_i Y_j^*)(I - S\Delta_j^*)^{-1}\}^{\wedge}(\Delta_i) \right)_{i,j=1}^{N}$$

Note that $\mathcal{H}$ act on ℓ_2^N and is also given by

$$(4.8) \qquad \mathcal{H} = \sum_{k=0}^{\infty} (\mathcal{A}S^{-1})^k \mathcal{B} J \mathcal{B}^* (S\mathcal{A}^*)^k,$$

where $\mathcal{A}$ and $\mathcal{B}$ are as in (4.1) and (4.2), respectively. Indeed, in terms of $\mathcal{A}$ and $\mathcal{B}$ the right hand side of (4.7) can be written as $\{\mathcal{B} J \mathcal{B}^* (I - S\mathcal{A}^*)^{-1}\}^{\wedge}(\mathcal{A})$. Now, we use that

$$\{\mathcal{B} J \mathcal{B}^* (I - S\mathcal{A}^*)^{-1}\}_{[k]} = S^{-k}\{\mathcal{B} J \mathcal{B}^* (S\mathcal{A}^*)^k\},$$

and apply the definition in (1.18) of the point evaluation map to get (4.8). From (4.8) we see that $\mathcal{H} \in \mathcal{D}^{N \times N}$.

THEOREM 4.2. *Let* $\{X_j, Y_j, \Delta_j : j = 1, \ldots, N\}$ *be an admissible TVNPI data set. Define block diagonal operators* $\mathcal{A} \in \mathcal{D}^{N \times N}$ *and* $\mathcal{B} \in \mathcal{D}^{N \times (m+r)}$ *as in (4.1) and*

(4.2), and let $\mathcal{H} \in \mathcal{D}^{N \times N}$ be the Hermitian operator defined by (4.7). Assume that $\mathcal{H}$ is invertible on ℓ_2^N and that the diagonal entries H_k of $\mathcal{H}$ have signature independent of k. Then there exists a J-unitary lower triangular matrix $\Theta \in \mathcal{L}^{(m+r) \times (m+r)}$ which satisfies (3.8). Moreover Θ is J-inner if and only if, in addition, the operator $\mathcal{H}$ given by (4.7) is positive definite.

The following corollary presents a recipe for the construction of a realization for the matrix Θ in Theorem 4.1.

COROLLARY 4.3. Let $\{X_j, Y_j, \Delta_j : j = 1, \ldots, N\}, \mathcal{A}, \mathcal{B}$ and $\mathcal{H}$ be as in Theorem 4.2, and assume that $\mathcal{H}$ is invertible with diagonal entries H_k having constant signature. Then $\mathcal{H} = \mathrm{diag}\,(H_k)_{k=-\infty}^{\infty}$ satisfies the following time-varying Stein equation:

$$(4.9) \qquad H_k - A_k H_{k+1} A_k^* = B_k J B_k^*, \qquad k \in \mathbf{Z},$$

where A_k and B_k are as in (4.1) and (4.2), respectively. Furthermore, one can find matrices $(C_k \quad D_k)$ such that $(C_k \quad D_k)$ are bounded in norm uniformly with respect to k and

$$(4.10) \qquad \begin{pmatrix} H_{k+1}^{-1} - A_k^* H_k^{-1} A_k & -A_k^* H_k^{-1} B_k \\ -B_k^* H_k^{-1} A_k & J - B_k^* H_k^{-1} B_k \end{pmatrix} = \begin{pmatrix} C_k^* \\ D_k^* \end{pmatrix} J (C_k^* \quad D_k^*).$$

Now, define matrices $\alpha_k, \beta_k, \gamma_k, \delta_k$ by

$$(4.11) \qquad \begin{pmatrix} \alpha_k & \beta_k \\ \gamma_k & \delta_k \end{pmatrix} = \begin{pmatrix} A_k^* & C_k^* J \\ J B_k^* & J D_k^* J \end{pmatrix},$$

and let Σ be the time-varying system given by

$$(4.12) \qquad \Sigma \begin{cases} x_{k+1} & = \alpha_k x_k + \beta_k u_k \\ y_k & = \gamma_k x_k + \delta_k u_k. \end{cases}$$

Then the input-output map

$$(4.13) \qquad \Theta = T_\Sigma = J \mathcal{D}^* J + J \mathcal{B}^* (I - S \mathcal{A}^*)^{-1} S \mathcal{C}^* J,$$

where $\mathcal{C} = \mathrm{diag}\,(C_k)_{k=-\infty}^{\infty}$, $\mathcal{D} = \mathrm{diag}\,(D_k)_{k=-\infty}^{\infty}$, and Θ is a lower triangular J-unitary operator satisfying (3.8).

PROOF OF THEOREM 4.2 AND COROLLARY 4.3. We seek a lower triangular J-unitary (or even J-inner) Θ satisfying (3.8). Condition (3.8) is a condition on an anticausal realization for $\Psi = \Theta^{-1} = J\Theta^* J$. Certainly Ψ is necessarily upper triangular J-unitary and, by Proposition 2.6, is anti-J-inner if and only if Θ is J-inner. Theorem 4.1 suggests that we seek an anticausal realization for $\Psi = J\Theta^* J$ of the form

$$(4.14) \qquad \sigma \begin{cases} x_k = A_k x_{k+1} + B_k u_k \\ y_k = C_k x_{k+1} + D_k u_k. \end{cases}$$

where (A_k, B_k) are given by the data as in (4.1) and (4.2), and C_k, D_k are to be determined. Note that A_k acts on $\mathbf{C}^N$, and hence the state spaces in (4.14) do not depend on time. In order for the input-output map of σ to be J-unitary, by Theorem 2.5 we are led to seek a sequence of invertible Hermitian matrices $(H_k^{-1})_{k=-\infty}^{\infty}$ so that

$$(4.15) \qquad \begin{pmatrix} A_k^* & C_k^* \\ B_k^* & D_k^* \end{pmatrix} \begin{pmatrix} H_k^{-1} & 0 \\ 0 & J \end{pmatrix} \begin{pmatrix} A_k & B_k \\ C_k & D_k \end{pmatrix} = \begin{pmatrix} H_{k+1}^{-1} & 0 \\ 0 & J \end{pmatrix}.$$

Since A_k, B_k are given and C_k, D_k are to be found, it is more convenient to work with the equivalent formulation

$$(4.16) \qquad \begin{pmatrix} A_k & B_k \\ C_k & D_k \end{pmatrix} \begin{pmatrix} H_{k+1} & 0 \\ 0 & J \end{pmatrix} \begin{pmatrix} A_k^* & C_k^* \\ B_k^* & D_k^* \end{pmatrix} = \begin{pmatrix} H_k & 0 \\ 0 & J \end{pmatrix}.$$

Equality of the (1,1)-blocks in (4.14) leads to the time-varying Stein equation (4.9).

Now, let $\mathcal{H}$ be as in (4.7). Since $\mathcal{H}$ also admits the representation (4.8), we have

$$(4.17) \qquad \mathcal{H} = (\mathcal{A}S^{-1})\mathcal{H}(S\mathcal{A}^*) + \mathcal{B}J\mathcal{B}^*.$$

By comparing the diagonal entries of the left and right hand side in (4.17) we see that the diagonal entries of $\mathcal{H}$ satisfy (4.9).

The next step is to find matrices $(C_k \quad D_k)$ satisfying (4.10) and such that $(C_k \quad D_k)$ are bounded in norm uniformly with respect to k. This problem may be viewed as the time-variant analogue of the embedding problem solved in [11]. To find the matrices $(C_k \quad D_k)$ set

$$\Pi_k = \begin{pmatrix} A_k^* \\ B_k^* \end{pmatrix} H_k^{-1} (A_k \quad B_k) \begin{pmatrix} H_{k+1} & 0 \\ 0 & J \end{pmatrix}.$$

Note that (4.9) can be rewritten as

$$(4.18) \qquad H_k = (A_k \quad B_k) \begin{pmatrix} H_{k+1} & 0 \\ 0 & J \end{pmatrix} \begin{pmatrix} A_k^* \\ B_k^* \end{pmatrix},$$

and hence we can use this identity to show that Π_k is a projection operator acting on the space $\mathbb{C}^{N+m+r}$. Since H_k is invertible, formula (4.18) also implies that

$$\mathrm{Im}\ \begin{pmatrix} A_k & B_k \end{pmatrix} = \mathbb{C}^N, \quad \mathrm{Ker}\ \begin{pmatrix} A_k^* \\ B_k^* \end{pmatrix} = (0),$$

and hence rank $\Pi_k = N$. Therefore the Hermitian matrix

$$(4.19) \qquad \Lambda_k := \begin{pmatrix} H_{k+1}^{-1} - A_k^* H_k^{-1} A_k & -A_k^* H_k^{-1} B_k \\ -B_k^* H_k^{-1} A_k & J - B_k^* H_k^{-1} B_k \end{pmatrix} = (I - \Pi_k) \begin{pmatrix} H_{k+1}^{-1} & 0 \\ 0 & J \end{pmatrix}$$

has rank $m + r$. Via the functional calculus for Hermitian matrices, one sees that Λ_k has a factorization

$$(4.20) \qquad \Lambda_k = \begin{pmatrix} C_k^* \\ D_k^* \end{pmatrix} j \begin{pmatrix} C_k & D_k \end{pmatrix}$$

for some $(m+r) \times (m+r)$ signature matrix j. In fact, one may choose $\begin{pmatrix} C_k & D_k \end{pmatrix}$ to have the form

$$(4.21) \qquad \begin{pmatrix} C_k & D_k \end{pmatrix} = U_k f(\Lambda_k),$$

where $f(t) = |t|^{\frac{1}{2}}$ and U_k is a partial isometry. Note that our assumptions on A_k, B_k and H_k guarantee that Λ_k in (4.19) is bounded in norm uniformly with respect to k. But then we see from (4.21) that the matrices $\begin{pmatrix} C_k & D_k \end{pmatrix}$ in (4.20) also may be chosen to be uniformly bounded relative to k.

Next, let us prove that in (4.20) we may take $j = J$. Note that (4.18) implies that

$$\begin{pmatrix} A_k & B_k \end{pmatrix} \begin{pmatrix} H_{k+1} & 0 \\ 0 & J \end{pmatrix} (I - \Pi_k) = 0,$$

and thus also

$$(4.22) \qquad \begin{pmatrix} A_k & B_k \end{pmatrix} \begin{pmatrix} H_{k+1} & 0 \\ 0 & J \end{pmatrix} \Lambda_k = 0.$$

From (4.20) we know that $\mathrm{Im}\ \Lambda_k = \mathrm{Im}\ \begin{pmatrix} C_k^* \\ D_k^* \end{pmatrix}$, and hence the equality (4.22) yields

$$(4.23) \qquad \begin{pmatrix} A_k & B_k \\ C_k & D_k \end{pmatrix} \begin{pmatrix} H_{k+1} & 0 \\ 0 & J \end{pmatrix} \begin{pmatrix} A_k^* & C_k^* \\ B_k^* & D_k^* \end{pmatrix} = \begin{pmatrix} H_k & 0 \\ 0 & C_k H_{k+1} C_k^* + D_k J D_k^* \end{pmatrix}.$$

On the other hand from $(I - \Pi_k)^2 = I - \Pi_k$ we get

$$\begin{pmatrix} C_k^* \\ D_k^* \end{pmatrix} j \, (\, C_k \quad D_k \,) \begin{pmatrix} H_{k+1} & 0 \\ 0 & J \end{pmatrix} \begin{pmatrix} C_k^* \\ D_k^* \end{pmatrix} j \, (\, C_k \quad D_k \,) \begin{pmatrix} H_{k+1} & 0 \\ 0 & J \end{pmatrix}$$

$$= \begin{pmatrix} C_k^* \\ D_k^* \end{pmatrix} j \, (\, C_k \quad D_k \,) \begin{pmatrix} H_{k+1} & 0 \\ 0 & J \end{pmatrix} .$$

Since $\begin{pmatrix} C_k^* \\ D_k^* \end{pmatrix}$ is injective and $(\, C_k \quad D_k \,) \begin{pmatrix} H_{k+1} & 0 \\ 0 & J \end{pmatrix}$ is surjective, this gives

$$(4.24) \qquad\qquad C_k H_{k+1} C_k^* + D_k J D_k^* = j^{-1} = j.$$

In particular, the right hand side of (4.23) is invertible. The latter implies that the first
and third term in the left hand side of (4.24) are invertible, and thus the matrices

$$\begin{pmatrix} H_{k+1} & 0 \\ 0 & J \end{pmatrix} , \begin{pmatrix} H_k & 0 \\ 0 & j \end{pmatrix}$$

have the same signature. According to our hypotheses, the signatures of H_{k+1} and H_k are
equal. It follows that J and j have the same signatures. Therefore in (4.20) we may take
$j = J$.

From (4.23) and (4.24) we get the identity (4.16), and hence the system σ
in (4.14) is (anticausal) anti-stable J-unitary. Furthermore, for the input-output map we
have

$$T_\sigma = \mathcal{D} + \mathcal{C}(S - \mathcal{A})^{-1}\mathcal{B}.$$

Now put $\Theta = J T_\sigma^* J$. Then Θ is the J-unitary input-output map of the system Σ in (4.12)
and Θ admits the representation (4.13). Since $\Theta^{-1} = J\Theta J = T_\sigma$, also (4.3) holds.

Next, let us check that condition (4.4) is fulfilled. The identity (4.15), which
is equivalent to (4.16), yields

$$\mathcal{H}^{-1} - (S\mathcal{A}^*)\mathcal{H}^{-1}(\mathcal{A}S^{-1}) = S\mathcal{C}^* J \mathcal{C} S^{-1},$$

and hence

$$\mathcal{H}^{-1} = \sum_{j=0}^{\infty} (S\mathcal{A}^*)^j S\mathcal{C}^* J \mathcal{C} S^{-1} (\mathcal{A}S^{-1})^j.$$

So, if $\vec{x}$ is a vector in the space defined by the left hand side of (4.4), then $\mathcal{H}^{-1}\vec{x}$ must be
zero, and therefore $\vec{x} = 0$. So condition (4.4) is fulfilled.

Thus, by Theorem 4.1, Θ satisfies (3.8). Note that, by definition, the system σ is anti-J-inner exactly when $\mathcal{H} > 0$. Hence, by Theorem 2.5, the same holds true for the associated input-output map T_σ. It follows (cf. Proposition 2.6) that Θ is J-inner if and only if $\mathcal{H} > 0$. So Θ meets all the requirements in Theorem 4.2. $\square$

Putting together the pieces we have the following more detailed form of Theorem 3.1.

THEOREM 4.4. *Let* $\{X_j, Y_j, \Delta_j : j = 1,\dots,N\}$ *be an admissible TVNPI data set, and let* $\mathcal{H}$ *be the block diagonal matrix in* $\mathcal{D}^{N \times N}$ *given by* (4.7). *Then solutions of the TVNPI problem exist if and only if* $\mathcal{H}$ *is positive definite. In this case any solution* F *of the TVNPI problem* (3.2) *and* (3.3) *is given by*

$$F = (\Theta_{11} G + \Theta_{12})(\Theta_{21} G + \Theta_{22})^{-1},$$

where G *is any lower triangular matrix in* $\mathcal{L}^{m \times r}$ *with* $\|G\| < 1$ *and where* $\Theta = T_\Sigma$ *is constructed as in Corollary 4.3.*

PROOF. Necessity of the condition $\mathcal{H} > 0$ has already been noted. Conversely, suppose $\mathcal{H} > 0$. Then we may construct Θ as in Corollary 4.3 satisfying all the conditions of Theorem 4.2. Now Theorem 3.3 gives that solutions F of the TVNPI problem exist and that the set of all such solutions is given by the linear fractional formula as described above. $\square$

5. AN ILLUSTRATIVE EXAMPLE

As an example which one may compute by hand, we consider the special case of one ($N = 1$) interpolation condition

$$(5.1) \qquad\qquad\qquad (X_1 F)^\wedge(\Delta_1) = Y_1,$$

where the unknown $F \in \mathcal{L}^{1 \times 1}$ as a scalar entries, and $X_1 = \text{diag}\,(X_1^{(k)})_{k=-\infty}^{\infty}$, $Y_1 = \text{diag}\,(Y_1^{(k)})_{k=-\infty}^{\infty}$ and $\Delta_1 = \text{diag}\,(\Delta_1^{(k)})_{k=-\infty}^{\infty}$ are given by

$$(5.2a) \qquad\qquad\qquad X_1^{(k)} = 1, \qquad k \in \mathbf{Z},$$

$$(5.2b) \qquad\qquad\qquad Y_1^{(k)} = y_k, \qquad k \in \mathbf{Z},$$

with $(y_k)_{k=-\infty}^{\infty}$ a bounded sequence of complex numbers, and

(5.2c)
$$\Delta_1^{(k)} = \begin{cases} 0 & \text{if } k \neq 0, \\ \omega & \text{if } k = 0. \end{cases}$$

In this case $(\Delta_1 S^{-1})^j = 0$ of each $j \geq 2$. Hence, for $F = \sum_{j=0}^{\infty} S^j F_{[j]} \in \mathcal{L}^{1\times1}$, the diagonal matrix $(X_1 F)^{\wedge}(\Delta_1)$ is given by

$$(X_1 F)^{\wedge}(\Delta_1) = F_{[0]} + \Delta_1 F_{[1]}.$$

Thus the interpolation condition (5.1) can be given explicitly in terms of the entries F_{ij} $(i \geq j)$ of F as follows:

(5.3)
$$\begin{cases} F_{ii} = y_i & 0 \neq i \in \mathbf{Z}, \\ F_{00} + \omega F_{10} = y_0. \end{cases}$$

We also want the interpolant $F \in \mathcal{L}^{1\times1}$ in (5.1) to be a strict contraction.

For a lower triangular matrix $F = (F_{ij})_{i,j=-\infty}^{\infty}$ to be a strict contraction, it is certainly necessary that all diagonal entries F_{ii} $(i \in \mathbf{Z})$ have modulus less than 1. We see from (5.3) immediately therefore that a necessary condition for (5.3) to be satisfied by a strict contraction F in $\mathcal{L}^{1\times1}$ is that $|y_i| < 1$ for all $i \neq 0$. If y_0 also has $|y_0| < 1$, clearly we may set $F = Y = \text{diag } (y_k)_{k=-\infty}^{\infty}$ to get a solution. It is not obvious from a casual glance what the precise necessary and sufficient condition for strictly contractive solutions to exist should be. Such a condition is easily computed by using the theory developed in the preceding sections. Note that in this case condition (3.1) is fulfilled.

PROPOSITION 5.1. *There exists* $F = (F_{ij})_{i,j=-\infty}^{\infty} \in \mathcal{L}^{1\times1}$ *with* $\|F\| < 1$ *satisfying the interpolation conditions (5.3) if and only if the following two conditions are fulfilled:*

    (i) *for some* $\varepsilon > 0$, *we have* $|y_i| \leq 1 - \varepsilon$ *for all* $0 \neq i \in \mathbf{Z}$,

    (ii) $|y_0|^2 < 1 + |\omega|^2(1 - |y_1|^2)$.

PROOF. By Theorem 4.4, solutions exist if and only if $\mathcal{H} = \text{diag } (H_k)_{k=-\infty}^{\infty}$ is positive definite on ℓ_2, where H_k $(k \in \mathbf{Z})$ is determined as the solution of the time-varying Stein equation

(5.4)
$$H_k - A_k H_{k+1} A_k^* = B_k J B_k^*.$$

For our case, $A_k = 0$ for $k \neq 0$, $A_0 = \omega$, $B_k = (1 \quad -y_k)$ for all k. Thus (5.4) becomes

$$H_k = 1 - |y_k|^2, \quad k \neq 0; \qquad H_0 - |\omega|^2 H_1 = 1 - |y_0|^2.$$

Solving gives

$$(5.5) \qquad H_k = \begin{cases} 1 - |y_k|^2, & 0 \neq k \in \mathbf{Z}, \\ 1 - |y_0|^2 + |\omega|^2(1 - |y_1|^2), & k = 0. \end{cases}$$

The condition $H_k \geq \varepsilon$ for some $\varepsilon > 0$ then leads to (i) and (ii) in the theorem. $\qquad \square$

Theorem 4.4 of course provides not only a necessary and sufficient condition for existence of solutions of the TVNPI problem, but also a linear fractional parametrization for the set of all solutions. For the specific data set (5.2) which we are discussing here, implementation of the algorithm for the construction of the linear fractional map Θ involves decisions at various steps as to whether a certain quantity is positive, negative or zero. The explicit formula for Θ as a result breaks out into five special cases. Here we present two of these cases explicitly for purposes of illustration.

PROPOSITION 5.2. *Suppose that the TVNPI data set given by (5.2) satisfies the following additional condition:*

(j) *for some $\varepsilon > 0$ we have $|y_i| \leq 1 - \varepsilon$ for all $i \in \mathbf{Z}$.*

Set

$$(5.6) \qquad \delta_i = 1 - |y_i|^2 \quad (i \in \mathbf{Z}), \qquad H_0 = \delta + |\omega|^2 \delta_1.$$

Then the interpolation problem (5.3) has a solution $F \in \mathcal{L}^{1 \times 1}$ with $\|F\| < 1$ and the block lower triangular matrix

$$\Theta = \begin{pmatrix} \Theta_{11} & \Theta_{12} \\ \Theta_{22} & \Theta_{22} \end{pmatrix} \in \mathcal{L}^{2 \times 2}$$

which parametrizes the set of all solutions is given by $\Theta = (\Theta^{(i,j)})_{i,j=-\infty}^{\infty}$, *where*

$$\Theta^{(0,0)} = \begin{pmatrix} -\omega \delta_0^{-\frac{1}{2}} \delta_1^{\frac{1}{2}} H_0^{-\frac{1}{2}} & \delta_0^{-\frac{1}{2}} y_0 \\ -\omega \delta_0^{-\frac{1}{2}} \delta_1^{\frac{1}{2}} H_0^{-\frac{1}{2}} \overline{y}_0 & \delta_0^{-\frac{1}{2}} \end{pmatrix}, \quad \Theta^{(i,i)} = \begin{pmatrix} 0 & y_i \delta_i^{-\frac{1}{2}} \\ 0 & \delta_i^{-\frac{1}{2}} \end{pmatrix}, \quad 0 \neq i \in \mathbf{Z},$$

$$\Theta^{(0,-1)} = \begin{pmatrix} H_0^{-\frac{1}{2}} & 0 \\ \overline{y}_0 H_0^{-\frac{1}{2}} & 0 \end{pmatrix}, \qquad \Theta^{(1,0)} = \begin{pmatrix} H_0^{-\frac{1}{2}} \delta_0^{\frac{1}{2}} \delta_1^{-\frac{1}{2}} & 0 \\ H_0^{-\frac{1}{2}} \delta_0^{\frac{1}{2}} \delta_1^{-\frac{1}{2}} \overline{y}_1 & 0 \end{pmatrix},$$

$$\Theta^{(i,i-1)} = \begin{pmatrix} \delta_i^{-\frac{1}{2}} & 0 \\ \delta_1^{-\frac{1}{2}} \overline{y}_i & 0 \end{pmatrix}, \quad i \in \mathbf{Z}, i \neq 0, 1, \qquad \Theta^{(1,-1)} = \begin{pmatrix} \overline{\omega} H_0^{-\frac{1}{2}} & 0 \\ \overline{\omega} \overline{y}_1 H_0^{-\frac{1}{2}} & 0 \end{pmatrix},$$

and $\Theta^{(i,j)} = 0$ for all other pairs (i,j). In particular, the central interpolant $\widetilde{F} = \Theta_{12}\Theta_{22}^{-1}$ is the diagonal matrix $\widetilde{F} = \mathrm{diag}\,(y_k)_{k=-\infty}^{\infty}$.

PROOF. Note that condition (j) implies that conditions (i) and (ii) in Theorem 5.1 are fulfilled. Hence, the interpolation problem (5.3) has a solution $F \in \mathcal{L}^{1\times 1}$ with $\|F\| < 1$. By Theorem 4.4 combined with Theorem 4.2 and Corollary 4.3, the desired Θ is given by

$$(5.7) \qquad \Theta = J\mathcal{D}^*J + J\mathcal{B}^*(I - S\mathcal{A}^*)^{-1}S\mathcal{C}^*J$$

where

$$\mathcal{A} = \mathrm{diag}\,(A_k)_{k=-\infty}^{\infty} \qquad \mathcal{B} = \mathrm{diag}\,(B_k)_{k=-\infty}^{\infty}$$

$$\mathcal{C} = \mathrm{diag}\,(C_k)_{k=-\infty}^{\infty} \qquad \mathcal{D} = \mathrm{diag}\,(D_k)_{k=-\infty}^{\infty}$$

with

$$A_k = 0 \quad (0 \neq k \in \mathbf{Z}), \qquad A_o = \omega, \qquad B_k = (1 \quad -y_k) \quad (k \in \mathbf{Z}),$$

where C_k is a 2×1 matrix and D_k is a 2×2 matrix, bounded in norm uniformly with respect to k, such that

$$\Lambda_k := \begin{pmatrix} H_{k+1}^{-1} - A_k^* H_k^{-1} A_k & -A_k^* H_k^{-1} B_k \\ -B_k^* H_k^{-1} A_k & J - B_k^* H_k^{-1} B_k \end{pmatrix} = \begin{pmatrix} C_k^* \\ D_k^* \end{pmatrix} J (C_k \quad D_k)$$

with $J = \begin{pmatrix} 1 & 0 \\ 0 & -1 \end{pmatrix}$. For our case here, H_k is given by (5.5). Thus

$$(5.8) \qquad \Lambda_k = \delta_k^{-1}\begin{pmatrix} \delta_k H_{k+1}^{-1} & 0 & 0 \\ 0 & -|y_k|^2 & y_k \\ 0 & \overline{y}_k & -1 \end{pmatrix}, \qquad k \neq 0.$$

By inspection we observe the factorization

$$\begin{pmatrix} -|y_k|^2 & y_k \\ y_k & -1 \end{pmatrix} = \begin{pmatrix} -y_k \\ 1 \end{pmatrix}(-1)(-\overline{y}_k \quad 1).$$

Since $\delta_k > 0$ by assumption, we conclude that a viable choice of $\begin{pmatrix} C_k \\ D_k \end{pmatrix}$ for $k \neq 0$ is

$$(5.9) \qquad C_k = \begin{pmatrix} H_{k+1}^{-\frac{1}{2}} \\ 0 \end{pmatrix}, \qquad D_k = \begin{pmatrix} 0 & 0 \\ -\delta_k^{-\frac{1}{2}}\overline{y}_k & \delta_k^{-\frac{1}{2}} \end{pmatrix}, \qquad k \neq 0.$$

For $k = 0$ a straigtforward computation gives us

$$(5.10) \qquad \Lambda_0 = H_0^{-1} \begin{pmatrix} \delta_1^{-1}\delta_0 & -\overline{\omega} & \overline{\omega}y_0 \\ -\omega & |\omega|^2\delta_1 - |y_0|^2 & y_0 \\ \overline{y}_0\omega & \overline{y}_0 & -1 - |\omega|^2\delta_1 \end{pmatrix}.$$

To factor Λ_0 we perform a sequence of row and symmetric column operations. The result is

$$(5.11) \qquad E_2^* E_1^* \Lambda_0 E_1 E_2 = \begin{pmatrix} \delta_1^{-1}\delta_0 H_0^{-1} & 0 & 0 \\ 0 & -\delta_0^{-1}|y_0|^2 & y_0 \\ 0 & \overline{y}_0 & -\delta_0 \end{pmatrix}$$

where

$$E_1 = \begin{pmatrix} 1 & 0 & 0 \\ 0 & 1 & y_0 \\ 0 & 0 & 1 \end{pmatrix}, \qquad E_2 = \begin{pmatrix} 1 & \overline{\omega}\delta_0^{-1}\delta_1 & 0 \\ 0 & 1 & 0 \\ 0 & 0 & 1 \end{pmatrix}.$$

Note that in the definition of E_2 we have already used that $\delta_0 = 1 - |y_0|^2 \neq 0$. Almost by inspection, where we now use the assumption that $\delta_0 > 0$, we get the factorization

$$(5.12) \qquad \begin{pmatrix} -\delta_0^{-1}|y_0|^2 & y_0 \\ \overline{y}_0 & -\delta_0 \end{pmatrix} = \begin{pmatrix} -\delta_0^{-\frac{1}{2}}y_0 \\ \delta_0^{\frac{1}{2}} \end{pmatrix} (-1) \begin{pmatrix} -\delta_0^{-\frac{1}{2}}\overline{y}_0 & \delta_0^{-\frac{1}{2}} \end{pmatrix}.$$

Putting the pieces together, we conclude that a viable choice for $(\, C_0 \quad D_0 \,)$ is

$$(5.13) \qquad \begin{aligned} (\, C_0 \quad D_0 \,) &= \begin{pmatrix} H_0^{-\frac{1}{2}}\delta_0^{\frac{1}{2}}\delta_1^{-\frac{1}{2}} & 0 & 0 \\ 0 & -\delta_0^{-\frac{1}{2}}\overline{y}_0 & \delta_0^{\frac{1}{2}} \end{pmatrix} E_2^{-1} E_1^{-1} \\ &= \begin{pmatrix} H_0^{-\frac{1}{2}}\delta_0^{\frac{1}{2}}\delta_1^{-\frac{1}{2}} & -\overline{\omega}H_0^{-\frac{1}{2}}\delta_0^{-\frac{1}{2}}\delta_1^{\frac{1}{2}} & H_0^{-\frac{1}{2}}\overline{\omega}\delta_0^{-\frac{1}{2}}\delta_1^{\frac{1}{2}}y_0 \\ 0 & -\delta_0^{-\frac{1}{2}}\overline{y}_0 & \delta_0^{-\frac{1}{2}} \end{pmatrix} \end{aligned}$$

Taking adjoints and multiplying by J as appropriate, in summary we have

$$(5.14) \qquad \begin{aligned} A_k^* &= \begin{cases} \overline{\omega}, & k = 0 \\ 0, & k \neq 0, \end{cases}; \qquad JB_k^* = \begin{pmatrix} 1 \\ \overline{y}_k \end{pmatrix} \\ C_k^* J &= \begin{cases} \left(H_{k+1}^{-\frac{1}{2}} \quad 0 \right), & k \neq 0 \\ \left(H_0^{-\frac{1}{2}}\delta_0^{\frac{1}{2}}\delta_1^{-\frac{1}{2}} \quad 0 \right), & k = 0 \end{cases} \\ JD_k^* J &= \begin{cases} \begin{pmatrix} 0 & y_k\delta_k^{-\frac{1}{2}} \\ 0 & \delta_k^{-\frac{1}{2}} \end{pmatrix}, & k \neq 0 \\ \begin{pmatrix} -\omega\delta_0^{-\frac{1}{2}}\delta_1^{\frac{1}{2}}H_0^{-\frac{1}{2}} & \delta_0^{-\frac{1}{2}}y_0 \\ -\omega\delta_0^{-\frac{1}{2}}\delta_1^{\frac{1}{2}}\overline{y}_0 H_0^{-\frac{1}{2}} & \delta_0^{-\frac{1}{2}} \end{pmatrix}, & k = 0. \end{cases} \end{aligned}$$

We are now ready to plug (5.14) into (5.7) to get Θ. To do this observe that $(I - S\mathcal{A}^*)^{-1}S$ is given by

$$(I - S\mathcal{A}^*)^{-1}S = \begin{pmatrix} \ddots & \ddots & \ddots & & & & \\ & 0 & 1 & 0 & & & \\ & & 0 & 1 & \boxed{0} & & \\ & & \varpi & 1 & 0 & & \\ & & & 0 & 1 & 0 & \\ & & & & 0 & 1 & 0 \\ & & & & \ddots & \ddots & \ddots \end{pmatrix}$$

Hence

(5.15a) $$\Theta^{(i,i)} = JD_i^*J$$

(5.15b) $$\Theta^{(i,i-1)} = JB_i^*C_{i-1}^*J$$

(5.15c) $$\Theta^{(1,-1)} = \varpi JB_{-1}^*C_{-1}^*J$$

and

(5.15d) $$\Theta^{(i,j)} = 0 \text{ otherwise.}$$

This leads to the formula for Θ stated in the theorem. From this formula for Θ we read off that Θ_{12} is diagonal with its diagonal entries given by

$$\Theta_{12}^{(i,i)} = \begin{cases} \delta_0^{-\frac{1}{2}}y_0, & i = 0, \\ y_i\delta_i^{-\frac{1}{2}}, & i \neq 0, \end{cases}$$

while Θ_{22} is also diagonal with diagonal entries

$$\Theta_{22}^{(i,i)} = \begin{cases} \delta_0^{-\frac{1}{2}}, & i = 0, \\ \delta_i^{-\frac{1}{2}}, & i \neq 0. \end{cases}$$

From this we read off that the central solution $\widetilde{F} = \Theta_{12}\Theta_{22}^{-1}$ is the diagonal solution $\widetilde{F} = \text{diag}\,(y_k)_{k=-\infty}^{\infty}$ which one can see by inspection (for the case where the strong sufficient condition (j) holds) without applying the theory. $\qquad\square$

We next present another special case where the structure of the solution is somewhat different.

PROPOSITION 5.3. *Suppose that the TVNPI data set given by (5.2) satisfies the following condition:*

(i) *for some $\varepsilon > 0$ we have $|y_i| < 1 - \varepsilon$ for all $0 \neq i \in \mathbf{Z}$,*

(ii) *$|y_0|^2 < 1 + |\omega|^2(1 - |y_1|^2)$,*

(iii) *$|y_0| > 1$.*

Again set

$$\delta_i = 1 - |y_i|^2 \quad (i \in \mathbf{Z}), \qquad H_0 = \delta_0 + |\omega|^2\delta_1.$$

Then the interpolation problem (5.3) has a solution $F \in \mathcal{L}^{1\times 1}$ with $\|F\| < 1$ and the block lower triangular matrix

$$\Theta = \begin{pmatrix} \Theta_{11} & \Theta_{12} \\ \Theta_{11} & \Theta_{22} \end{pmatrix} \in \mathcal{L}^{2\times 2}$$

which parametrizes the set of all solutions is given by $\Theta = (\Theta^{(i,j)})_{i,j=-\infty}^{\infty}$, where

$$\Theta^{(0,0)} = \begin{pmatrix} |\delta_0|^{-\frac{1}{2}}y_0 & -\omega|\delta_0|^{-\frac{1}{2}}\delta_1^{\frac{1}{2}}H_0^{-\frac{1}{2}} \\ |\delta_0|^{-\frac{1}{2}} & -\omega|\delta_0|^{-\frac{1}{2}}\delta_1^{\frac{1}{2}}\overline{y}_0 H_0^{-\frac{1}{2}} \end{pmatrix}, \qquad \Theta^{(i,i)} = \begin{pmatrix} 0 & y_i\delta_i^{-\frac{1}{2}} \\ 0 & \delta_i^{-\frac{1}{2}} \end{pmatrix}, \quad 0 \neq i \in \mathbf{Z},$$

$$\Theta^{(0,-1)} = \begin{pmatrix} H_0^{-\frac{1}{2}} & 0 \\ \overline{y}_0 H_0^{-\frac{1}{2}} & 0 \end{pmatrix}, \qquad \Theta^{(1,0)} = \begin{pmatrix} 0 & -H_0^{-\frac{1}{2}}|\delta_0|^{\frac{1}{2}}\delta_1^{-\frac{1}{2}} \\ 0 & -\overline{y}_1 H_0^{-\frac{1}{2}}|\delta_0|^{\frac{1}{2}}\delta_1^{-\frac{1}{2}} \end{pmatrix},$$

$$\Theta^{(i,i-1)} = \begin{pmatrix} \delta_i^{-\frac{1}{2}} & 0 \\ \overline{y}_i\delta_i^{-\frac{1}{2}} & 0 \end{pmatrix}, \quad i \in \mathbf{Z}, i \neq 0,1, \qquad \Theta^{(1,-1)} = \begin{pmatrix} \overline{\omega}H_0^{-\frac{1}{2}} & 0 \\ \overline{\omega}\overline{y}_1 H_0^{-\frac{1}{2}} & 0 \end{pmatrix},$$

with $\Theta^{(i,j)} = 0$ for all other pairs (i,j). In particular, the central solution $\widetilde{F} - \Theta_{12}\Theta_{22}^{-1}$ is given by

$$\widetilde{F} = \begin{pmatrix} \ddots & & & & & \\ & y_{-2} & & & & \\ & & y_{-1} & & & \\ & & & \boxed{\overline{y}_0^{-1}} & & \\ & & x & y_1 & & \\ & & & & y_2 & \\ & & & & & \ddots \end{pmatrix}$$

where unspecified entries are all zero and where $x = -\omega^{-1}(1 - |y_0|^2)\overline{y}_0^{-1}$.

PROOF. Since (i), (ii) hold, Theorem 5.1 implies that the associated TVNPI problem has a solution. To get the parametrization of all solutions we proceed as in the proof of Proposition 5.2. The first step is to find uniformly bounded matrices $(\,C_k \quad D_k\,)$ which solve the factorization problem

$$\Lambda_k = \begin{pmatrix} C_k^* \\ D_k^* \end{pmatrix} J (\,C_k \quad D_k\,),$$

where Λ_k is again given by (5.8) and (5.10). For $k \neq 0$ the situation is exactly the same as in the proof of Proposition 5.2; a viable choice for $(\,C_k \quad D_k\,)$ is given by (5.9) for $k \neq 0$. For $k = 0$, all the details are the same up to formula (5.11). Since in our present setting $\delta_0 < 0$, we factorize

$$\delta_1^{-1}\delta_0 = (|\delta_0|^{\frac{1}{2}}\delta_1^{-\frac{1}{2}})(-1)(|\delta_0|^{\frac{1}{2}}\delta_1^{-\frac{1}{2}})$$

and

$$\begin{pmatrix} -\delta_0^{-1}|y_0|^2 & y_0 \\ \overline{y}_0 & -\delta_0 \end{pmatrix} = \begin{pmatrix} |\delta_0|^{-\frac{1}{2}}y_0 \\ |\delta_0|^{\frac{1}{2}} \end{pmatrix} (\,|\delta_0|^{-\frac{1}{2}}\overline{y}_0 \quad |\delta_0|^{\frac{1}{2}}\,).$$

We conclude that a viable choice of $(\,C_0 \quad D_0\,)$ is

$$
\begin{aligned}
(\,C_0 \quad D_0\,) &= \begin{pmatrix} 0 & |\delta_0|^{-\frac{1}{2}}\overline{y}_0 & |\delta_0|^{\frac{1}{2}} \\ H_0^{-\frac{1}{2}}|\delta_0|^{\frac{1}{2}}\delta_1^{-\frac{1}{2}} & 0 & 0 \end{pmatrix} E_2^{-1} E_1^{-1} \\
&= \begin{pmatrix} 0 & |\delta_0|^{-\frac{1}{2}}\overline{y}_0 & -|\delta_0|^{-\frac{1}{2}} \\ H_0^{-\frac{1}{2}}|\delta_0|^{\frac{1}{2}}\delta_1^{-\frac{1}{2}} & \varpi H_0^{-\frac{1}{2}}|\delta_0|^{-\frac{1}{2}}\delta_1^{\frac{1}{2}} & -\varpi H_0^{-\frac{1}{2}}|\delta_0|^{-\frac{1}{2}}\delta_1^{\frac{1}{2}}y_0 \end{pmatrix}.
\end{aligned}
$$

Next, taking asjoints and multiplying by J as appropoiate, in summary we have

$$
\begin{aligned}
A_k^* &= \begin{cases} \varpi, & k = 0 \\ 0, & k \neq 0, \end{cases} \qquad\qquad JB_k^* = \begin{pmatrix} 1 \\ \overline{y}_k \end{pmatrix} \\[2mm]
C_k^* J &= \begin{cases} (\,H_{k+1}^{-\frac{1}{2}} \quad 0\,), & k \neq 0 \\ (\,0 \quad -H_0^{-\frac{1}{2}}|\delta_0|^{\frac{1}{2}}\delta_1^{-\frac{1}{2}}\,), & k = 0 \end{cases} \\[2mm]
JD_k^* J &= \begin{cases} \begin{pmatrix} 0 & y_k\delta_k^{-\frac{1}{2}} \\ 0 & \delta_k^{-\frac{1}{2}} \end{pmatrix}, & k \neq 0 \\[4mm] \begin{pmatrix} |\delta_0|^{-\frac{1}{2}}y_0 & -\omega|\delta_0|^{-\frac{1}{2}}\delta_1^{\frac{1}{2}}H_0^{-\frac{1}{2}} \\ |\delta_0|^{-\frac{1}{2}} & -\omega|\delta_0|^{-\frac{1}{2}}\delta_1^{\frac{1}{2}}\overline{y}_0 H_0^{-\frac{1}{2}} \end{pmatrix}, & k = 0. \end{cases}
\end{aligned}
$$

(5.16)

As before the entries $\Theta^{(i,j)}$ are given by (5.15a) - (5.15d). Use of (5.16) then leads to the formulas for $\Theta^{(i,j)}$ as stated in Proposition 5.3. In this case Θ_{12} and Θ_{22} are not diagonal.

Indeed, we have

$$\Theta_{22} = \begin{pmatrix} \ddots & & & & & \\ & \delta_{-2}^{-\frac{1}{2}} & & & & \\ & & \delta_{-1}^{-\frac{1}{2}} & & & \\ & & & \boxed{-\omega|\delta_0|^{-\frac{1}{2}}\delta_1^{\frac{1}{2}}\overline{y}_0 H_0^{-\frac{1}{2}}} & & \\ & & & -\overline{y}_1 H_0^{-\frac{1}{2}}|\delta_0|^{\frac{1}{2}}\delta_1^{-\frac{1}{2}} & \delta_1^{-\frac{1}{2}} & \\ & & & & & \delta_2^{-\frac{1}{2}} \\ & & & & & & \ddots \end{pmatrix}$$

and

$$\Theta_{12} = \begin{pmatrix} \ddots & & & & & \\ & y_{-2}\delta_{-2}^{-\frac{1}{2}} & & & & \\ & & y_{-1}\delta_{-1}^{-\frac{1}{2}} & & & \\ & & & \boxed{-\omega|\delta_0|^{-\frac{1}{2}}\delta_1^{\frac{1}{2}}H_0^{-\frac{1}{2}}} & & \\ & & & -H_0^{-\frac{1}{2}}|\delta_0|^{\frac{1}{2}}\delta_1^{-\frac{1}{2}} & y_1\delta_1^{-\frac{1}{2}} & \\ & & & & & y_2\delta_2^{-\frac{1}{2}} \\ & & & & & & \ddots \end{pmatrix}$$

Since Θ_{22} has only one nonzero off diagonal entry, its inverse is easily computed; the result is

$$\Theta_{22}^{-1} = \begin{pmatrix} \ddots & & & & & \\ & \delta_{-2}^{\frac{1}{2}} & & & & \\ & & \delta_{-1}^{\frac{1}{2}} & & & \\ & & & \boxed{-\omega^{-1}|\delta_0|^{\frac{1}{2}}\delta_1^{-\frac{1}{2}}\overline{y}_0^{-1}H_0^{\frac{1}{2}}} & & \\ & & & \omega^{-1}\overline{y}_1\overline{y}_0^{-1}\delta_0\delta_1^{-\frac{1}{2}} & \delta_1^{\frac{1}{2}} & \\ & & & & & \delta_2^{\frac{1}{2}} \\ & & & & & & \ddots \end{pmatrix}$$

Multiplying out now gives that $\widetilde{F} = \Theta_{12}\Theta_{22}^{-1}$ as is specified in the theorem. $\qquad\square$

Of course it is also possible to verify directly that $\widetilde{F}$ as in Proposition 5.3 is a solution of the TVNPI problem associated with the data set (5.2). The interpolation conditions (5.3) are clearly fulfilled when $i \neq 0$. When $i = 0$, we have

$$\widetilde{F}_{00} + \omega\widetilde{F}_{10} = \overline{y}_0^{-1} + \omega x = \overline{y}_0^{-1} - (1 - |y_0|^2)\overline{y}_0^{-1} = y_0$$

as required. To show that $\widetilde{F}$ has norm less than 1, since $|y_k| \leq 1 - \varepsilon < 1$ for all $k \neq 0$ for some $\varepsilon > 0$, it suffices to show that

$$\left\| \begin{pmatrix} \overline{y}_0^{-1} & 0 \\ x & y_1 \end{pmatrix} \right\| < 1.$$

By assumption we know $|\overline{y}_0^{-1}| < 1$, $|y_1| < 1$. In general one can show that $\left\| \begin{pmatrix} \overline{y}_0^{-1} & 0 \\ x & y_1 \end{pmatrix} \right\| < 1$ if and only if

$$|x|^2 < (1 - |y_0|^{-2})(1 - |y_1|^2).$$

With $x = -\omega^{-1}(1 - |y_0|^2)\overline{y}_0^{-1}$, the condition to be verified is

$$(5.17) \qquad |\omega|^{-2}(|y_0|^2 - 1)^2|y_0|^{-2} < (1 - |y_0|^{-2})(1 - |y|_1)^2.$$

Multiply both sides by $|\omega|^2|y_0|^2$ and divide by $|y_0|^2 - 1$ to convert (5.17) to

$$|y_0|^2 - 1 < |\omega|^2(1 - |y_1|^2).$$

This condition in turn is exactly equivalent to $H_0 = \delta_0 + |\omega|^2\delta_1 > 0$. Thus, given that $|y_0| > 1$, $\widetilde{F}$ is a strict contraction exactly when the necessary and sufficient condition for solutions of the TVNPI problem holds.

A similar analysis can be done for the case $|y_0| = 1$. In this case the explicit formula for Θ breaks into three cases depending on whether $|\omega| > 1$, $|\omega| = 1$ or $|\omega| < 1$. We invite the interested reader to explore the details of this case for his or her self. For more complicated examples, of course, one would want to automate the algorithm on a computer.

REFERENCES

[1]   D. Alpay and P. Dewilde, Time-varying signal approximation and estimation, in: *Signal processing, scattering and operator theory, and numerical methods*, Proceedings of the international symposium MTNS-89, Volume III (Eds. M.A. Kaashoek, J.H. van Schuppen and A.C.M. Ran), Birkhäuser Verlag, Boston, 1990.

[2]   D. Alpay, P. Dewilde and H. Dym, Lossless inverse scattering and reproducing kernels for upper triangular operators, in: *Extension and interpolation of linear operators and matrix functions*, OT 47 (Ed. I. Gohberg), Birkhäuser Verlag, Basel, 1990, pp. 61-135.

[3]   T. Azizov and I.S. Yokhvidov, *Foundations of the theory of linear operators in spaces with an indefinite metric*, Wiley, New York, 1989.

[4]   J.A. Ball, I. Gohberg and M.A. Kaashoek, Time-varying systems: Nevanlinna-Pick interpolation and sensitivity minimization, Proceedings MTNS-91, submitted.

[5]   J.A. Ball, I. Gohberg and M.A. Kaashoek, Nevanlinna-Pick interpolation for time-varying input-output maps: the continuous time case, In this Volume.

[6]   J.A. Ball, I. Gohberg and L. Rodman, Realization and interpolation of rational matrix functions, in: *Topics in interpolation theory of rational matrix-valued functions*, OT 33 (Ed. I. Gohberg), Birkhäuser Verlag, Basel, 1988, pp. 1-72.

[7]   J.A. Ball, I. Gohberg and L. Rodman, *Interpolation of rational matrix functions*, OT 45 , Birkhäuser Verlag, Basel, 1990

[8]   P. Dewilde, A course on the algebraic Schur and Nevanlinna-Pick interpolation problems, in *Algorithms and Parallel VLSI Architectures*, Volume A: Tutorials (Eds. E.F. Deprettere and A.-J. van der Veen), Elsevier, Amsterdam, 1991.

[9]   P. Dewilde and H. Dym, Interpolation for upper triangular operators, in this Volume.

[10]   M.A. Drischel and J. Rovnyak, Extension theorem for contraction operators on Krein spaces, in: *Extension and interpolation of linear operators and matrix functions*, OT 47 (Ed. I. Gohberg), Birkhäuser Verlag, Basel, 1990, pp. 221-305.

[11]   Y. Genin, P. Van Dooren, T. Kailath, M. Delosme and M. Morf, On Σ-lossless transfer functions and related questions, *Linear Algebra Appl.* 50 (1983), 251-275.

[12]   I. Gohberg, M.A. Kaashoek and L.Lerer, Minimality and realization of discrete time-varying systems, in this Volume.

J.A. Ball

Department of Mathematics, Virginia Tech

Blacksburg, VA 24061, U.S.A.

I. Gohberg

Raymond and Beverly Sackler Faculty of Exact Sciences

School of Mathematical Sciences, Tel-Aviv University

Ramat-Aviv, Israel.

M.A. Kaashoek

Faculteit Wiskunde en Informatica, Vrije Universiteit

Amsterdam, The Netherlands

52

Operator Theory:
Advances and Applications, Vol. 56
© 1992 Birkhäuser Verlag Basel

NEVANLINNA-PICK INTERPOLATION FOR TIME-VARYING INPUT-OUTPUT MAPS: THE CONTINUOUS TIME CASE

J.A. Ball, I. Gohberg and **M.A. Kaashoek**

In this paper the tangential Nevanlinna-Pick interpolation problem for time-varying continuous time input-output maps is introduced and solved. The conditions of solvability are derived and all solutions are described via a linear fractional representation.

0. INTRODUCTION

For functions analytic on the open right half plane $\mathbb{C}^+$ the simplest Nevanlinna-Pick interpolation problem reads as follows. Given N different points $z_1, \ldots, z_N$ in $\mathbb{C}^+$ and arbitrary complex numbers $y_1, \ldots, y_N$, determine a function F, analytic on $\mathbb{C}^+$, such that

(i) $F(z_j) = y_j$, $j = 1, \ldots, N$,

(ii) $\sup\{|F(\lambda)| \mid \lambda \in \mathbb{C}^+\} < 1$.

Let us assume that we look for solutions F of the form

$$(0.1) \qquad F(\lambda) = d + \int_0^\infty e^{-\lambda t} f(t)dt,$$

where d is a complex number and f is in $L_1(\mathbb{R})$ with $\operatorname{supp} f \subset [0, \infty)$. Then (i) can be rewritten as

$$(0.2) \qquad d + \int_0^\infty e^{-z_j t} f(t)dt = y_j, \qquad j = 1, \ldots, N,$$

and the above interpolation problem can be restated as a problem involving operators on $L_2(\mathbb{R})$. To see this, note that for the function F in (0.1) the operator of multiplication by F on $L^2(i\mathbb{R})$ is unitarily equivalent via the bilateral Laplace transform to the convolution operator T on $L_2(\mathbb{R})$ given by

$$(0.3) \qquad (T\varphi)(t) = d\varphi(t) + \int_{-\infty}^t f(t - s)\varphi(s)ds, \qquad t \in \mathbb{R}.$$

The number z_j we view as the operator of multiplication by z_j on $L_2(\mathbb{R})$. Since $z_j \in \mathbb{C}^+$, the maximal operator on $L_2(\mathbb{R})$ associated with the differential expression $\frac{d}{dt} - z_j$ is invertible and its inverse is given by

$$\left(\left(\frac{d}{dt} - z_j\right)^{-1}\varphi\right)(t) = -\int_t^\infty e^{z_j(t-s)}\varphi(s)ds, \qquad t \in \mathbb{R}.$$

Thus $\left(\frac{d}{dt} - z_j\right)^{-1}T$ is an integral operator on $L_2(\mathbb{R})$ with kernel function

$$k_j(t,s) = -de^{z_j(t-s)} - \int_{\max(t,s)}^\infty e^{z_j(t-\alpha)}f(\alpha - s)d\alpha,$$

and the interpolation condition (0.2) is equivalent to the requirement that

$$k_j(t,t) = -y_j, \qquad j = 1,\ldots,N.$$

The interpolation problem (i), (ii) mentioned above can now be reformulated as a problem involving operators on $L_2(\mathbb{R})$, namely find a lower triangular Wiener-Hopf operator T on $L_2(\mathbb{R})$ of the form (0.3) such that

(i)$'$ the kernel function of the integral operator $-\left(\frac{d}{dt} - z_j\right)^{-1}T$ evaluated at $t = s$ is equal to y_j, $j = 1,\ldots,N$,

(ii)$'$ $\|T\| < 1$.

In this form the problem can be extended in a natural way to an interpolation problem for operators that are not of Wiener-Hopf type and in which the interpolation data $z_1,\ldots,z_N$ and $y_1,\ldots,y_N$ are L_∞-functions on $\mathbb{R}$.

More precisely, in the present paper we study the following problem. Let $z_1,\ldots,z_N$ and $y_1,\ldots,y_N$ be in $L_\infty(\mathbb{R})$. Assume that the maximal operator on $L_2(\mathbb{R})$ associated with the differential expression $\frac{d}{dt} - z_j(t)$ is invertible, and let its inverse be the upper triangular integral operator

$$(0.4) \qquad \left(\left(\frac{d}{dt} - z_j(\cdot)\right)^{-1}\varphi\right)(t) = -\int_t^\infty Z_j(t,s)\varphi(s)ds, \qquad t \in \mathbb{R}.$$

Furthermore, assume that

$$(0.5) \qquad \left(\int_t^\infty Z_i(t,\alpha)\overline{Z_j(t,\alpha)}d\alpha\right)_{i,j=1}^N \geq \varepsilon I_N, \qquad t \in \mathbb{R},$$

where ε is a positive number independent of t and I_N is the $N \times N$ identity matrix. In the classical case condition (0.4) means that the points where the interpolation takes place are in $\mathbb{C}^+$ and (0.5) is equivalent to the requirement that these points are different. Now the problem is to find a lower triangular integral operator T on $L_2(\mathbb{R})$ of the form

$$(0.6) \qquad (T\varphi)(t) = d(t)\varphi(t) + \int_{-\infty}^t f(t,s)\varphi(s)ds,$$

such that $\|T\| < 1$ and for $j = 1, \ldots, N$ the following interpolation requirements are fulfilled:

$$(0.7) \qquad d(t) + \int_0^\infty Z_j(t, t+\alpha) f(t+\alpha, t) d\alpha = y_j(t), \qquad t \in \mathbb{R}.$$

The function $d(\cdot)$ in (0.6) is required to be an L_∞-function on $\mathbb{R}$, and the kernel function f in (0.6) has to be measurable on $\mathbb{R} \times \mathbb{R}$ and such that

$$\int_0^\infty \left(\sup_{t-s=\alpha} |f(t,s)| \right) d\alpha < \infty.$$

The class of operators T as in (0.6) for which d and f satisfy these conditions can be viewed as the time-varying analogue of the Wiener algebra on the line. We shall prove that this generalized continuous-time Nevanlinna-Pick interpolation problem is solvable if and only if for some $\varepsilon > 0$

$$(0.8) \qquad \left(\int_t^\infty Z_i(t, \alpha) \{ 1 - y_i(\alpha)\overline{y_j(\alpha)} \} \overline{Z_j(t, \alpha)} d\alpha \right)_{i,j=1}^N \geq \varepsilon I_N, \qquad t \in \mathbb{R}.$$

In what follows we also treat the matrix version of this problem.

In the classical Nevanlinna-Pick interpolation problem there is a special interest in solutions (0.1) that are rational, i.e., a ratio of polynomials. Using the realization theorem from systems theory the latter can be interpreted to mean that the operator T in (0.2) is the input-output map of a causal stable time-invariant system, i.e., the relation $T\varphi = g$ is given in the following way:

$$\begin{cases} x'(t) = Ax(t) + B\varphi(t), & t \in \mathbb{R}, \\ g(t) = Cx(t) + D\varphi(t), \end{cases}$$

where A, B, C and D are matrices of appropriate sizes. In the general problem the requirement that the solution is rational is replaced by the condition that the operator T in (0.6) is the input-output map of a time-variant system. In other words the action of the operator T in (0.6) is given by

$$\begin{cases} x'(t) = A(t)x(t) + B(t)\varphi(t), & t \in \mathbb{R}, \\ g(t) = C(t)x(t) + D(t)\varphi(t), \end{cases}$$

where now the state, input, output and feedthrough matrices $A(t)$, $B(t)$, $C(t)$ and $D(t)$, respectively, may vary in time.

The left hand side of (0.7) will be denoted by $\widehat{T}(z_j)(t)$. The map $z \mapsto \widehat{T}(z)$, which assigns to an L_∞-function z the function $\widehat{T}(z)$, is the continuous analogue of the generalized point evaluation map for lower triangular doubly infinite matrices appearing in [Dew], [ADeDy], [DeDy], and [BallGK]. The generalized Nevanlinna-Pick interpolation problem introduced above is the natural continuous version of the interpolation problem appearing in [DewDy] and [BallGK].

In the present paper we develop the continuous analogue of the method of solution employed in [BallGK]. The latter is based on reduction to a homogeneous interpolation problem and uses systematically the system theoretic point of view. We derive the conditions of solvability and describe all input-output maps that are solutions of the matrix version of the generalized Nevanlinna-Pick interpolation problem stated above. The coefficients of the linear fractional transformation describing the solutions are given explicitly in terms of the original data. The results are illustrated by computing in detail the solutions for one example, namely when $N = 1$ and

$$z_1(t) = \begin{cases} 1 & , \quad t > 0 \\ 1 + i, & t < 0 \end{cases}; \qquad y_1(t) = \begin{cases} 2, & 0 \leq t \leq c, \\ 0, & \text{otherwise,} \end{cases}$$

where c is an arbitrary positive number.

A few words about notation. By $L_2^m(\mathbb{R})$ we denote the Hilbert space of all square integrable $\mathbb{C}^m$-valued functions on $\mathbb{R}$. We write $L_\infty^{r \times m}(\mathbb{R})$ for the set of all $r \times m$ matrices whose entries are L_∞-functions on $\mathbb{R}$. Each $A \in L_\infty^{r \times m}(\mathbb{R})$ defines in a natural way, via multiplication, on operator acting from $L_2^m(\mathbb{R})$ into $L_2^r(\mathbb{R})$, which we shall denote by M_A. Thus

$$(0.9) \qquad (M_A\varphi)(t) = A(t)\varphi(t), \qquad t \in \mathbb{R}.$$

An identity operator is denoted by I; from the context it should be clear on which space it acts. For an invertible matrix or Hilbert space operator A we let A^{-*} denote the adjoint of A^{-1}. The norm of a matrix will always be its spectral norm, i.e., the largest singular value of the matrix.

1. GENERALIZED POINT EVALUATION

1.1. Time-varying points. By Δ we denote the operator of differentiation on $L_2^m(\mathbb{R})$. The domain $\mathcal{D}(\Delta)$ of Δ is the set of functions in $L_2^m(\mathbb{R})$ that are absolutely continuous on finite intervals and such that $\Delta f = f'$ belongs to $L_2^m(\mathbb{R})$. For $z \in L_\infty^{m \times m}(\mathbb{R})$ we write M_z for the operator of multiplication by z on $L_2^m(\mathbb{R})$ (cf., formula (0.9)). Note that $\Delta - M_z$ is the (unbounded) operator on $L_2^m(\mathbb{R})$ defined by

$$\mathcal{D}(\Delta - M_z) = \mathcal{D}(\Delta),$$
$$((\Delta - M_z)\varphi)(t) = \varphi'(t) - z(t)\varphi(t), \qquad t \in \mathbb{R}.$$

Let $z \in L_\infty^{m \times m}(\mathbb{R})$. By $Z(t, s)$ we denote the transition matrix associated with the differential equation

$$(1.1) \qquad x'(t) = z(t)x(t), \qquad -\infty < t < \infty.$$

Thus $Z(t, s)$ is an $m \times m$ matrix function, absolutely continuous in t on each finite interval, and

$$\frac{\partial}{\partial t} Z(t, s) = z(t)Z(t, s), \qquad Z(s, s) = I.$$

We call z an *anti-stable* (or *right half plane*) *time-varying point* (for short: astv-point) if there exists constants $M > 0$ and $0 < a < 1$ such that

$$(1.2) \qquad\qquad \|Z(t,s)\| \le M a^{s-t}, \qquad s \ge t.$$

In that case the differential operator $\Delta - M_z$ is invertible and

$$(1.3) \qquad\qquad ((\Delta - M_z)^{-1}\varphi)(t) = -\int_t^\infty Z(t,s)\varphi(s)ds, \qquad t \in \mathbb{R}.$$

To illustrate this notion of anti-stable time-varying points, assume that $z(t) = \omega \in \mathbb{C}$ for all $t \in \mathbb{R}$. Then the corresponding transition function is $e^{\omega(t-s)}$, and in this case (1.2) is equivalent to the requirement that ω is a point in the open right half plane.

For later purposes we mention the following lemma.

LEMMA 1.1. *Let* $z \in L_\infty^{m \times m}(\mathbb{R})$ *be an astv-point with associated transition matrix* $z(t,s)$, *and let* $y \in L_2^{p \times m}(\mathbb{R})$. *Then*

$$k(t,s) = \begin{cases} y(t)Z(t,s), & t \le s, \\ 0 & , \quad t > s, \end{cases}$$

is a Hilbert-Schmidt kernel on $\mathbb{R} \times \mathbb{R}$.

PROOF. By using (1.2) we see that

$$\int_{-\infty}^\infty \int_{-\infty}^\infty \|k(t,s)\|^2 dt ds = \int_{-\infty}^\infty \left(\int_t^\infty \|k(t,s)\|^2 ds \right) dt$$

$$\le \int_{-\infty}^\infty \|y(t)\|^2 \left(\int_t^\infty \|Z(t,s)\|^2 ds \right) dt$$

$$\le \int_{-\infty}^\infty \|y(t)\|^2 \left(\int_t^\infty M^2 a^{2(s-t)} ds \right) dt$$

$$= \left(\int_{-\infty}^\infty \|y(t)\|^2 dt \right) \left(\int_0^\infty M^2 a^{2\alpha} d\alpha \right) < \infty. \quad \square$$

1.2. The nonstationary Wiener algebra and point evaluation. By $\mathcal{W}^{m \times r}$ we denote the set of all operators $T \colon L_2^r(\mathbb{R}) \to L_2^m(\mathbb{R})$ that admit a representation of the form

$$(1.4) \qquad\qquad (T\varphi)(t) = D(t)\varphi(t) + \int_{-\infty}^\infty k(t,s)\varphi(s)ds, \qquad t \in \mathbb{R},$$

where $D(\cdot)$ is an $m \times r$ matrix function whose entries are in $L_\infty(\mathbb{R})$ and the kernel function k is an $m \times r$ matrix function whose entries are measurable functions on $\mathbb{R} \times \mathbb{R}$ such that

$$(1.5) \qquad\qquad \sup_{t-s=\alpha} \|k(t,s)\| \le \ell_k(\alpha)$$

for some $\ell_k \in L_1(\mathbb{R})$. We write $\mathcal{W}$ in place of $\mathcal{W}^{1\times 1}$. By identifying $L_2^n(\mathbb{R})$ with the Hilbert space direct sum of n copies of $L_2(\mathbb{R})$ each $T \in \mathcal{W}^{m\times r}$ may be written as an $m \times r$ operator matrix of which the entries are in $\mathcal{W}$. The set $\mathcal{W}$, which is closed under taking sums and products of operators, will be referred to as the *nonstationary Wiener algebra* (on the real line).

LEMMA 1.2. *Each* $T \in \mathcal{W}^{m\times r}$ *defines a bounded linear operator from* $L_2^r(\mathbb{R})$ *into* $L_2^m(\mathbb{R})$, *and the representation* (1.4) *is unique modulo changes of* $D(\cdot)$ *and* $k(\cdot,\cdot)$ *on sets of measure zero.*

PROOF. We may assume without loss of generality that $m = r = 1$. Take $\varphi, \psi \in L_2(\mathbb{R})$. Note that $\overline{\psi(t)}k(t,s)\varphi(s)$ is measurable on $\mathbb{R} \times \mathbb{R}$. Let $\ell_k(\cdot) \in L_1(\mathbb{R})$ be as in (1.5). Since $\ell_k(\cdot)$ is an L_1-function, the integral

$$\int_{-\infty}^{\infty} \ell_k(t - s)|\varphi(s)|ds$$

exists for almost all $t \in \mathbb{R}$ and the resulting function is in $L_2(\mathbb{R})$ (see [HeS], item 21.32, page 397). Now

(1.6)
$$|\overline{\psi(t)}k(t,s)\varphi(s)| \le |\psi(t)|\ell_k(t - s)|\varphi(s)|,$$

and by the preceding remark the right hand side of (1.6) is integrable on $\mathbb{R} \times \mathbb{R}$. It follows that the same is true for the left hand side of (1.6). Thus $T\varphi \in L_2(\mathbb{R})$, and T is a well-defined bounded linear operator on $L_2(\mathbb{R})$.

Next, we prove the statement about the uniqueness of the representation (1.4). By taking the difference of two representations of the type (1.4), one sees that we have to establish the following property. If for each $\varphi \in L_2(\mathbb{R})$ we have

(1.7)
$$d(t)\varphi(t) = \int_{-\infty}^{\infty} k(t,s)\varphi(s)ds, \qquad t \in \mathbb{R}, \text{ a.e.},$$

then d and k are zero almost everywhere on $\mathbb{R}$ and $\mathbb{R} \times \mathbb{R}$, respectively. Now, let $[a,b]$ and $[c,d]$ be disjoint intervals, and assume that $\operatorname{supp}\varphi \subset [a,b]$. Then (1.7) implies that

$$\int_c^d \int_a^b \overline{\psi(t)}k(t,s)\varphi(s)dsdt = 0$$

for each $\psi \in L_2([c,d])$. It follows that $k(\cdot,\cdot)$ is zero almost everywhere on $[c,d] \times [a,b]$. By varying $[a,b]$ and $[c,d]$ in an appropriate way we see that $k(\cdot,\cdot)$ is zero almost everywhere on $\mathbb{R} \times \mathbb{R}$. But then (1.7) shows that the same holds for $d(\cdot)$ on $\mathbb{R}$. $\square$

An operator $T \in \mathcal{W}^{m\times r}$ is called *lower triangular* (notation: $T \in \mathcal{LW}^{m\times r}$) if the kernel function k in (1.4) is zero a.e. on $s \ge t$. In this case

(1.8)
$$(T\varphi)(t) = D(t)\varphi(t) + \int_{-\infty}^{t} k(t,s)\varphi(s)ds, \qquad t \in \mathbb{R}.$$

By $\mathcal{U}W^{m\times r}$ we denote the set of $T \in W^{m\times r}$ for which the kernel function k in (1.4) is zero a.e. on $s \leq t$. An operator $T \in \mathcal{U}W^{m\times r}$ is said to be *upper triangular*.

Let $z \in L_\infty^{m\times m}(\mathbb{R})$ be an astv-point with associated transition matrix $Z(t,s)$, and let $T \in \mathcal{L}W^{m\times r}$ be given by (1.8). From (1.2) and (1.3) we see that $(\Delta - M_z)^{-1} \in \mathcal{U}W^{m\times m}$. Note that $(\Delta - M_z)^{-1}T \in W^{m\times r}$ and its kernel function is given by

$$(1.9) \qquad -Z(t,s)D(s) - \int_{\max(t,s)}^{\infty} Z(t,u)k(u,s)du.$$

By taking $t = s$ in (1.9) and changing the sign we arrive at the following function:

$$(1.10) \qquad y(t) := D(t) + \int_0^\infty Z(t,t+\alpha)k(t+\alpha,t)d\alpha, \qquad t \in \mathbb{R}.$$

LEMMA 1.3. *The function y in (1.10) is in $L_\infty^{m\times r}$ and y is uniquely determined by T.*

PROOF. It suffices to consider the case when $m = r = 1$. Note that

$$|Z(t,t+\alpha)k(t+\alpha,t)| \leq Ma^\alpha \left(\sup_{t-s=\alpha} |k(t,s)| \right), \qquad \alpha \geq 0,$$

for some $0 < a < 1$ and $M > 0$. Thus, by (1.5) and the properties of $D(\cdot)$, we obtain

$$|y(t)| \leq |D(t)| + \int_0^\infty |Z(t,t+\alpha)k(t+\alpha,t)|d\alpha$$

$$\leq \sup_{t\in\mathbb{R}} |D(t)| + M\left(\int_0^\infty \ell_k(\alpha)d\alpha \right) = C < \infty,$$

where C does not depend on t. Thus $y \in L_\infty(\mathbb{R})$.

To prove that y is uniquely determined by T, consider a second representation of T,

$$(T\varphi)(t) = \tilde{D}(t)\varphi(t) + \int_{-\infty}^t \tilde{k}(t,s)\varphi(s)ds, \qquad t \in \mathbb{R},$$

and put

$$\tilde{y}(t) := \tilde{D}(t) + \int_0^\infty Z(t,t+\alpha)\tilde{k}(t+\alpha,t)d\alpha, \qquad t \in \mathbb{R}.$$

We have to show $y = \tilde{y}$ a.e. on $\mathbb{R}$. From Lemma 1.2 we know that $D = \tilde{D}$ and $k = \tilde{k}$ almost everywhere on $\mathbb{R}$ and $\mathbb{R} \times \mathbb{R}$, respectively. Integrating over $a \leq t \leq b$ yields

$$\int_a^b y(t)dt = \int_a^b D(t)dt + \int_a^b \left(\int_0^\infty Z(t,t+\alpha)k(t+\alpha,t)d\alpha \right)dt,$$

$$\int_a^b \tilde{y}(t)dt = \int_a^b \tilde{D}(t)dt + \int_a^b \left(\int_0^\infty Z(t,t+\alpha)\tilde{k}(t+\alpha,t)d\alpha \right)dt.$$

Since $D = \widetilde{D}$ and $k = \widetilde{k}$ a.e., it follows that

$$\int_a^b y(t)dt = \int_a^b \widetilde{y}(t)dt$$

for each $[a, b]$. Thus y and $\widetilde{y}$ have equal primitives, and hence $y = \widetilde{y}$ a.e. $\square$

In what follows we write $\widehat{T}(z)$ for the function y defined by (1.10), and we refer to $z \mapsto \widehat{T}(z)$ as the *generalized point evaluation map* on $\mathcal{L}\mathcal{W}^{m \times r}$. For the time-invariant case this map reduces to the usual point evaluation in the frequency domain. Indeed, let $z(t) \equiv \omega$ with $\mathrm{Re}\,\omega > 0$ and

$$(F\varphi)(t) = D\varphi(t) + \int_{-\infty}^t f(t - s)\varphi(s)ds, \qquad t \in \mathbb{R},$$

where $f \in L_1^{m \times m}(\mathbb{R})$ and $\mathrm{supp}\, f \subset [0, \infty)$. Then z is an astv-point, the associated transition function is $e^{\omega(t-s)}$, and thus

$$\widehat{F}(z)(t) = D + \int_0^\infty e^{-\omega\alpha} f(\alpha)d\alpha = D + \widehat{f}(\omega),$$

where $\widehat{f}$ is the Laplace transform of f.

PROPOSITION 1.4. *Let* $T \in \mathcal{L}\mathcal{W}^{m \times r}$, *and let* $z \in L_\infty^{m \times m}$ *be an astv-point. Put* $y = \widehat{T}(z)$. *Then*

$$(1.11) \qquad\qquad (\Delta - M_z)^{-1}(T - M_y) \in \mathcal{L}\mathcal{W}^{m \times r},$$

and y *is the unique function in* $L_\infty^{m \times r}(\mathbb{R})$ *such that* (1.11) *holds.*

PROOF. Both $(\Delta - M_z)^{-1}T$ and $(\Delta - M_z)^{-1}M_y$ are in $\mathcal{W}^{m \times r}$. Let us denote their kernel functions by $h(\cdot, \cdot)$ and $\ell(\cdot, \cdot)$, respectively. Then (cf., formula (1.9)) we have for $t < s$

$$h(t, s) = -Z(t, s)D(s) - \int_s^\infty Z(t, u)k(u, s)du,$$

$$\ell(t, s) = -Z(t, s)D(s) - Z(t, s)\int_s^\infty Z(s, u)k(u, s)du.$$

Here $k(\cdot, \cdot)$ is the kernel function of T. Now use that $Z(t, s)$ is a transition matrix. So $Z(t, s)Z(s, u) = Z(t, u)$, which implies that $h(t, s) - \ell(t, s) = 0$ for $t < s$, and thus (1.11) holds.

Next, let $\widetilde{y} \in L_\infty^{m \times r}(\mathbb{R})$, and assume that (1.11) holds with $\widetilde{y}$ in place of y. Then $(\Delta - M_z)^{-1}(M_y - M_{\widetilde{y}})$ is lower triangular, and thus

$$Z(t, s)\big(y(s) - \widetilde{y}(s)\big) = 0, \qquad t < s.$$

Since $\det Z(t, s) \neq 0$ for all t and s, we conclude that $y = \widetilde{y}$. $\square$

COROLLARY 1.5. *If $T \in \mathcal{L}W^{m \times r}$ and $S \in \mathcal{L}W^{r \times p}$, then for each astv-point $z \in L_\infty^{m \times m}(\mathbb{R})$ we have*

$$(1.12) \qquad \widehat{TS}(z) = (M_{\widehat{T}(z)} S)\widehat{}(z).$$

PROOF. Put $y = \widehat{T}(z)$ and $x = \widehat{TS}(z)$. From Proposition 1.4 we know that $(\Delta - M_z)^{-1}(T - M_y)$ is lower triangular. Since S is lower triangular, we conclude that $(\Delta - M_z)^{-1}(TS - M_y S)$ is lower triangular. Also, again by Proposition 1.4, $(\Delta - M_z)^{-1}(TS - M_x)$ is lower triangular. By taking the difference we see that

$$(\Delta - M_z)^{-1}(M_y S - M_x) \in \mathcal{L}W^{m \times p}.$$

Now use the uniqueness part of Proposition 1.4, and we are done. $\quad\square$

1.3. Point evaluation for Hilbert-Schmidt operators. By $S_2^{m \times r}$ we denote the set of all Hilbert-Schmidt integral operators $K: L_2^r(\mathbb{R}) \to L_2^m(\mathbb{R})$. Note that $K \in S_2^{m \times r}$ admits a representation of the form

$$(1.13) \qquad (K\varphi)(t) = \int_{-\infty}^{\infty} k(t,s)\varphi(s)ds, \qquad t \in \mathbb{R},$$

where the kernel function k belongs to $L_2^{m \times r}(\mathbb{R} \times \mathbb{R})$. We write $K \in \mathcal{L}S_2^{m \times r}$ if $k(t,s) = 0$ for $t < s$ (and hence K is lower triangular).

Let $z \in L_\infty^{m \times m}(\mathbb{R})$ be an astv-point, and let $Z(t,s)$ be the associated transition matrix. For $K \in \mathcal{L}S_2^{m \times r}$ we let $\widehat{K}(z)$ be the function defined by

$$(1.14) \qquad \widehat{K}(z)(t) = \int_0^{\infty} Z(t, t+\alpha)k(t+\alpha, t)d\alpha.$$

By integrating over an arbitrary finite interval one sees (cf., the proof of Lemma 1.3) that the right hand side of (1.14) is uniquely determined by K (and does not depend on the choice of the kernel function k in (1.13)). We claim that $\widehat{K}(z) \in L_2^{m \times r}(\mathbb{R})$. Indeed, by (1.2),

$$\|\widehat{K}(z)(t)\| \leq \int_0^{\infty} Ma^{\alpha}\|k(t+\alpha, t)\|d\alpha$$

$$\leq \left(\int_0^{\infty} M^2 a^{2\alpha} d\alpha\right)^{1/2} \left(\int_0^{\infty} \|k(t+\alpha, t)\|^2 d\alpha\right)^{1/2}.$$

Put $C = \int_0^{\infty} M^2 a^{2\alpha} d\alpha$. Then

$$\int_{-\infty}^{\infty} \|\widehat{K}(z)(t)\|^2 dt \leq C \int_{-\infty}^{\infty} \int_0^{\infty} \|k(t+\alpha, t)\|^2 d\alpha dt < \infty.$$

In the next lemma we use that $S_2^{m \times r}$ has a natural inner product, namely $\langle T, S \rangle = \operatorname{tr} S^* T$.

LEMMA 1.6. *Let $K \in \mathcal{L}S_2^{m \times 1}$ and $x \in L_\infty^{1 \times m}(\mathbb{R})$. If $z \in L_\infty(\mathbb{R})$ is an astv-point, then*

$$(1.15) \qquad \langle (M_x K)^\wedge(z), d \rangle = \langle K, M_x^*(\Delta - M_z)^{-*} M_d \rangle, \qquad d \in L_2(\mathbb{R}) \cap L_\infty(\mathbb{R}).$$

PROOF. The inner product in the right hand side of (1.15) is that of Hilbert-Schmidt operators referred to above. Note that $M_x^*(\Delta - M_z)^{-1} M_d$ is a Hilbert-Schmidt operator because of Lemma 1.1. The inner product in the left hand side of (1.15) is that of $L_2(\mathbb{R})$. Thus

$$\langle (M_x K)^\wedge(z), d \rangle = \int_{-\infty}^\infty \overline{d(t)} \left(\int_0^\infty Z(t, t+\alpha) x(\alpha+t) k(\alpha+t, t) d\alpha \right) dt$$

$$= \int_{-\infty}^\infty \int_0^\infty \left\{ \overline{d(t)} Z(t, t+\alpha) x(\alpha+t) \right\} k(\alpha+t, t) d\alpha dt$$

$$= \langle K, M_x^*(\Delta - M_z)^{-*} M_d \rangle. \quad \square$$

Let $F \in \mathcal{L}W^{m \times r}$. With F we associate the operator of left multiplication on $\mathcal{L}S_2^{r \times 1}$ as follows:

$$(1.16) \qquad L_F : \mathcal{L}S_2^{r \times 1} \to \mathcal{L}S_2^{m \times 1}, \qquad L_F K = FK.$$

Note that $\mathcal{L}S_2^{m \times 1}$ and $\mathcal{L}S_2^{r \times 1}$ are Hilbert spaces. Thus L_F^* is a well-defined operator from $\mathcal{L}S_2^{m \times 1}$ into $\mathcal{L}S_2^{r \times 1}$.

LEMMA 1.7. *Let $F \in \mathcal{L}W^{m \times r}$, $x \in L_\infty^{1 \times m}(\mathbb{R})$, and let $z \in L_\infty(\mathbb{R})$ be an astv-point. Then $(M_x F)^\wedge(z) = y$ implies that*

$$L_F^* M_x^*(\Delta - M_z)^{-*} M_d = M_y^*(\Delta - M_z)^{-*} M_d, \qquad d \in L_2(\mathbb{R}) \cap L_\infty(\mathbb{R}).$$

PROOF. Let $K \in \mathcal{L}S_2^{r \times 1}$. We apply Lemma 1.6 to get

$$\langle K, L_F^* M_x^*(\Delta - M_z)^{-*} M_d \rangle = \langle FK, M_x^*(\Delta - M_z)^{-*} M_d \rangle$$
$$= \langle (M_x FK)^\wedge(z), d \rangle.$$

Similarly,

$$\langle K, M_y^*(\Delta - M_z)^{-*} M_d \rangle = \langle (M_y K)^\wedge(z), d \rangle.$$

Since $(M_x F)^\wedge(z) = y$, Corollary 1.5 shows that $(M_x FK)^\wedge(z) = (M_y K)^\wedge(z)$, and therefore

$$\langle K, L_F^* M_x^*(\Delta - M_z)^{-*} M_d \rangle = \langle K, M_y^*(\Delta - M_z)^{-*} M_d \rangle,$$

which proves the lemma. $\square$

2. BOUNDED INPUT-OUTPUT MAPS

In what follows we shall employ linear time-varying input-output systems of the following type:

$$\Sigma \begin{cases} x'(t) = A(t)x(t) + B(t)\varphi(t), & t \in \mathbb{R}, \\ g(t) = C(t)x(t) + D(t)\varphi(t), \end{cases}$$

where $A \in L_\infty^{n \times n}(\mathbb{R})$, $B \in L_\infty^{n \times r}(\mathbb{R})$, $C \in L_\infty^{m \times n}(\mathbb{R})$, and $D \in L_\infty^{m \times r}(\mathbb{R})$. Furthermore, we shall assume that the differential equation

$$(2.1) \qquad x'(t) = A(t)x(t), \qquad -\infty < t < \infty,$$

admits a dichotomy P.

Let us recall (see [DK], [MS], also [C]) the definition of a dichotomy. By $U(t)$ we denote the evolution matrix associated with (2.1), i.e., $U(\cdot)$ is absolutely continuous on finite intervals and

$$(2.2) \qquad \frac{d}{dt}U(t) = A(t)U(t) \quad (t \in \mathbb{R}), \qquad U(0) = I.$$

In other words, in the terminology of the previous section, $U(t)U(s)^{-1}$ is the transition matrix of (2.1). The equation (2.1) is said to have a *dichotomy* P if P is a projection of $\mathbb{C}^n$ (where n is the order of $A(t)$) and there exist constants $M > 0$ and $0 < a < 1$ such that

$$(2.3a) \qquad \|U(t)PU(s)^{-1}\| \leq Ma^{t-s}, \quad t \geq s,$$
$$(2.3b) \qquad \|U(t)(I - P)U(s)^{-1}\| \leq Ma^{s-t}, \quad s \geq t.$$

If a dichotomy for (2.1) exists, then it is unique. Since $A \in L_\infty^{n \times n}$, we may consider the operator $\Delta - M_A$ (see subsection 1.1). In [BenG] it is proved that (2.1) admits a dichotomy P if and only if the operator $\Delta - M_A$ is invertible, and in that case

$$(2.4) \qquad ((\Delta - M_A)^{-1}\varphi)(t) = \int_{-\infty}^{\infty} \gamma_A(t,s)\varphi(s)ds, \qquad t \in \mathbb{R},$$

where

$$(2.5) \qquad \gamma_A(t,s) = \begin{cases} U(t)PU(s)^{-1} & , \ t > s, \\ -U(t)(I - P)U(s)^{-1}, & t < s. \end{cases}$$

Now let us return to the system Σ. Since (2.1) is assumed to have a dichotomy P, it follows (cf., [GKvS], Section I.2) that the input-output map of Σ extends to a bounded linear operator T_Σ acting from $L_2^r(\mathbb{R})$ into $L_2^m(\mathbb{R})$. In fact,

$$(2.6) \qquad T_\Sigma = M_D + M_C(\Delta - M_Z)^{-1}M_B,$$

or more explicitly

$$(2.7) \qquad (T_\Sigma\varphi)(t) = D(t)\varphi(t) + \int_{-\infty}^{\infty} C(t)\gamma_A(t,s)B(s)\varphi(s)ds, \qquad t \in \mathbb{R},$$

where $\gamma_A(t,s)$ is defined by (2.5). From the dichotomy inequalities (2.3a,b) and the boundedness of the coefficients $B(\cdot)$ and $C(\cdot)$ we see that there are constants $\widetilde{M} > 0$ and $0 < a < 1$ such that

$$\|C(t)\gamma_A(t,s)B(s)\| \le \widetilde{M}a^{|t-s|},$$

and hence the kernel function of T_Σ satisfies (1.5). Since the feedthrough coefficient $D(\cdot)$ is also (essentially) bounded, we conclude that $T_\Sigma \in \mathcal{W}^{m\times r}$.

By $\mathcal{R}^{m\times r}$ we denote the set of all operators $T: L_2^r(\mathbb{R}) \to L_2^m(\mathbb{R})$ that appear as the input-output operator of a system Σ of the type considered in the first paragraph of this section. In that case we call Σ or the right hand side of (2.6) a *realization* of T. (The symbol $\mathcal{R}$ stands here for rational.) We write $T \in \mathcal{LR}^{m\times r}$ if $T \in \mathcal{R}^{m\times r}$ and T is *lower triangular* (i.e., the kernel function of T is zero a.e. on $s \ge t$). Obviously, $\mathcal{LR}^{m\times r} \subset \mathcal{LW}^{m\times r}$. By $\mathcal{UR}^{m\times r}$ we denote the set of all $T \in \mathcal{R}^{m\times r}$ that are *upper triangular* (i.e., the kernel function of T is zero a.e. on $s \le t$). A system Σ with dichotomy $P = I$ will be called *causal*; in this case the corresponding input-output map T_Σ is in $\mathcal{LR}^{m\times r}$. If Σ has dichotomy $P = 0$, then Σ is said to be *anti-causal*, and this implies $T \in \mathcal{UR}^{m\times r}$.

We shall need the following inversion theorem.

THEOREM 2.1. *Let* $T \in \mathcal{R}^{m\times m}$ *with realization*

$$(2.8) \qquad\qquad T = I + M_C(\Delta - M_A)^{-1}M_B.$$

Put $A^\times(t) = A(t) - B(t)C(t)$. *Then* T *is invertible if and only if the equation*

$$(2.9) \qquad\qquad x'(t) = A^\times(t)x(t), \qquad -\infty < t < \infty,$$

has a dichotomy, and in that case

$$(2.10) \qquad\qquad T^{-1} = I - M_C(\Delta - M_{A^\times})^{-1}M_B.$$

PROOF. Note that $A^\times \in L_\infty^{n\times n}(\mathbb{R})$. Thus, by Theorem 1.1 in [BenG], the differential operator $\Delta - M_{A^\times}$ is invertible if and only if (2.9) has a dichotomy. From the realization in (2.8) we see that T is invertible if and only if $I + M_B M_C(\Delta - M_A)^{-1}$ is invertible. Now

$$(2.11) \qquad I + M_B M_C(\Delta - M_A)^{-1} = (\Delta - M_{A^\times})(\Delta - M_A)^{-1}.$$

Note that $\mathcal{D}(\Delta - M_{A^\times}) = \mathcal{D}(\Delta) = \mathcal{D}(\Delta - M_A)$. Thus $\Delta - M_{A^\times}$ maps $\mathcal{D}(\Delta - M_{A^\times})$ in a one-one way onto $L_2^n(\mathbb{R})$ if and only if the left hand side of (2.11) is invertible. With these remarks the proof is completed. $\square$

The analogue of Theorem 2.1 for input-output operators acting on l_p-spaces with $p = 1$ or $p = \infty$ appears in [GKvS], Section 1.3.

Let $z \in L_\infty^{m\times m}(\mathbb{R})$ be an astv-point. Then $(\Delta - M_z)^{-1} \in \mathcal{UR}^{m\times m}$. In fact, $(\Delta - M_z)^{-1}$ is the input-output operator of the system

$$(2.12) \qquad\qquad \begin{cases} x'(t) = z(t)x(t) + \varphi(t), & t \in \mathbb{R}, \\ g(t) = x(t). \end{cases}$$

Note that (2.12) has dichotomy $P = 0$.

PROPOSITION 2.2. *Let $z \in L_\infty^{m \times m}(\mathbb{R})$ be an astv-point, and let $Z(t,s)$ be the associated transition matrix. Let $T \in \mathcal{LR}^{m \times r}$ with causal realization*

$$T = M_D + M_C(\Delta - M_A)^{-1} M_B.$$

Then $(\Delta - M_z)^{-1} T$ is the input-output operator of the system

$$(13) \quad \begin{cases} \begin{pmatrix} x_1(t) \\ x_2(t) \end{pmatrix}' = \begin{pmatrix} A(t) & 0 \\ 0 & z(t) \end{pmatrix} \begin{pmatrix} x_1(t) \\ x_2(t) \end{pmatrix} + \begin{pmatrix} B(t) \\ D(t) + R(t)B(t) \end{pmatrix} \varphi(t), \quad t \in \mathbb{R}, \\ g(t) = \begin{pmatrix} R(t) & I \end{pmatrix} \begin{pmatrix} x_1(t) \\ x_2(t) \end{pmatrix}, \end{cases}$$

which has dichotomy $P = \begin{pmatrix} I & 0 \\ 0 & 0 \end{pmatrix}$. Here

$$(2.14) \qquad R(t) = \int_t^\infty Z(t,s)C(s)U(s)U(t)^{-1}\,ds, \qquad t \in \mathbb{R}.$$

In particular, $(\Delta - M_z)^{-1} T \in \mathcal{R}^{m \times r}$ and

$$(2.15) \qquad ((\Delta - M_z)^{-1} T\varphi)(t) = \int_{-\infty}^\infty \ell(t,s)\varphi(s)\,ds, \qquad t \in \mathbb{R},$$

with

$$(2.16) \qquad \ell(t,s) = \begin{cases} R(t)U(t)U(s)^{-1}B(s) & , \quad t > s, \\ -Z(t,s)(D(s) + R(s)B(s)), & t < s. \end{cases}$$

PROOF. The operator $(\Delta - M_z)^{-1} T$ is the input-output operator of the cascade connection of the system (2.12) with the system Σ, where Σ is as in the first paragraph of this section with $P = I$. This cascade connection is obtained by taking the input φ in (2.12) to be the output of Σ. Thus $(\Delta - M_z)^{-1} T$ is the input-output operator of

$$\begin{cases} \begin{pmatrix} \widetilde{x}_1(t) \\ \widetilde{x}_2(t) \end{pmatrix} = \begin{pmatrix} A(t) & 0 \\ C(t) & z(t) \end{pmatrix} \begin{pmatrix} \widetilde{x}_1(t) \\ \widetilde{x}_2(t) \end{pmatrix} + \begin{pmatrix} B(t) \\ D(t) \end{pmatrix} \varphi(t), \quad t \in \mathbb{R}, \\ g(t) = \begin{pmatrix} 0 & I \end{pmatrix} \begin{pmatrix} \widetilde{x}_1(t) \\ \widetilde{x}_2(t) \end{pmatrix}. \end{cases}$$

Now apply a time-varying state space similarity by setting

$$(2.17) \qquad \begin{pmatrix} x_1(t) \\ x_2(t) \end{pmatrix} := \begin{pmatrix} I & 0 \\ R(t) & I \end{pmatrix} \begin{pmatrix} \widetilde{x}_1(t) \\ \widetilde{x}_2(t) \end{pmatrix}, \qquad t \in \mathbb{R},$$

where $R(t)$ is given by (2.14). This similarity does not change the input-output operator of the system. Since

$$\frac{d}{dt}R(t) = -C(t) + z(t)R(t) - R(t)A(t), \qquad t \in \mathbb{R},$$

one checks that with the left hand side of (2.17) as the new state the resulting system is (2.13).

The evolution matrix associated with the differential equation

$$\begin{pmatrix} x_1(t) \\ x_2(t) \end{pmatrix}' = \begin{pmatrix} A(t) & 0 \\ 0 & z(t) \end{pmatrix} \begin{pmatrix} x_1(t) \\ x_2(t) \end{pmatrix}$$

is given by

$$\begin{pmatrix} U(t) & 0 \\ 0 & Z(t) \end{pmatrix},$$

where $Z(t) = Z(t,0)$. It follows that (2.13) has dichotomy $P = \begin{pmatrix} I & 0 \\ 0 & 0 \end{pmatrix}$. Thus $(\Delta - M_z)^{-1}T$ belongs to $\mathcal{R}^{m \times r}$, and one sees that $(\Delta - M_z)^{-1}T$ is the integral operator described by (2.15) and (2.16). □

For z and T as in Proposition 2.2 we have

$$(2.18) \qquad \widehat{T}(z) = D(t) + R(t)B(t) = \lim_{\alpha \uparrow 0} -\ell(t + \alpha, t),$$

where $R(\cdot)$ is given by (2.14) and ℓ by (2.16). The first identity in (2.18) follows from the definition of $\widehat{T}(z)$, and the second follows by inspection from (2.16).

The fact that $(\Delta - M_z)^{-1}T$ in Proposition 2.2 belongs to $\mathcal{R}^{m \times r}$ is a particular case of the following proposition.

PROPOSITION 2.3. *If $T \in \mathcal{R}^{m \times r}$ and $S \in \mathcal{R}^{p \times m}$, then $ST \in \mathcal{R}^{p \times r}$.*

PROOF. One can use the same arguments as for the proof of Theorem I.2.4 in [GKvS]. One has only to add that the state matrix of the cascade connection is again essentially bounded. □

3. RESIDUE CALCULUS AND DIAGONAL EXPANSION

Let $f \in L_1^{m \times m}(\mathbb{R})$, and consider its two-sided Laplace transform

$$(3.1) \qquad F(\lambda) = \int_{-\infty}^{\infty} e^{-\lambda t} f(t)\,dt, \qquad \lambda \in i\mathbb{R}.$$

Assume that F has a meromorphic extension in the open right half plane, and let us denote by $\mathcal{R}es^+(F)$ the sum of the residues of F in the open right half plane $\mathbb{C}^+$. It can be shown that

$$(3.2) \qquad \mathcal{R}es^+(F) = -\lim_{t \uparrow 0} f(t).$$

Furthermore, if $\operatorname{supp} f \subset [0, \infty)$ and $\omega \in \mathbb{C}^+$, then

$$(3.3) \qquad \mathcal{R}es^+\left(\frac{1}{\cdot - \omega}F(\cdot)\right) = F(\omega).$$

In this section we discuss the time-varying version of these results. The aim is to illuminate further the generalized point evaluation map.

Let $T \in \mathcal{W}^{m \times r}$, and assume that T is given by (1.4). Since the kernel function k of T satisfies (1.5), it follows that for almost each α the function $k(\cdot + \alpha, \cdot)$ is an $m \times r$ matrix function with entries in $L_\infty(\mathbb{R})$. We write $M(k; \alpha)$ for the operator of multiplication by $k(\cdot + \alpha, \cdot)$, i.e.,

$$\big(M(k, \alpha)\varphi\big)(t) = k(t + \alpha, t)\varphi(t), \qquad t \in \mathbb{R}.$$

By S_α we denote the operator of translation by $-\alpha$. Thus

$$(S_\alpha\varphi)(t) = \varphi(t - \alpha), \qquad t \in \mathbb{R}.$$

This allows us to rewrite (1.4) in the form

$$(3.4) \qquad\qquad T = M_D + \int_{-\infty}^{\infty} S_\alpha M(k, \alpha)d\alpha.$$

LEMMA 3.1. *The integral in (3.4) converges in the strong operator topology to $T - M_D$.*

PROOF. We may assume without loss of generality that $m = r = 1$. Take $\varphi, \psi \in L_2(\mathbb{R})$. Then $\overline{\psi(t)}k(t, s)\varphi(s)$ is measurable on $\mathbb{R} \times \mathbb{R}$ and we know from the proof of Lemma 1.2 (see formula (1.6)) that

$$(3.5) \qquad\qquad \overline{\psi(t)}k(t, s)\varphi(s)$$

is integrable on $\mathbb{R} \times \mathbb{R}$. So we can apply Fubini's theorem to show that

$$
\begin{aligned}
\int_{-\infty}^{\infty}\int_{-\infty}^{\infty} \overline{\psi(t)}k(t, s)\varphi(s)dsdt &= \int_{-\infty}^{\infty}\left(\int_{-\infty}^{\infty} \overline{\psi(s + \alpha)}k(s + \alpha, s)\varphi(s)ds\right)d\alpha \\
&= \int_{-\infty}^{\infty} \langle M(k, \alpha)\varphi, S_{-\alpha}\psi\rangle d\alpha \\
&= \int_{-\infty}^{\infty} \langle S_\alpha M(k, \alpha)\varphi, \psi\rangle d\alpha,
\end{aligned}
$$

(3.6)

and $\langle S_\alpha M(k, \alpha)\varphi, \psi\rangle$ is integrable as a function of α. In particular, $\langle S_\alpha M(k, \alpha)\varphi, \psi\rangle$ is measurable as a function of α. This holds for each ψ in $L_2(\mathbb{R})$. Now, let $\psi_1, \psi_2, \ldots$ be an orthonormal basis of $L_2(\mathbb{R})$. Then

$$S_\alpha M(k, \alpha)\varphi = \sum_{j=1}^{\infty}\langle S_\alpha M(k, \alpha)\varphi, \psi_j\rangle \psi_j,$$

which implies that the vector function $S_\alpha M(k, \alpha)\varphi$ is measurable (in α). On the other hand

$$\int_{-\infty}^{\infty} \|S_\alpha M(k, \alpha)\varphi\|d\alpha \le \left(\int_{-\infty}^{\infty} \ell_k(\alpha)d\alpha\right)\|\varphi\| < \infty,$$

where $\ell_k(\cdot)$ is as in (1.5), and hence the (Bochner) integral

$$(3.7) \qquad \int_{-\infty}^{\infty} S_\alpha M(k,\alpha)\varphi \, d\alpha$$

exists for each $\varphi \in L_2(\mathbb{R})$. From (3.6) we see that

$$\langle T\varphi, \psi \rangle = \langle M_D\varphi, \psi \rangle + \int_{-\infty}^{\infty} \int_{-\infty}^{\infty} \overline{\psi(t)} k(t,s)\varphi(s) \, ds \, dt$$

$$= \langle M_D\varphi, \psi \rangle + \int_{-\infty}^{\infty} \langle S_\alpha M(k,\alpha)\varphi, \psi \rangle \, d\alpha$$

$$= \langle M_D\varphi + \int_{-\infty}^{\infty} S_\alpha M(k,\alpha)\varphi \, d\alpha, \psi \rangle,$$

and thus (3.4) holds pointwise. $\square$

We shall refer to (3.4) as the *left diagonal expansion* of T. It is a time-varying analogue of (3.1). To make the connection more transparent, let $V(\lambda)$ be the operator of multiplication by $e^{\lambda t}$, where $\lambda \in i\mathbb{R}$. Then $V(\lambda)$ is unitary, $V(\lambda)^* = V(-\lambda)$, and

$$V(\lambda) T V(\lambda)^* = M_D + \int_{-\infty}^{\infty} e^{-\lambda\alpha} S_\alpha M(k,\alpha) \, d\alpha.$$

Formula (3.4) suggests the following definition for the total right half plane residue to T. For $h > 0$ put

$$\rho_h(t) = \frac{-1}{h} \int_{-h}^{0} k(t+\alpha, t) \, d\alpha.$$

Note that $\rho_h \in L_\infty^{m \times r}(\mathbb{R})$ because of (1.5). By integrating ρ_h over an arbitrary finite interval we see that ρ_h is uniquely determined by T and does not depend on the particular choice of the kernel function k in (1.4). We say that the *total right half plane residue* of T, $\mathcal{R}es^+(T)$ exists and is equal to ρ, whenever $\rho_h \to \rho$ in the weak-$*$ topology of $L_\infty^{m \times r}(\mathbb{R})$ for $h \downarrow 0$. In other words we set

$$(3.8) \qquad \mathcal{R}es^+(T) = \lim_{h \downarrow 0} \frac{-1}{h} \int_{-h}^{0} k(\cdot + \alpha, \cdot) \, d\alpha$$

with convergence in the weak-$*$ topology of $L_\infty^{m \times r}(\mathbb{R})$.

PROPOSITION 3.2. *Let* $z \in L_\infty^{m \times m}(\mathbb{R})$ *be an astv-point, and let* $T \in \mathcal{L}W^{m \times r}$. *Then the total right half plane residue of* $(\Delta - M_z)^{-1} T$ *exists and*

$$(3.9) \qquad \mathcal{R}es^+\big((\Delta - M_z)^{-1} T\big) = \widehat{T}(z).$$

PROOF. We know that $(\Delta - M_z)^{-1} T \in W^{m \times r}$. Let g be its kernel function. Since $(\Delta - M_z)^{-1}$ is upper triangular, Proposition 1.3 shows that

$$g(t,s) = -Z(t,s)y(s), \qquad s > t,$$

where $Z(t,s)$ is the transition matrix of the equation (1.1) and $y = \widehat{T}(z)$. Put $Z(t) = Z(t,0)$. Then $Z(t)$ is the evolution matrix of (1.1) and $Z(t,s) = Z(t)Z(s)^{-1}$. So for $h > 0$ we have

$$y(t) - \frac{1}{h}\int_{-h}^{0} Z(t+\alpha,t)y(t)d\alpha = \frac{1}{h}\int_{-h}^{0} y(t)d\alpha - \frac{1}{h}\int_{-h}^{0} Z(t+\alpha)Z(t)^{-1}y(t)d\alpha$$
$$= \left(\frac{1}{h}\int_{-h}^{0}\{I - Z(t+\alpha)Z(t)^{-1}\}d\alpha\right)y(t).$$

Now use (see formula (2.5) in [BenG]) that

$$\|Z(t)Z(s)^{-1} - I\| \le e^{|t-s|\|z\|_\infty} - 1, \qquad t,s \in \mathbb{R}.$$

Here $\|z\|_\infty$ is the norm of z as an element of $L_\infty^{m\times m}(\mathbb{R})$. It follows that

$$\left|y(t) - \frac{1}{h}\int_{-h}^{0} Z(t+\alpha,t)y(t)d\alpha\right| \le \left(\frac{1}{h}\int_{-h}^{0}\{1 - e^{\alpha\|z\|_\infty}\}\right)\|y\|_\infty \to 0 \quad (h \downarrow 0),$$

which convergence in the norm of $L_\infty^{m\times r}(\mathbb{R})$. $\square$

4. J-UNITARY AND J-INNER OPERATORS

In what follows we assume the reader to be familiar with the contents of Section 2.1 in [BallGK]. In particular, we shall use freely the notions of J-isometric, J-unitary, J-contractive and J-bicontractive operators.

4.1. J-inner lower triangular operators. In this section J is the signature operator on $L_2^{m+r}(\mathbb{R})$ defined by

$$(4.1) \qquad\qquad J = \begin{pmatrix} I_{L_2^m(\mathbb{R})} & 0 \\ 0 & -I_{L_2^r(\mathbb{R})} \end{pmatrix}.$$

Here we identify $L_2^{m+r}(\mathbb{R})$ with $L_2^m(\mathbb{R}) \oplus L_2^r(\mathbb{R})$, that is, any $f \in L_2^{m+r}(\mathbb{R})$ is written as $f = \binom{g}{h}$ with $g \in L_2^m(\mathbb{R})$ and $h \in L_2^r(\mathbb{R})$. The projection of $L_2^{m+r}(\mathbb{R})$ onto $L_2^r(\mathbb{R})$ along $L_2^m(\mathbb{R})$ is denoted by Π.

By P_τ, $\tau \in \mathbb{R}$, we denote the projection on $L_2^{m+r}(\mathbb{R})$ defined by

$$(P_\tau f)(t) = \begin{cases} f(t), & t \ge \tau, \\ 0, & t < \tau. \end{cases}$$

The complementary projection $I - P_\tau$ is denoted by Q_τ. Note that P_τ commutes with J. We write J_τ for the restriction of J to $\operatorname{Im} P_\tau$, and $\widetilde{J}_\tau$ will denote the restriction of J to $\operatorname{Im} Q_\tau$. The projection P_τ and Π also commute, and Π_τ (resp., $\widetilde{\Pi}_\tau$) denotes the restriction of Π to $\operatorname{Im} P_\tau$ (resp., $\operatorname{Im} Q_\tau$).

In what follows T is a bounded linear operator on $L_2^{m+r}(\mathbb{R})$. We say that T is *J-inner* if T is J-unitary (see [BallGK], Section 2.1) and

$$(4.2) \qquad Q_\tau^* T^* J Q_\tau T Q_\tau \le J Q_\tau, \qquad \tau \in \mathbb{R}.$$

We call T *lower triangular* (notation: $T \in \mathcal{L}$) if

$$(4.3) \qquad Q_\tau T = Q_\tau T Q_\tau, \qquad \tau \in \mathbb{R}.$$

Note that this definition of lower triangularity agrees with the one we have used in Section 1.2 for operators in the nonstationary Wiener algebra. Consider the following partitioning of T:

$$(4.4) \qquad T = \begin{pmatrix} A_\tau & B_\tau \\ C_\tau & D_\tau \end{pmatrix} : \operatorname{Im} Q_\tau \oplus \operatorname{Im} P_\tau \to \operatorname{Im} Q_\tau \oplus \operatorname{Im} P_\tau.$$

In terms of this partitioning T is lower triangular if and only if $B_\tau = 0$ for each $\tau \in \mathbb{R}$.

LEMMA 4.1. *Assume T is J-unitary. Then T is J-inner if and only if A_τ in (4.4) is $\widetilde{J}_\tau$-contractive for each $\tau \in \mathbb{R}$.*

PROOF. We have

$$\langle Q_\tau T^* J Q_\tau T Q_\tau f, f \rangle = \langle \widetilde{J}_\tau A_\tau (Q_\tau f), A_\tau (Q_\tau f) \rangle$$

and

$$\langle J Q_\tau f, f \rangle = \langle \widetilde{J}_\tau (Q_\tau f), (Q_\tau f) \rangle.$$

Thus (4.2) holds if and only if A_τ is $\widetilde{J}_\tau$-contractive. $\square$

Assume T is J-unitary and lower triangular. Thus

$$(4.5) \qquad T = \begin{pmatrix} A_\tau & 0 \\ C_\tau & D_\tau \end{pmatrix},$$

and

$$(4.6a) \qquad \begin{pmatrix} A_\tau^* & C_\tau^* \\ 0 & D_\tau^* \end{pmatrix} \begin{pmatrix} \widetilde{J}_\tau & 0 \\ 0 & J_\tau \end{pmatrix} \begin{pmatrix} A_\tau & 0 \\ C_\tau & D_\tau \end{pmatrix} = \begin{pmatrix} \widetilde{J}_\tau & 0 \\ 0 & J_\tau \end{pmatrix},$$

$$(4.6b) \qquad \begin{pmatrix} A_\tau & 0 \\ C_\tau & D_\tau \end{pmatrix} \begin{pmatrix} \widetilde{J}_\tau & 0 \\ 0 & J_\tau \end{pmatrix} \begin{pmatrix} A_\tau^* & C_\tau^* \\ 0 & D_\tau^* \end{pmatrix} = \begin{pmatrix} \widetilde{J}_\tau & 0 \\ 0 & J_\tau \end{pmatrix}.$$

Note that (4.6a) implies that $D_\tau^* J_\tau D_\tau = J_\tau$, and thus D_τ is a J_τ-isometry. In particular, D_τ is a J_τ-contraction.

We need another partitioning of T, namely

$$(4.7) \qquad T = \begin{pmatrix} T_{11} & T_{12} \\ T_{21} & T_{22} \end{pmatrix} : \operatorname{Ker} \Pi \oplus \operatorname{Im} \Pi \to \operatorname{Ker} \Pi \oplus \operatorname{Im} \Pi.$$

If T is J-unitary, then T is a J-bicontraction, and hence by Theorem 2.1 in [BallGK1], the operator T_{22} is invertible and

$$(4.8) \qquad \|T_{22}^{-1}T_{21}\| < 1.$$

We shall prove the following theorem.

THEOREM 4.2. *Let T be J-unitary and lower triangular. Then the following are equivalent:*

(i) *T is J-inner,*

(ii) *$TJP_\tau T^* \leq JP_\tau$, $\tau \in \mathbb{R}$,*

(iii) *T_{22}^{-1} is lower triangular.*

PROOF. (i) $\Rightarrow$ (ii). Since $T \in \mathcal{L}$, we have $T^*JQ_\tau T \leq JQ_\tau$. Now replace Q_τ by $I - P_\tau$ and use that $T^*JT = J$. It follows that $T^*JP_\tau T \geq JP_\tau$. But then

$$JP_\tau = TJ(T^*JP_\tau T)JT^*$$
$$\geq TJ(JP_\tau)JT^* = TP_\tau JT^*.$$

(ii) $\Rightarrow$ (iii). By (4.3) we have $P_\tau T P_\tau = TP_\tau$. Hence

$$\langle TJP_\tau T^* f, f \rangle = \langle P_\tau T P_\tau J P_\tau T^* P_\tau f, f \rangle$$
$$= \langle J_\tau D_\tau^*(P_\tau f), D_\tau^*(P_\tau f) \rangle,$$

and

$$\langle JP_\tau f, f \rangle = \langle J(P_\tau f), (P_\tau f) \rangle.$$

From these two identities we see that (ii) is equivalent to the requirement that D_τ^* is a J_τ-contraction. We already know that D_τ is a J_τ-contraction. Thus D_τ is a J_τ-bicontraction. Now consider the following partitionings:

$$(4.9a) \qquad A_\tau = \begin{pmatrix} (A_\tau)_{11} & (A_\tau)_{12} \\ (A_\tau)_{21} & (A_\tau)_{22} \end{pmatrix} : \operatorname{Ker}\widetilde{\Pi}_\tau \oplus \operatorname{Im}\widetilde{\Pi}_\tau \to \operatorname{Ker}\widetilde{\Pi}_\tau \oplus \operatorname{Im}\widetilde{\Pi}_\tau,$$

$$(4.9b) \qquad C_\tau = \begin{pmatrix} (C_\tau)_{11} & (C_\tau)_{12} \\ (C_\tau)_{21} & (C_\tau)_{22} \end{pmatrix} : \operatorname{Ker}\widetilde{\Pi}_\tau \oplus \operatorname{Im}\widetilde{\Pi}_\tau \to \operatorname{Ker}\Pi_\tau \oplus \operatorname{Im}\Pi_\tau,$$

$$(4.9c) \qquad D_\tau = \begin{pmatrix} (D_\tau)_{11} & (D_\tau)_{12} \\ (D_\tau)_{21} & (D_\tau)_{22} \end{pmatrix} : \operatorname{Ker}\Pi_\tau \oplus \operatorname{Im}\Pi_\tau \to \operatorname{Ker}\Pi_\tau \oplus \operatorname{Im}\Pi_\tau.$$

Since D_τ is a J_τ-bicontraction, we know (from Theorem 2.1 in [BallGK1]) that $(D_\tau)_{22}$ is invertible. Now note that

$$(4.10) \qquad T_{22} = \begin{pmatrix} (A_\tau)_{22} & 0 \\ (C_\tau)_{22} & (D_\tau)_{22} \end{pmatrix} : L_2^r((-\infty,\tau]) \oplus L_2^r([\tau,\infty))$$
$$\to L_2^r((-\infty,\tau]) \oplus L_2^r([r,\infty)).$$

Thus $(D_\tau)_{22}$ invertible implies that

$$(4.11) \qquad T_{22}^{-1} = \begin{pmatrix} * & 0 \\ * & * \end{pmatrix} : L_2^r((-\infty,\tau]) \oplus L_2^r([\tau,\infty)) \to L_2^r((-\infty,\tau]) \oplus L_2^r([\tau,\infty)).$$

This holds for each τ. Hence $T_{22}^{-1} \in \mathcal{L}$.

(iii) $\Rightarrow$ (i). Since $T_{22}^{-1} \in \mathcal{L}$, we know that T_{22}^{-1} has the form (4.11) for each $\tau \in \mathbb{R}$. Since T_{22} is as in (4.10), it follows that $(D_\tau)_{22}$ is invertible. But then (apply again Theorem 2.1 in [BallGK1]) we have that D_τ is a J_τ-bicontraction. In particular, D_τ^* is a J_τ-contraction. But the latter statement is equivalent with (ii). From (ii) follows (i) in the same way as (i) follows from (ii). $\square$

4.2. J-unitary input-output maps. This section concerns lower triangular J-unitary operators that appear as input-output maps of systems. We shall consider linear time-varying input-output systems Σ,

$$(4.12) \qquad \Sigma \begin{cases} x'(t) = A(t)x(t) + B(t)\varphi(t), & t \in \mathbb{R}, \\ g(t) = C(t)x(t) + \varphi(t), \end{cases}$$

such that $A \in L_\infty^{n \times n}(\mathbb{R})$, $B \in L_\infty^{n \times (m+r)}(\mathbb{R})$ and $C \in L_\infty^{(m+r) \times n}(\mathbb{R})$. The feedthrough coefficient is assumed to be equal to the $(m+r) \times (m+r)$ identity matrix. We shall assume that Σ is *causal*, i.e., the differential equation

$$(4.13) \qquad x'(t) = A(t)x(t), \qquad -\infty < t < \infty,$$

has dichotomy $P = I$. It follows (see Section 2) that the input-output map T_Σ is lower triangular.

In this section the symbol J (see (4.1)) will also be used for the signature matrix

$$(4.14) \qquad J = \begin{pmatrix} I_m & 0 \\ 0 & -I_r \end{pmatrix}.$$

From the context it will be clear which J we are dealing with. By a *causal J-unitary time-varying system* we shall mean a causal system Σ of the type described above with the additional property that the time-varying Lyapunov equation,

$$(4.15) \qquad R'(t) = A(t)R(t) + R(t)A(t)^* + B(t)JB(t)^*, \qquad t \in \mathbb{R},$$

has a solution $R(\cdot)$ such that

(a) $R(t)^* = R(t)$, $t \in \mathbb{R}$,

(b) $\sup_{t \in \mathbb{R}}\{\|R(t)\|, \|R(t)^{-1}\|\} < \infty$,

(c) $C(t)R(t) = -JB(t)^*$, $t \in \mathbb{R}$.

Here $R(\cdot)$ is said to be a solution of (4.15) if $R(\cdot)$ is absolutely continuous on each finite interval and the identity (4.15) holds true almost everywhere on $\mathbb{R}$. A causal J-unitary time-varying system Σ is *J-inner* if, in addition, the Hermitian matrix $R(t)$ is positive definite for each $t \in \mathbb{R}$.

THEOREM 4.3. *The input-output map of a causal J-unitary time-varying system is a lower triangular J-unitary operator on $L_2^{m+r}(\mathbb{R})$.*

PROOF. Let $T = T_\Sigma$ with Σ as in (4.12) such that Σ is causal and J-unitary. First let us prove that T is invertible. To do this we pass to the inverse system

$$\Sigma^\times \begin{cases} x'(t) = \big(A(t) - B(t)C(t)\big)x(t) + B(t)g(t), & t \in \mathbb{R}, \\ \varphi(t) = -C(t)x(t) + g(t), \end{cases}$$

and apply the similarity $z(t) = R(t)^{-1}x(t)$. This transforms $\Sigma^\times$ into the system

$$\Sigma^\sim \begin{cases} z'(t) = -A(t)^* z(t) - C(t)^* Jg(t), & t \in \mathbb{R}, \\ \varphi(t) = JB(t)^* z(t) + g(t). \end{cases}$$

To see this note that

$$(4.16)\quad \begin{aligned} \frac{d}{dt} R(t)^{-1} &= -R(t)^{-1} R(t)' R(t)^{-1} \\ &= -R(t)^{-1}A(t) - A(t) - A(t)^* R(t)^{-1} - R(t)^{-1}B(t)JB(t)^* R(t)^{-1}, \qquad t \in \mathbb{R}. \end{aligned}$$

Thus

$$\begin{aligned} z'(t) = &-R(t)^{-1}A(t)x(t) - A(t)^* z(t) - R(t)^{-1}B(t)JB(t)^* R(t)^{-1}x(t) \\ &+ R(t)^{-1}\big(A(t) - B(t)C(t)\big)x(t) + R(t)^{-1}B(t)g(t). \end{aligned}$$

Next use that $R(t)^{-1}B(t) = -C(t)^* J$, because of (c), and we obtain the first equation in $\Sigma^\sim$. As

$$C(t)x(t) = C(t)R(t)z(t) = -JB(t)^* z(t),$$

we also obtain the second. Now note that the evolution matrix of the differential equation

$$(4.17)\qquad\qquad x'(t) = -A(t)^* x(t), \qquad -\infty < t < \infty,$$

is equal to $U(t)^{-*}$, where $U(t)$ is the evolution matrix of (4.13). According to our hypothesis (4.13) has dichotomy $P = I$. Thus (4.17) has dichotomy $P = 0$. It follows that T is invertible and $T^{-1} = T_{\Sigma^\sim}$, the input-output operator of $\Sigma^\sim$. Thus

$$(T\varphi)(t) = \varphi(t) + \int_{-\infty}^{t} C(t)U(t)U(s)^{-1}B(s)\varphi(s)ds,$$

$$(T^{-1}g)(t) = g(t) + \int_{t}^{\infty} JB(t)^* U(t)^{-*}U(s)^* C(s)Jg(s)ds,$$

and therefore $T^{-1} = JT^* J$. $\quad\square$

LEMMA 4.4. *Assume that the system Σ in (4.12) is J-unitary, and let $R(\cdot)$ be the corresponding solution of (4.15). Then*

$$(4.18) \qquad \frac{d}{dt}\langle R(t)^{-1}x(t), x(t)\rangle = \langle J\varphi(t), \varphi(t)\rangle - \langle Jg(t), g(t)\rangle, \qquad t \in \mathbb{R}.$$

PROOF. By using (4.17) and the first equation in (4.12) we get

$$\begin{aligned}
\frac{d}{dt}\langle R(t)^{-1}x(t), x(t)\rangle &= \langle -R(t)^{-1}A(t)x(t) - A(t)^*R(t)^{-1}x(t), x(t)\rangle \\
&\quad + \langle -R(t)^{-1}B(t)JB(t)^*R(t)^{-1}x(t), x(t)\rangle \\
&\quad + \langle R(t)^{-1}A(t)x(t) + R(t)^{-1}B(t)\varphi(t), x(t)\rangle \\
&\quad + \langle x(t), R(t)^{-1}A(t)x(t) + R(t)^{-1}B(t)\varphi(t)\rangle \\
&= \langle -R(t)^{-1}B(t)JB(t)^*R(t)^{-1}x(t), x(t)\rangle + \langle R(t)^{-1}B(t)\varphi(t), x(t)\rangle \\
&\quad + \langle x(t), R(t)^{-1}B(t)\varphi(t)\rangle.
\end{aligned}$$

Since $C(t) = -JB(t)^*R(t)^{-1}$, the second equation in (4.12) yields:

$$\begin{aligned}
\frac{d}{dt}\langle R(t)^{-1}x(t), x(t)\rangle &= \langle R(t)^{-1}B(t)g(t), x(t)\rangle + \langle -JC(t)x(t), \varphi(t)\rangle \\
&= \langle Jg(t), -C(t)x(t)\rangle + \langle -JC(t)x(t), \varphi(t)\rangle \\
&= \langle Jg(t), \varphi(t) - g(t)\rangle + \langle J\varphi(t) - Jg(t), \varphi(t)\rangle \\
&= \langle J\varphi(t), \varphi(t)\rangle - \langle Jg(t), g(t)\rangle. \qquad \square
\end{aligned}$$

The matrix function $R(t)^{-1}$ in Lemma 4.4 is the so-called storage function appearing in the papers [HiM] and [W].

THEOREM 4.5. *Let T be the input-output map of the causal J-unitary time-varying system Σ. Then T is J-inner if and only if the system Σ is J-inner.*

PROOF. Since Σ is J-unitary we have a solution $R(\cdot)$ of (4.15) with the properties (a)–(c). Fix $\tau \in \mathbb{R}$ and let $\tau > t$. By (4.18) we have

$$\begin{aligned}
&\langle R(\tau)^{-1}x(\tau), x(\tau)\rangle - \langle R(t)^{-1}x(t), x(t)\rangle \\
&= \int_t^\tau \langle J\varphi(s), \varphi(s)\rangle ds - \int_t^\tau \langle Jg(s), g(s)\rangle ds,
\end{aligned} \qquad (4.19)$$

where x, φ and g are as in (4.12). Both φ and g are square integrable on $\mathbb{R}$. Thus for $t \to -\infty$ the right hand side of (4.19) has a limit. It follows that

$$(4.20) \qquad \lim_{t \to -\infty} \langle R(t)^{-1}x(t), x(t)\rangle$$

exists. Since x is square integrable over $\mathbb{R}$, property (b) of $R(\cdot)$ implies that

$$\int_{-\infty}^\infty |\langle R(t)^{-1}x(t), x(t)\rangle| dt$$

exists. But then the limit in (4.20) must be zero, and hence we have

$$\langle R(\tau)^{-1}x(\tau), x(\tau)\rangle = \int_{-\infty}^{\tau} \langle J\varphi(s), \varphi(s)\rangle ds - \int_{-\infty}^{\tau} \langle Jg(s), g(s)\rangle ds.$$

Now recall that $T\varphi = g$, and let us use the projection $Q_\tau = I - P_\tau$. It follows that

$$\langle R(\tau)^{-1}x(\tau), x(\tau)\rangle = \int_{-\infty}^{\infty} \langle J(Q_\tau\varphi)(s), (Q_\tau\varphi)(s)\rangle ds - \int_{-\infty}^{\infty} \langle J(Q_\tau T\varphi)(s), (Q_\tau T\varphi)(s) ds$$
$$= \langle JQ_\tau\varphi, Q_\tau\varphi\rangle - \langle JQ_\tau T\varphi, Q_\tau T\varphi\rangle$$
$$= \langle JQ_\tau\varphi, \varphi\rangle - \langle T^*JQ_\tau T\varphi, \varphi\rangle.$$

Here we use that J and Q_τ commute. It follows that Σ is J-inner if and only if

$$(4.21) \qquad\qquad T^*JQ_\tau T \le JQ_\tau, \qquad \tau \in \mathbb{R}.$$

Since T is lower triangular, $T^*Q_\tau = Q_\tau T^*Q_\tau$. Thus (4.21) means that T is J-inner, which proves the theorem. $\square$

Theorem 4.3 (with the words causal and lower triangular deleted) and Lemma 4.4 hold true for arbitrary J-unitary systems. Also the part of the proof of Theorem 4.5 that precedes (4.21) does not use causality.

For later purposes (see Section 5.4 below) we mention the following proposition, which may be viewed as a supplement to Theorem 2.1.

PROPOSITION 4.6. *Let $T \in \mathcal{R}^{m\times m}$ be invertible, and assume that both T and T^{-1} are lower triangular. Then $T^{-1} \in \mathcal{R}^{m\times m}$.*

PROOF. Fix $-\infty < a < b < \infty$, and let us decompose $L_2^m(\mathbb{R})$ as

$$(4.22) \qquad\qquad L_2^m(\mathbb{R}) = H_1 \oplus H_0 \oplus H_2,$$

where

$$H_1 = \{\varphi \in L_2^m(\mathbb{R}) \mid \operatorname{supp}\varphi \subset (-\infty, a)\},$$
$$H_0 = \{\varphi \in L_2^m(\mathbb{R}) \mid \operatorname{supp}\varphi \subset [a, b]\},$$
$$H_2 = \{\varphi \in L_2^m(\mathbb{R}) \mid \operatorname{supp}\varphi \subset (b, \infty)\}.$$

Since T and T^{-1} are both lower triangular, $\operatorname{Im} P_\tau$ is invariant under T and T^{-1} for each τ. It follows that the spaces H_2 and $H_0 \oplus H_2$ are invariant under T and T^{-1}. Hence with respect to the decomposition (4.22) the operators T and T^{-1} admit the following partitioning:

$$(4.23) \qquad T = \begin{bmatrix} T_{11} & 0 & 0 \\ * & T_{00} & 0 \\ * & * & T_{22} \end{bmatrix}, \qquad T^{-1} = \begin{bmatrix} T_{11}^\times & 0 & 0 \\ * & T_{00}^\times & 0 \\ * & * & T_{22}^\times \end{bmatrix}.$$

Now use that TT^{-1} and $T^{-1}T$ are equal to the identity operator on $L_2^m(\mathbb{R})$. It follows that T_{00} is invertible and $T_{00}^{-1} = T_{00}^\times$.

By our hypotheses $T \in \mathcal{R}^{m \times m}$, and hence T admits a realization, i.e.,

$$(4.24) \qquad T = M_D + M_C(\Delta - M_A)^{-1} M_B.$$

It follows that the action of T_{00} is given by

$$(T_{00}\varphi)(t) = D(t)\varphi(t) + \int_a^b k(t,s)\varphi(s)ds, \qquad a \le t \le b,$$

where

$$(4.25) \qquad k(t,s) = C(t)\gamma_A(t,s)B(s),$$

with γ_A being given by (2.5). From (4.25) we see that $k(\cdot,\cdot)$ is a Hilbert-Schmidt kernel on $[a,b] \times [a,b]$. Since T_{00} is invertible, it follows that M_D acts as a Fredholm operator on $L_2^m([a,b])$. But M_D is an operator of multiplication. So we may conclude that M_D restricted to $L_2^m([a,b])$ is invertible. This implies that $D(t)$ is non-singular for almost all $a \le t \le b$ and

$$(4.26) \qquad \operatorname{ess\,sup}\{\|D(t)^{-1}\| \mid a \le t \le b\} < \infty.$$

Furthermore, we see that

$$(4.27) \qquad T_{00}^{\times} = T_{00}^{-1} = M_E + L,$$

where $E(t) = D(t)^{-1}$ for $a \le t \le b$ and L is a Hilbert-Schmidt operator.

Next, we show that

$$(4.28) \qquad \|M_E + L\| \ge \|M_E\|.$$

Put $E^*(t) = E(t)^*$ for $a \le t \le b$. Then

$$\|M_E + L\|^2 = \|(M_E + L)^*(M_E + L)\| = \|M_{E^*E} + \widetilde{L}\|$$

with $\widetilde{L}$ being a selfadjoint Hilbert-Schmidt operator. Note that $\lambda - M_{E^*E} = M_F$, where $F(t) = \lambda - E(t)^*E(t)$ for $a \le t \le b$. So, if $\lambda \in \sigma(M_{E^*E})$, the spectrum of M_{E^*E}, then $\lambda - M_{E^*E}$ is not Fredholm, and therefore $\lambda - M_{E^*E} - \widetilde{L}$ is not Fredholm, because $\widetilde{L}$ is compact. It follows that $\sigma(M_{E^*E}) \subset \sigma(M_{E^*E} + L)$. Now, use that M_{E^*E} is selfadjoint and apply Theorem V.2.1. We conclude that

$$\|M_E\|^2 = \|M_{E^*E}\| \le \|M_{E^*E} + \widetilde{L}\|,$$

and hence (4.28) is proved.

From (4.28) it follows that

$$(4.29) \qquad \operatorname{ess\,sup}\{\|D(t)^{-1}\| \mid t \in \mathbb{R}\} < \infty.$$

Indeed, from the partitioning of T^{-1} we see that $\|T_{00}^{\times}\| \leq \|T^{-1}\|$. So, by (4.27) and (4.28), the number in the left hand side of (4.26) has $\|T^{-1}\|$ as an upper bound. This holds for each compact interval $[a, b]$. So the number in the left hand side of (4.29) is less than or equal to $\|T^{-1}\|$. In particular, (4.29) holds.

From (4.29) it follows that M_D is invertible as operator on $L_2^m(\mathbb{R})$. Thus

$$(4.30) \qquad S := M_D^{-1}T = I + M_{\widetilde{C}}(\Delta - M_A)^{-1}M_B$$

is invertible. Here $\widetilde{C}(t) = D(t)^{-1}C(t)$ for $t \in \mathbb{R}$, and $A(\cdot)$, $B(\cdot)$ and $C(\cdot)$ are given by the realization (4.24). Note that S is invertible. So we may apply Theorem 2.1 to the right hand side of (4.30). This yields $S^{-1} \in \mathcal{R}^{m \times m}$, and hence $T^{-1} \in \mathcal{R}^{m \times m}$. $\square$

5. TIME-VARYING NEVANLINNA-PICK INTERPOLATION

5.1. Problem and main theorem. In this section we introduce the time-varying tangential Nevanlinna-Pick interpolation problem and state its solution. The given data are row vector valued functions

$$x_1, \ldots, x_N \in L_\infty^{1 \times m}(\mathbb{R}),$$
$$y_1, \ldots, y_N \in L_\infty^{1 \times r}(\mathbb{R}),$$

and anti-stable time-varying points

$$z_1, \ldots, z_N \in L_\infty(\mathbb{R}).$$

Let $Z_1(t, s), \ldots, Z_N(t, s)$ be the transition matrices associated with the astv-points $z_1, \ldots, z_N$. For $i, j = 1, \ldots, N$ put

$$(5.1) \qquad \omega_{ij}(t) = \int_t^\infty Z_i(t, \alpha)x_i(\alpha)x_j(\alpha)^* Z_j(t, \alpha)^* d\alpha.$$

In what follows we require additionally that for some $\varepsilon_0 > 0$ we have

$$(5.2) \qquad \begin{pmatrix} \omega_{11}(t) & \cdots & \omega_{1N}(t) \\ \vdots & & \vdots \\ \omega_{N1}(t) & \cdots & \omega_{NN}(t) \end{pmatrix} \geq \varepsilon_0 I_N, \qquad t \in \mathbb{R}.$$

Here I_N stands for the $N \times N$ identity matrix. If the condition (5.2) is fulfilled, then we call

$$(5.3) \qquad \{x_i, y_i, z_i : i = 1, \ldots, N\}$$

an *admissible* CTNPI *data set*. The problem we want to consider is the following. Find all lower triangular operators $F \in \mathcal{L}\mathcal{R}^{m \times r}$ with $\|F\| < 1$ such that

$$(5.4) \qquad (M_{x_j}F)^{\widehat{\ }}(z_j) = y_j, \qquad j = 1, \ldots, N.$$

Here the norm is the usual operator norm. Note that we seek interpolants F that are input-output maps of systems of the type considered in the first paragraph of Section 2. In the time-invariant case the latter means that the interpolants are required to be rational matrix functions.

For the solution of the above problem we need the functions:

$$(5.5) \qquad h_{ij}(t) = \int_t^\infty Z_i(t,\alpha)\{x_i(\alpha)x_j(\alpha)^* - y_i(\alpha)y_j(\alpha)^*\}Z_j(t,\alpha)^* d\alpha, \qquad t \in \mathbb{R}.$$

The following theorem is our main result.

THEOREM 5.1. *Let* $\{x_i, y_i, z_i : i = 1, \ldots, N\}$ *be an admissible* CTNPI *data set, and let* $h_{ij}(t)$ *be as in* (5.5). *Then there exists* $F \in \mathcal{LW}^{m \times r}$ *such that*

(i) $(M_{x_j} F)^\wedge(z_j) = y_j$, $j = 1, \ldots, N$,

(ii) $\|F\| < 1$,

if and only if for some $\varepsilon > 0$ *we have*

$$(5.6) \qquad \begin{pmatrix} h_{11}(t) & \cdots & h_{N1}(t) \\ \vdots & & \vdots \\ h_{1N}(t) & \cdots & h_{NN}(t) \end{pmatrix} \geq \varepsilon I_N, \qquad t \in \mathbb{R}.$$

In this case all operators $F \in \mathcal{LR}^{m \times r}$ *satisfying* (i) *and* (ii) *are given by*

$$(5.7) \qquad F = (\theta_{11} G + \theta_{12})(\theta_{21} G + \theta_{22})^{-1},$$

where G *is an arbitrary operator in* $\mathcal{LR}^{m \times r}$ *such that* $\|G\| < 1$. *Furthermore,*

$$(5.8) \qquad \begin{pmatrix} \theta_{11} & \theta_{12} \\ \theta_{21} & \theta_{22} \end{pmatrix} = I + J M_B^* (\Delta - M_A)^{-*} M_H^{-1} M_B,$$

with

$$A(t) = \begin{pmatrix} z_1(t) & 0 & \cdots & 0 \\ 0 & z_2(t) & \cdots & 0 \\ \vdots & \vdots & \cdots & \vdots \\ 0 & 0 & \cdots & z_N(t) \end{pmatrix}, \qquad t \in \mathbb{R},$$

$$B(t) = \begin{pmatrix} x_1(t) & -y_1(t) \\ x_2(t) & -y_2(t) \\ \vdots & \vdots \\ x_N(t) & -y_N(t) \end{pmatrix}, \qquad t \in \mathbb{R},$$

and J *is the operator on* $L_2^{m \times r}(\mathbb{R})$ *of multiplication on the left by*

$$J(t) = J = \begin{pmatrix} I_m & 0 \\ 0 & -I_r \end{pmatrix}, \qquad t \in \mathbb{R}.$$

For the time-invariant case the above theorem reduces to the tangential Nevanlinna-Pick interpolation theorem of [BGR1]. To see this let us take $x_j(t) \equiv x_j$, $y_j(t) \equiv y_j$ and $z_j(t) \equiv z_j$ for $j = 1, \ldots, N$. We know (see Section 1.1) that $z_1, \ldots, z_N$ are ordinary points in the open right half plane and $Z_j(t,s) = e^{z_j(t-s)}$ for $j = 1, \ldots, N$. Thus

$$
\omega_{ij}(t) = \int_t^\infty e^{z_i(t-\alpha)} x_i x_j^* e^{\overline{z}_j(t-\alpha)} \, d\alpha
$$

$$
= \int_0^\infty e^{-(z_i+\overline{z}_j)\alpha} x_i x_j^* \, d\alpha = \frac{x_i x_j^*}{z_i + \overline{z}_j},
$$

and hence

$$
\begin{pmatrix} \omega_{11}(t) & \cdots & \omega_{1N}(t) \\ \vdots & & \vdots \\ \omega_{N1}(t) & \cdots & \omega_{NN}(t) \end{pmatrix} = \begin{pmatrix} x_1 & & \\ & \ddots & \\ & & x_N \end{pmatrix} \left(\frac{1}{z_i + \overline{z}_j} \right)_{i,j=1}^N \begin{pmatrix} x_1^* & & \\ & \ddots & \\ & & x_N^* \end{pmatrix}.
$$

It follows that for this case condition (5.2) reduces to the requirement that the points $z_1, \ldots, z_N$ are different and the vectors $x_1, \ldots, x_N$ are nonzero. Furthermore, we see that

$$
\left(h_{ij}(t) \right)_{i,j=1}^N = \left(\frac{x_i x_j^* - y_i y_j^*}{z_i + \overline{z}_j} \right)_{i,j=1}^N,
$$

and thus (5.6) is the positive definiteness condition on the Pick matrix.

Returning to the general case we note that the functions ω_{ij} and h_{ij} from (5.1) and (5.5) may be rewritten as

$$
(5.9) \qquad \omega_{ij} = -\left[M_{x_i} M_{x_j}^* (\Delta - M_{z_j})^{-*} \right]^{\widehat{\,}}(z_i),
$$

$$
(5.10) \qquad h_{ij} = -\left[(M_{x_i} M_{x_j}^* - M_{y_i} M_{y_j}^*)(\Delta - M_{z_j})^{-*} \right]^{\widehat{\,}}(z_i).
$$

From these identities it follows that ω_{ij} and h_{ij} are in $L_\infty(\mathbb{R})$.

5.2. Proof of necessity. In this section we proof the necessity of the condition (5.6). So in what follows we assume that we have an operator $F \in \mathcal{L}W^{m \times r}$ such that the conditions (i) and (ii) in Theorem 5.1 are fulfilled.

The norm constraint (ii) implies that the operator L_F defined by (1.16) has norm strictly less than one. By duality, the same holds for L_F^*. Now apply Lemma 1.7. Let $d_1, \ldots, d_N$ be continuous complex-valued functions on $\mathbb{R}$ with compact support, and put

$$
\gamma := \left\| \sum_{j=1}^N M_{x_j}^* (\Delta - M_{z_j})^{-*} M_{d_j} \right\|_2^2,
$$

$$
\eta := \left\| \sum_{j=1}^N M_{y_j}^* (\Delta - M_{z_j})^{-*} M_{d_j} \right\|_2^2,
$$

where $\| \cdot \|_2$ denotes the Hilbert-Schmidt norm. Since $d_1, \ldots, d_N$ are in $L_2(\mathbb{R}) \cap L_\infty(\mathbb{R})$ and $\|L_F^*\| < 1$, we know from Lemma 1.7 that

$$(5.11) \qquad \gamma - \eta \geq \left(1 - \|L_F^*\|\right)\gamma.$$

Now $\gamma = \sum_{i=1}^N \sum_{j=1}^N \operatorname{tr} K_{ij}$, where K_{ij} is the trace class integral operator with kernel function k_{ij} given by

$$k_{ij}(t,s) = \overline{d_i(t)}\left(\int_{\max(t,s)}^\infty Z_i(t,\alpha)x_i(\alpha)x_j(\alpha)^* Z_j(s,\alpha)^* d\alpha\right)d_j(s).$$

The function k_{ij} is continuous on $\mathbb{R} \times \mathbb{R}$. It follows (see [GGK], Theorem VII.2.3) that

$$\operatorname{tr} K_{ij} = \int_{-\infty}^\infty k_{ij}(t,t)dt = \int_{-\infty}^\infty \overline{d_i(t)}\omega_{ij}(t)d_j(t)dt.$$

But then

$$\begin{aligned}
\gamma &= \sum_{i=1}^N \sum_{j=1}^N \int_{-\infty}^\infty \overline{d_i(t)}\omega_{ij}(t)d_j(t)dt \\
&= \int_{-\infty}^\infty \left\langle \begin{pmatrix} \omega_{11}(t) & \cdots & \omega_{1N}(t) \\ \vdots & & \vdots \\ \omega_{N1}(t) & \cdots & \omega_{NN}(t) \end{pmatrix} \begin{pmatrix} d_1(t) \\ \vdots \\ d_N(t) \end{pmatrix}, \begin{pmatrix} d_1(t) \\ \vdots \\ d_N(t) \end{pmatrix} \right\rangle dt \\
&\geq \varepsilon_0 \sum_{i=1}^N \|d_i\|^2,
\end{aligned}$$

where ε_0 is as in (5.2). In a similar way as for γ one shows that

$$\gamma - \eta = \sum_{i=1}^N \sum_{j=1}^N \int_{-\infty}^\infty \overline{d_i(t)}h_{ij}(t)d_j(t)dt.$$

It follows (use (5.11)) that

$$(5.12) \qquad \begin{aligned}
&\int_{-\infty}^\infty \left\langle \begin{pmatrix} h_{11}(t) & \cdots & h_{1N}(t) \\ \vdots & & \vdots \\ h_{N1}(t) & \cdots & h_{NN}(t) \end{pmatrix} \begin{pmatrix} d_1(t) \\ \vdots \\ d_N(t) \end{pmatrix}, \begin{pmatrix} d_1(t) \\ \vdots \\ d_N(t) \end{pmatrix} \right\rangle dt \\
&\geq \left(1 - \|L_F^*\|^2\right)\varepsilon_0 \left(\sum_{i=1}^N \|d_i\|^2\right).
\end{aligned}$$

Finally we use that the continuous functions with compact support are dense in $L_2(\mathbb{R})$. Thus (5.12) holds for all $d_1, \ldots, d_N$ in $L_2(\mathbb{R})$, and hence we have (5.6) with $\varepsilon = \left(1 - \|L_F^*\|^2\right)\varepsilon_0$.

5.3. Auxiliary results and homogeneous interpolation. In this section we derive some auxiliary results which we shall need for the proof of the sufficiency part of Theorem 5.1. Throughout this section we assume that condition (5.6) is fulfilled.

Put

$$(5.13) \qquad H(t) = \begin{pmatrix} h_{11}(t) & \cdots & h_{1N}(t) \\ \vdots & & \vdots \\ h_{N1}(t) & \cdots & h_{NN}(t) \end{pmatrix}, \qquad t \in \mathbb{R},$$

where h_{ij} is defined by (5.5). We already know (see (5.10)) that the functions h_{ij} are in $L_\infty(\mathbb{R})$. Also, $h_{ij}(t)$ depends continuously on t. Thus condition (5.6) may be summarized in the following way:

(a)$'$ $H(t)$ is positive definite for each $t \in \mathbb{R}$,

(b)$'$ $\sup_{t \in \mathbb{R}}\{\|H(t)\|, \|H(t)^{-1}\|\} < \infty$.

From the definition of h_{ij} in (5.5) we also see that the entries of $H(t)$ are absolutely continuous on finite intervals on $\mathbb{R}$ and

$$(5.14) \qquad \frac{d}{dt}H(t) = A(t)H(t) + H(t)A(t)^* - B(t)JB(t)^*, \qquad t \in \mathbb{R},$$

where $A(t)$, $B(t)$ and J are as in Theorem 5.1. Since $H(t)$ is invertible, we can compute $\frac{d}{dt}H(t)^{-1}$ via

$$\frac{d}{dt}H(t)^{-1} = -H(t)^{-1}\left(\frac{d}{dt}H(t)\right)H(t)^{-1}$$

which yields:
(5.15)
$$\frac{d}{dt}H(t)^{-1} = -A(t)^*H(t)^{-1} - H(t)^{-1}A(t) + H(t)^{-1}B(t)JB(t)^*H(t)^{-1}, \qquad t \in \mathbb{R}.$$

From our conditions on $z_1, \ldots, z_N$ it follows that $A \in L_\infty^{N \times N}(\mathbb{R})$ and A is an astv-point. Furthermore, $B \in L_\infty^{N \times (m+r)}$, and from (5.10) we see that

$$(5.16) \qquad H = -\{M_B J M_B^*(\Delta - M_A)^{-*}\}\hat{\ }(A),$$

where J is now the operator on $L_2^{m+r}(\mathbb{R})$ defined by left multiplication with

$$(5.17) \qquad J(t) = J = \begin{pmatrix} I_m & 0 \\ 0 & -I_r \end{pmatrix}, \qquad t \in \mathbb{R}.$$

LEMMA 5.2. *The operator*

$$(5.18) \qquad \theta = I + JM_B^*(\Delta - M_A)^{-*}M_H^{-1}M_B$$

is a lower triangular J-inner operator on $L_2^{m+r}(\mathbb{R})$.

PROOF. Consider the system

$$(5.19) \qquad \Sigma \begin{cases} x'(t) = \mathcal{A}(t)x(t) + \mathcal{B}(t)\varphi(t), & t \in \mathbb{R}, \\ g(t) = \mathcal{C}(t)x(t) + \varphi(t), \end{cases}$$

where $\mathcal{A}(t) = -A(t)^*$, $\mathcal{B}(t) = H(t)^{-1}B(t)$ and $\mathcal{C}(t) = -JB(t)^*$. The differential equation

$$(5.20) \qquad x'(t) = \mathcal{A}(t)x(t), \qquad -\infty < t < \infty,$$

has dichotomy $P = I$. Indeed, the evolution matrix of (5.20) is $U(t)^{-*}$, where $U(t)$ is the evolution matrix of $x'(t) = A(t)x(t)$. From this observation it also follows that

$$(5.21) \qquad (\Delta - M_{\mathcal{A}})^{-1} = -(\Delta - M_A)^{-*},$$

and thus $\theta = T_\Sigma$, the input-output operator of Σ.

Put $R(t) = H(t)^{-1}$. From (5.15) we see that

$$(5.22) \qquad \frac{d}{dt}R(t)^{-1} = A(t)R(t) + R(t)A(t)^* + B(t)JB(t)^*, \qquad t \in \mathbb{R}.$$

Furthermore, $\mathcal{C}(t)R(t) = -JB(t)^*H(t)^{-1} = -JB(t)^*$. Thus, conditions (a), (b) and (c) in Section 4.2 hold true for Σ. But then we can apply Theorems 4.3 and 4.5 to show that θ has the desired properties. $\square$

LEMMA 5.3. *Let θ be as in (5.18). Then*

$$(5.23) \qquad (M_B\theta)\hat{\ }(A) = 0.$$

PROOF. From the representation (5.18) it follows that the left hand side of (5.23) is equal to

$$B + \big\{ M_B J M_B^*(\Delta - M_A)^{-*} M_H^{-1} M_B \big\}\hat{\ }(A).$$

Now apply formula (1.12) with $T = M_B J M_B^*(\Delta - M_A)^{-*}$ and $S = M_H^{-1}M_B$, and use (5.16). It follows that

$$\begin{aligned} (M_B\theta)\hat{\ }(A) &= B + (-M_H M_H^{-1} M_B)\hat{\ }(A) \\ &= B - B = 0. \quad \square \end{aligned}$$

THEOREM 5.4. *Let θ be as in (5.18). Then $\theta\mathcal{L}\mathcal{W}^{(m+r)\times 1}$ is precisely the set of all $L \in \mathcal{L}\mathcal{W}^{(m+r)\times 1}$ such that $(M_B L)\hat{\ }(A) = 0$.*

PROOF. Since θ is the input-output operator of the system Σ in (5.19) we know that $\theta \in \mathcal{L}\mathcal{R}^{(m+r)\times(m+r)}$. Now let $L = \theta S$ with $S \in \mathcal{L}\mathcal{W}^{(m+r)\times 1}$. Then L is the product of two lower triangulars, and hence L is lower triangular. From $\mathcal{R} \subset \mathcal{W}$ and $\mathcal{W}$ is an algebra we conclude that $L \in \mathcal{L}\mathcal{W}^{(m+r)\times 1}$. Furthermore, by formula (1.12) and Lemma 5.3, we have

$$\begin{aligned} (M_B L)\hat{\ }(A) &= (M_B\theta S)\hat{\ }(A) \\ &= (M_{\widehat{M_B\theta(A)}} S)\hat{\ }(A) = 0. \end{aligned}$$

To prove the reverse implication, take L in $\mathcal{L}\mathcal{W}^{(m+r)\times 1}$ such that $(M_B L)^{\wedge}(A)$ $= 0$. By Proposition 1.4 this implies that $(\Delta - M_A)^{-1} M_B L$ is lower triangular. Now $\theta^{-1} = J\theta^* J$, and thus

$$(5.24) \qquad \theta^{-1} = I + J M_B^* M_H^{-1}(\Delta - M_A)^{-1} M_B.$$

So we may conclude that $S := \theta^{-1} L$ is lower triangular. Note that $\theta^{-1} \in \mathcal{R}^{(m+r)\times(m+r)}$. Since $\mathcal{R} \subset \mathcal{W}$ and $\mathcal{W}$ is an algebra, we see that $S \in \mathcal{W}^{(m+r)\times 1}$. So $L = \theta S$ with $S \in \mathcal{L}\mathcal{W}^{(m+r)\times 1}$. $\square$

5.4. Proof of sufficiency and construction of all solutions. In what follows we assume that condition (5.6) holds for some $\varepsilon > 0$. We have to show that the interpolation problem (i), (ii) is solvable and that the solutions in $\mathcal{L}\mathcal{R}^{m\times r}$ have the desired parametrization. This will be done in a number of steps.

Step (a). By Lemma 5.2 the operator

$$\theta = \begin{pmatrix} \theta_{11} & \theta_{12} \\ \theta_{21} & \theta_{22} \end{pmatrix}$$

appearing in Theorem 5.1 is in $\mathcal{L}\mathcal{R}^{(m+r)\times(m+r)}$ and θ is J-inner. By Theorem 4.2, the operator θ_{22}^{-1} exists and is lower triangular. Hence, we may apply Proposition 4.6 to show that $\theta_{22}^{-1} \in \mathcal{L}\mathcal{R}^{m\times m}$.

Step (b). Let $G \in \mathcal{L}\mathcal{R}^{m\times r}$ with $\|G\| < 1$. Then

$$X(G) := (\theta_{22}^{-1}\theta_{21}G + I)^{-1}\theta_{22}^{-1}$$

is a lower triangular operator on $L_2^m(\mathbb{R})$. To see this, note that $\theta \in \mathcal{L}\mathcal{R}^{(m+r)\times(m+r)}$. So θ_{21} is lower triangular. By the result proved under Step (a), the operator θ_{22}^{-1} is well-defined and lower triangular. It follows that the same holds for $\theta_{22}^{-1}\theta_{21}G$. Note that $\|\theta_{22}^{-1}\theta_{21}G\| < 1$ because $\|\theta_{22}^{-1}\theta_{21}\| < 1$ by Theorem 2.1 in [BallGK1]. Thus

$$(5.25) \qquad (\theta_{22}^{-1}\theta_{21}G + I)^{-1} = \sum_{\nu=0}^{\infty}(-\theta_{22}^{-1}\theta_{21}G)^{\nu}$$

is lower triangular and hence so is $X(G)$.

Since $G \in \mathcal{R}^{m\times r}$, the operator $T := \theta_{22}^{-1}\theta_{21}G + I$ is in $\mathcal{R}^{m\times m}$. We also know that T is invertible as an operator on $L_2^m(\mathbb{R})$. Furthermore, from what has been proved in the previous paragraph we know that T and T^{-1} are lower triangular. Hence, by Proposition 4.6, we have $T^{-1} \in \mathcal{R}^{m\times m}$, and therefore $X(G) \in \mathcal{L}\mathcal{R}^{m\times m}$.

Step (c). Let $G \in \mathcal{L}\mathcal{R}^{m\times r}$ with $\|G\| < 1$, and define F by (5.7). We shall prove that F is a solution of our interpolation problem. Note that

$$\begin{pmatrix} F \\ I \end{pmatrix} = \theta \begin{pmatrix} G \\ I \end{pmatrix} X(G).$$

Thus, by the result of Step (b), the operator F is in $\mathcal{LR}^{m \times r}$. By Theorem 5.5

$$\left(M_B \begin{pmatrix} F \\ I \end{pmatrix}\right)^{\widehat{}}(A) = 0,$$

and thus the interpolation condition (i) is fulfilled. It remains to show that $\|F\| < 1$. For this purpose, we make the following computation:

$$F^* F - I = (F^* \quad I) J \begin{pmatrix} F \\ I \end{pmatrix}$$

$$= X(G)^* (G^* \quad I) \theta^* J \theta \begin{pmatrix} G \\ I \end{pmatrix} X(G)$$

$$= X(G)^* (G^* \quad I) J \begin{pmatrix} G \\ I \end{pmatrix} X(G)$$

$$= X(G)^* (G^* G - I) X(G) < 0,$$

which yields $\|F\| < 1$.

Step (d). In this part we assume that $F \in \mathcal{LR}^{m \times r}$ is a solution of our interpolation problem and we show that F admits the representation (5.7). By identifying $L_2^r(\mathbb{R})$ with the Hilbert space direct sum,

$$L_2(\mathbb{R}) \oplus L_2(\mathbb{R}) \oplus \cdots \oplus L_2(\mathbb{R}),$$

of r copies of $L_2(\mathbb{R})$, we may write

$$\begin{pmatrix} F \\ I \end{pmatrix} = (L_1 \ L_2 \ \cdots \ L_r).$$

Since F satisfies the interpolation condition (i), we have $(M_B L_j)^{\widehat{}}(A) = 0$, and thus, by Theorem 5.4, each $L_j \in \theta \mathcal{LW}^{(m+r) \times 1}$. Recall (see formula (5.24)) that $\theta^{-1} \in \mathcal{R}^{(m+r) \times (m+r)}$. Since $F \in \mathcal{LR}^{m \times r}$, we have $L_j \in \mathcal{LR}^{(m+r) \times 1}$, and we can find $G_1 \in \mathcal{LR}^{m \times r}$ and $G_2 \in \mathcal{LR}^{r \times r}$ such that

$$(5.26) \qquad\qquad\qquad \begin{pmatrix} F \\ I \end{pmatrix} = \theta \begin{pmatrix} G_1 \\ G_2 \end{pmatrix}.$$

A computation similar to the one made in Step (c) shows that

$$0 > F^* F - I = G_1^* G_1 - G_2^* G_2,$$

and thus

$$(5.27) \qquad\qquad\qquad G_2^* G_2 - G_1^* G_1 > 0.$$

From (5.27) it follows that G_2 is invertible and G_2^{-1} is lower triangular. To see this, take $\varphi \in L_2^r(\mathbb{R})$ with $G_2 \varphi = 0$. Then (5.27) forces $G_1 \varphi = 0$. But then we can use (5.26) to show that

$$\varphi = (0 \quad I) \begin{pmatrix} F \\ I \end{pmatrix} \varphi = (0 \quad I) \theta \begin{pmatrix} G_1 \\ G_2 \end{pmatrix} \varphi = 0,$$

and hence $\operatorname{Ker} G_2 = \{0\}$. Next, since θ is J-unitary, θ^{-1} is a bounded operator on $L_2^{m+r}(\mathbb{R})$, and thus

$$
(G_1^* \quad G_2^*)\begin{pmatrix} G_1 \\ G_2 \end{pmatrix} = (F^* \quad I)\theta^{-*}\theta^{-1}\begin{pmatrix} F \\ I \end{pmatrix}
$$
$$
\geq \gamma(F^*F + I) \geq \gamma I,
$$

for some constant $\gamma > 0$. But then we can use (5.27) to show that $2G_2^*G_2 \geq \gamma I$, and so G_2 has a closed range. To show that G_2 is invertible with inverse lower triangular, it remains to prove that $G_2 L_2^r([\tau, \infty))$ is a dense subset of $L_2^r([\tau, \infty))$ for every $\tau \in \mathbb{R}$. Therefore, suppose that $\varphi \in L_2^r([\tau, \infty))$ is orthogonal to $G_2 L_2^r([\tau, \infty))$ for some τ. Set

$$
\begin{pmatrix} \varphi_1 \\ \varphi_2 \end{pmatrix} := \Theta \begin{pmatrix} 0 \\ \varphi \end{pmatrix} \in L_2^{m+r}([\tau, \infty)).
$$

Then for all $\psi \in L_2^r([\tau, \infty))$ we have

$$
\left\langle J\begin{pmatrix} F \\ I \end{pmatrix}\psi, \begin{pmatrix} \varphi_1 \\ \varphi_2 \end{pmatrix}\right\rangle = \left\langle J\Theta\begin{pmatrix} G_1 \\ G_2 \end{pmatrix}\psi, \Theta\begin{pmatrix} 0 \\ \varphi \end{pmatrix}\right\rangle
$$
$$
= \left\langle J\begin{pmatrix} G_1 \\ G_2 \end{pmatrix}\psi, \begin{pmatrix} 0 \\ \varphi \end{pmatrix}\right\rangle
$$
$$
= -\langle G_2\psi, \varphi\rangle = 0.
$$

In particular, this holds with $\psi = \varphi_2$. So

$$
\|\varphi_1 - F\varphi_2\|^2 = \left\langle J\left\{\begin{pmatrix} \varphi_1 \\ \varphi_2 \end{pmatrix} - \begin{pmatrix} F \\ I \end{pmatrix}\varphi_2\right\}, \begin{pmatrix} \varphi_1 \\ \varphi_2 \end{pmatrix} - \begin{pmatrix} F \\ I \end{pmatrix}\varphi_2\right\rangle
$$
$$
= \left\langle J\begin{pmatrix} \varphi_1 \\ \varphi_2 \end{pmatrix}, \begin{pmatrix} \varphi_1 \\ \varphi_2 \end{pmatrix}\right\rangle + \left\langle J\begin{pmatrix} F \\ I \end{pmatrix}\varphi_2, \begin{pmatrix} F \\ I \end{pmatrix}\varphi_2\right\rangle
$$
$$
\leq \left\langle J\Theta\begin{pmatrix} 0 \\ \varphi \end{pmatrix}, \Theta\begin{pmatrix} 0 \\ \varphi \end{pmatrix}\right\rangle
$$
$$
= \left\langle J\begin{pmatrix} 0 \\ \varphi \end{pmatrix}, \begin{pmatrix} 0 \\ \varphi \end{pmatrix}\right\rangle = -\|\varphi\|^2 \leq 0,
$$

where the first inequality comes from $\|F\| < 1$. We conclude that $F\varphi_2 = \varphi_1$. Thus

$$
\Theta\begin{pmatrix} 0 \\ \varphi \end{pmatrix} = \begin{pmatrix} \varphi_1 \\ \varphi_2 \end{pmatrix} = \begin{pmatrix} F \\ I \end{pmatrix}\varphi_2 = \Theta\begin{pmatrix} G_1 \\ G_2 \end{pmatrix}\varphi_2,
$$

and thus $G_2\varphi_2 = \varphi$. As $\varphi \in L_2^r([\tau, \infty))$ was chosen to be orthogonal to $G_2 L_2^r([\tau, \infty))$, we see that $\varphi = 0$ as needed. Thus G_2^{-1} exists and is lower triangular. Since $G_2 \in \mathcal{LR}^{r \times r}$, we know from Proposition 4.6 that $G_2^{-1} \in \mathcal{R}^{r \times r}$, and so $G_2^{-1} \in \mathcal{LR}^{r \times r}$.

Since $G_2^{-1} \in \mathcal{LR}^{r \times r}$, we see that $G := G_1 G_2^{-1} \in \mathcal{LR}^{m \times r}$. Moreover (5.27) implies that $\|G\| < 1$. Finally, from (5.26) we see that

$$
F = \theta_{11}G_1 + \theta_{12}G_2 = (\theta_{11}G + \theta_{12})G_2,
$$
$$
I = \theta_{21}G_1 + \theta_{22}G_2 = (\theta_{21}G + \theta_{22})G_2.
$$

From Step (c) we know that $\theta_{21} G + \theta_{22}$ is invertible. Thus F has the desired linear fractional representation (5.7). $\square$

6. AN EXAMPLE

In this section we illustrate the general theory on a simple example. Let $z \in L_\infty(\mathbb{R})$ be given by

$$(6.1) \qquad z(t) = \begin{cases} 1 & , \quad t > 0, \\ 1+i, & t < 0. \end{cases}$$

We claim that z is an anti-stable time-varying point. To see this, note that the transition function $Z(t,s)$ of the associated differential equation is given by

$$(6.2) \qquad Z(t,s) = \begin{cases} e^{t-s} & , \quad s \geq t \geq 0, \\ e^{it} e^{t-s} & , \quad s \geq 0 > t, \\ e^{i(t-s)} e^{t-s}, & 0 > s \geq t. \end{cases}$$

In particular, $|Z(t,s)| = e^{t-s}$, and hence inequality (1.2) holds true with $a = e^{-1}$ and $M = 1$. Thus (6.1) is an astv-point.

Next, let $y \in L_\infty(\mathbb{R})$ be given by

$$(6.3) \qquad y(t) = \begin{cases} \sqrt{2}, & 0 \leq t \leq c, \\ 0 & , \quad \text{otherwise.} \end{cases}$$

Here $c > 0$ is a parameter which we shall specify later. We consider the following interpolation problem: find $F \in \mathcal{LR}$ such that

(i) $\widehat{F}(z) = y$,

(ii) $\|F\| < 1$.

If $d(\cdot)$ is the external coefficient of F and $\sigma(t,s)$ its kernel function, then (i) can be rewritten as

$$d(t) + \int_0^{-t} e^{-(1+i)\alpha} \sigma(t+\alpha,t)d\alpha + e^{it} \int_{-t}^\infty e^{-\alpha} \sigma(t+\alpha,t)d\alpha = 0, \qquad t < 0;$$

$$d(t) + \int_0^\infty e^{-\alpha} \sigma(t+\alpha,t)d\alpha = \begin{cases} \sqrt{2}, & 0 \leq t \leq c, \\ 0 & , \quad t > c. \end{cases}$$

To find the condition of solvability we have to compute

$$h(t) = \int_t^\infty Z(t,\alpha)\{1 - |y(\alpha)|^2\}\overline{Z(t,\alpha)}d\alpha.$$

Since $|Z(t,s)| = e^{t-s}$ we have

$$h(t) = \int_0^\infty e^{-2\alpha}\{1 - |y(t+\alpha)|^2\}\,d\alpha.$$

An elementary computation yields:

$$(6.4) \qquad h(t) = \begin{cases} \frac{1}{2} & , \quad t \geq c, \\ -\frac{1}{2} + e^{2t}e^{-2c} & , \quad 0 \leq t \leq c, \\ \frac{1}{2} - e^{2t}(1 - e^{-2c}), & t \leq 0. \end{cases}$$

Note that $h(t) \geq h(0)$ for all t. It follows that our interpolation problem is solvable if and only if

$$(6.5) \qquad 0 < c < \log\sqrt{2}.$$

Assume (6.5) is fulfilled. Then, according to Theorem 5.1, the operator

$$\theta = \begin{pmatrix} \theta_{11} & \theta_{12} \\ \theta_{21} & \theta_{22} \end{pmatrix} \in \mathcal{LR}^2$$

that parametrizes the set of all solutions is the integral operator

$$\left(\theta\begin{pmatrix} \varphi_1 \\ \varphi_2 \end{pmatrix}\right)(t) = \begin{pmatrix} \varphi_1(t) \\ \varphi_2(t) \end{pmatrix} - \int_{-\infty}^t \begin{pmatrix} 1 \\ y(t) \end{pmatrix} \overline{Z(s,t)} h(s)^{-1}(\varphi_1(s) - y(s)\varphi_2(s))\,ds,$$

where $Z(t,s)$ is defined by (6.2) and $h(s)$ is given by (6.4).

The central solution $\widetilde{F}$ is equal to $\theta_{12}\theta_{22}^{-1}$. To compute θ_{22}^{-1}, note (cf., (5.19)) that θ_{22} is the input-output map of the system:

$$(6.6) \qquad \begin{cases} x'(t) = -\overline{z(t)}x(t) + h(t)^{-1}y(t)\varphi(t), \\ g(t) = y(t)x(t) + \varphi(t). \end{cases}$$

The system (6.6) has dichotomy $P = I$. In fact, the transition function $\widetilde{Z}(t,s)$ associated with the differential equation $x'(t) = -\overline{z(t)}x(t)$ is equal to $\overline{Z(s,t)}$. From (6.6) we see that θ_{22}^{-1} is the input-output map of the system

$$(6.7) \qquad \begin{cases} x'(t) = -(\overline{z(t)} + h(t)^{-1}|y(t)|^2) + h(t)^{-1}y(t)g(t), \\ \varphi(t) = -y(t)x(t) + g(t). \end{cases}$$

Put $z^\times(t) = -\overline{z(t)} - h(t)^{-1}|y(t)|^2$. Then

$$z^\times(t) = \begin{cases} -1 & , \quad t \geq c \\ -1 - 2h(t)^{-1}, & 0 \leq t \leq c \\ -(1 - i) & , \quad t < 0. \end{cases}$$

Put

$$(6.8) \qquad u^\times(t) = \begin{cases} -t + c + 3c + 2\log 2 , & t \ge c, \\ 3t - 2\log h(t) & , \quad 0 \le t \le c, \\ -(1-i)t - 2\log h(0), & t \le 0. \end{cases}$$

The function $u^\times(t)$ is a primitive of $z^\times(t)$. Thus the transition function $Z^\times(t,s)$ associated with (6.7) is given by

$$(6.8) \qquad Z^\times(t,s) = \exp\big(u^\times(t) - u^\times(s)\big).$$

It follows that

$$(\theta_{22}^{-1}g)(t) = g(t) - \int_{-\infty}^{t} y(t) Z^\times(t,s) h(s)^{-1} y(s) g(s) ds.$$

We also know that

$$(\theta_{12}\varphi)(t) = \int_{-\infty}^{t} \overline{Z(s,t)} h(s)^{-1} y(s) g(s) ds.$$

Thus the central solution $\widetilde{F} = \theta_{12}\theta_{22}^{-1}$ is given by

$$(\widetilde{F}g)(t) = \int_{-\infty}^{t} \widetilde{f}(t,s) g(s) ds, \qquad t \in \mathbb{R},$$

where for $s \le t$

$$\widetilde{f}(t,s) = \overline{Z(s,t)} h(s)^{-1} y(s) - \int_{s}^{t} \overline{Z(\alpha,t)} h(\alpha)^{-1} |y(\alpha)|^2 Z^\times(\alpha,s) h(s)^{-1} y(s) d\alpha.$$

Note that $\widetilde{f}(t,s) = 0$ for $s < 0$ or $s > c$. Furthermore,

$$Z^\times(t,s) = e^{3(t-s)} \left(\frac{h(s)}{h(t)} \right)^2, \qquad 0 \le s \le t \le c.$$

It follows that for $0 \le s \le c$ and $s \le t$ the kernel function $\widetilde{f}$ of the central solution is equal to

$$\widetilde{f}(t,s) = \frac{\sqrt{2}}{h(s)} e^{s-t} - 2\sqrt{2} h(s) e^{-t-3s} \int_{s}^{\min(t,c)} e^{4\alpha} \frac{1}{h(\alpha)^3} d\alpha,$$

where $h(s) = -\frac{1}{2} + e^{2(s-c)}$ for $0 \le s \le c$.

REFERENCES

[BallGK]   J.A. Ball, I. Gohberg and M.A. Kaashoek, Nevanlinna-Pick interpolation for time-varying input-output maps: the discrete case, *Operator Theory: Advances and Applications*, this volume.

[BallGR1]   J.A. Ball, I. Gohberg and L. Rodman, Tangential interpolation problems for rational matrix functions, in *Proceedings Symposia Applied Mathematics* **40** (1990), 59–86.

[BallGR2]   J.A. Ball, I. Gohberg and L. Rodman, *Interpolation of rational matrix functions*, OT45, Birkhäuser Verlag, Basel, 1990.

[BenG]   A. Ben-Artzi and I. Gohberg, Dichotomy of systems and invertibility of linear ordinary differential operators, *Operator Theory: Advances and Applications*, this volue.

[C]   W.A. Coppel, *Dichotomies in stability theory*, Lecture Notes in Mathematics **629**, Springer-Verlag, Berlin, 1978.

[DK]   Ju.L. Daleckii and M.G. Krein, *Stability of solutions of differential equations in Banach space*, Transl. Math. Monographs **43**, Amer. Math. Soc., Providence, Rhode Island, 1974.

[Dew]   P. Dewilde, A course on the algebraic Schur and Nevanlinna-Pick interpolation problems, in *Algorithms and Parallel VLSI Architectures*, Volume A: *Tutorials* (Eds. E.F. Deprettere and A.-J. van der Veen), Elsevier, Amsterdam, 1991.

[DewDy]   P. Dewilde and H. Dym, Interpolation for upper triangular operators, *Operator Theory: Advances and Applications*, this volume.

[GGK]   I. Gohberg, S. Goldberg and M.A. Kaashoek, *Classes of Linear Operators*, Vol. I, *Operator Theory: Advances and Applications* **49**, Birkhäuser Verlag, Basel, 1990.

[GKvS]   I. Gohberg, M.A. Kaashoek and F. van Schagen, Non-compact integral operators with semi-separable kernels and their discrete analogues: inversion and Fredholm properties, *Integral Equations Operator Theory* **7** (1984), 642–703.

[HeS]   E. Hewitt and K. Stromberg, *Real and Abstract Analysis*, Springer-Verlag, Heidelberg, 1965.

[HiM]   D.J. Hill and P.J. Moylan, Dissipative dynamical systems: basic input-output properties, *J. Franklin Inst.* **309** (1980), 327–357.

[MS]   J.L. Massera and J.J. Schäffer, *Linear Differential Equations and Function Spaces*, Academic Press, New York, 1966.

[W]   J.C. Willems, Dissipative dynamical systems, Part I: General theory, *Arch. Rat. Mech. Anal.* **45** (1972), 321–351.

J.A. Ball
Department of Mathematics
Virginia Polytechnic Institute
Blacksburg, VA 24061, U.S.A.

I. Gohberg
School of Mathematical Sciences
The Raymond and Beverly Sackler Faculty of Exact Sciences
Tel-Aviv University
Tel-Aviv, Ramat-Aviv 69989, Israel

M.A. Kaashoek
Faculteit Wiskunde en Informatica
Vrije Universiteit
Amsterdam, The Netherlands

Operator Theory:
Advances and Applications, Vol. 56
© 1992 Birkhäuser Verlag Basel

DICHOTOMY OF SYSTEMS AND INVERTIBILITY OF LINEAR ORDINARY DIFFERENTIAL OPERATORS

Asher Ben-Artzi and **Israel Gohberg**

A linear ordinary differential operator of first order with bounded coefficients is shown to be invertible in $L_2^n(-\infty, \infty)$ or Fredholm in $L_2^n(0, \infty)$ if and only if the underlying system of homogeneous differential equations has a dichotomy. In that case the operator is proved to be a direct sum of two generators of C_0-semigroups, one of which has support on the negative half line and the other on the positive half line.

1. INTRODUCTION

The aim of this paper is to give conditions for the invertibility of the following operator T in $L_2^n(-\infty, \infty)$

$$(Tf)(t) = f'(t) - A(t)f(t) \qquad (-\infty < t < \infty).$$

The domain of T is the set of functions f in $L_2^n(-\infty, \infty)$, which are absolutely continuous on finite intervals, and such that f' (defined componentwise almost everywhere) belongs to $L_2^n(-\infty, \infty)$. The given coefficient $A(t)$ is an $n \times n$ complex matrix valued function defined on $(-\infty, \infty)$, whose entries are bounded and measurable, and $L_2^n(-\infty, \infty)$ is the Hilbert space of square integrable functions with values in $\mathbb{C}^n$.

An essential role is played by a dichotomy of the ordinary differential equation

$$(1.1) \qquad \frac{dx}{dt} = A(t)x(t) \qquad (-\infty < t < \infty).$$

A projection P in $\mathbb{C}^n$ is called a *dichotomy* for (1.1), see [DK] and see also [MS], if there exist two positive numbers a and M, with $a < 1$ and such that the following inequalities hold

$$(1.2) \qquad \|U(t)PU(s)^{-1}\| \leq Ma^{t-s} \qquad (t \geq s),$$

and

$$(1.3) \qquad \|U(s)(I_n - P)U(t)^{-1}\| \leq Ma^{t-s} \qquad (t \geq s).$$

Here, $I_n = (\delta_{ij})_{ij=1}^n$ and $U(t)$ is the evolution operator of (1.1), namely the unique matrix valued function which is absolutely continuous on finite intervals and satisfies $dU/dt = A(t)U(t)$ and $U(0) = I_n$. We refer to a pair of positive numbers a and M with $a < 1$ and such that (1.2)–(1.3) hold as a *pair of constants* for the dichotomy. It is easy to see that if a dichotomy for (1.1) exists then it is unique (see also [C]). One of our main results concerning the operator T is the following.

THEOREM 1.1. *Let $A(t)$ be a bounded and measurable $n \times n$ complex matrix valued function defined for $-\infty < t < \infty$. The operator T defined in $L_2^n(-\infty, \infty)$ by the expression*

$$(Tf)(t) = f'(t) - A(t)f(t) \qquad (-\infty < t < \infty),$$

with the domain

$$\mathrm{Dom}\, T = \{f \in L_2^n(-\infty, \infty) : f \text{ is locally absolutely continuous and } f' \in L_2^n(-\infty, \infty)\},$$

is invertible in $L_2^n(-\infty, \infty)$ if and only if the ordinary differential equation

$$(1.4) \qquad \frac{dx}{dt} = A(t)x(t) \qquad (-\infty < t < \infty)$$

admits a dichotomy. Moreover, if T is invertible then its inverse T^{-1} is given by the formula

$$(1.5) \qquad (T^{-1}f)(t) = \int_{-\infty}^{t} U(t)PU(s)^{-1}f(s)ds - \int_{t}^{\infty} U(t)(I_n - P)U(s)^{-1}f(s)ds$$

for $-\infty < t < \infty$ and $f \in L_2^n(-\infty, \infty)$, where P and $U(\cdot)$ denote respectively the dichotomy and the evolution operator of (1.4). Furthermore, a pair of constants for the dichotomy P is given by

$$(1.6) \qquad a = e^{-[2\|T^{-1}\|(1+\|A\|_\infty)]^{-2}} \qquad \text{and} \qquad M = e^{5[(1+\|T^{-1}\|)(1+\|A\|_\infty)]^4},$$

and for any constants a and M of the dichotomy of (1.4) we have

$$(1.7) \qquad \|T^{-1}\| \leq 2M/\log a^{-1}.$$

For related results, under the assumption that the closure $H(A)$ of the set of translates of $A(t)$ is compact, see [SaS1–3] and [CS]. In these papers the results are obtained by using skew products or bundle flows over the compact space $H(A)$, see [SaS1] pp. 432–433 or [CS] pp. 160–161.

The Fredholm properties of T are related to a dichotomy of an ordinary differential equation of type

$$\frac{dx}{dt} = A(t)x(t) \qquad (t \geq 0),$$

where A is as above, but t is restricted to the positive half line. The dichotomy is defined in analogy with the full line case. The difference being that the numbers t and s appearing in (1.2)–(1.3) are now assumed to be nonnegative. Let us remark that in this case the dichotomy is no longer unique and only $\mathrm{Im}\, P$ is uniquely determined (see [C]).

THEOREM 1.2. *Let $A(t)$ be a bounded and measurable complex matrix valued function defined for $-\infty < t < \infty$. The operator T defined in $L_2^n(-\infty, \infty)$, by the expression*

$$(Tf)(t) = f'(t) - A(t)f(t) \qquad (-\infty < t < \infty)$$

with the domain

$$\operatorname{Dom} T = \left\{ f \in L_2^n(-\infty, \infty) : f \text{ is locally absolutely continuous and } f' \in L_2^n(-\infty, \infty) \right\}$$

is Fredholm if and only if the following two ordinary differential equations admit dichotomies

$$(1.8) \qquad \frac{dx}{dt} = A(t)x(t) \qquad (t \geq 0)$$

and

$$(1.9) \qquad \frac{dx}{dt} = -A(-t)x(t) \qquad (t \geq 0).$$

Furthermore, if P and P_1 are dichotomies for (1.8) and (1.9) respectively, then

$$(1.10) \qquad \operatorname{Ker} T = \left\{ f : f(t) = U(t)x, \ x \in \operatorname{Im} P \cap \operatorname{Im} P_1 \right\},$$

$$(1.11) \qquad \operatorname{Im} T = \left\{ f : f(t) = U(t)^{*-1}x, \ x \perp (\operatorname{Im} P + \operatorname{Im} P_1) \right\}^{\perp},$$

and

$$(1.12) \qquad \operatorname{index} T = \operatorname{Rank} P + \operatorname{Rank} P_1 - n,$$

where $U(\cdot)$ is the evolution operator of $dx/dt = A(t)x(t)$ $(-\infty < t < \infty)$.

It turns out that the differential equation (1.2) admits a dichotomy if and only if T is *exponentially dichotomous* in the sense of [BGK]. For sake of completeness, we now give the corresponding definitions. We shall use the basic facts of the theory of strongly continuous semigroups of operators (also called C_0-semigroups) as explained for instance in [DS], [HP], [P] and [Y]. Besides ordinary semigroups defined on the positive half line, henceforth to be called *right semigroups*, we also consider semigroups defined on the negative semiaxis. These will be termed *left semigroups*.

Let $T(t)$ be a strongly continuous right or left semigroup. As is well-known there exist constant M and ω such that

$$\|T(t)\| \leq M e^{\omega |t|}, \qquad t \in J.$$

Here J is the half line $[0, \infty)$ or $(-\infty, 0]$ according to $T(t)$ being a right or left semigroup. If the above inequality is satisfied for a given real number ω and some positive constant M, we say that $T(t)$ is of *exponential type ω*. Semigroups of negative exponential type will be called *exponentially decaying*.

Next we introduce the concept of an *exponentially dichotomous* operator. Let X be a complex Banach space, and let $G(X \to X)$ be a linear operator acting in X. So the domain $\operatorname{Dom} G$ of G is a linear subspace of X and $G: \operatorname{Dom} G \to X$ is a linear operator (not necessarily bounded). Further, let $\mathcal{P}: X \to X$ be a (bounded) projection of X commuting with G, i.e., $\mathcal{P}$ maps $\operatorname{Dom} G$ into itself and $\mathcal{P} G x = G \mathcal{P} x$ for all $x \in \operatorname{Dom} G$. Put $X_- = \operatorname{Im} \mathcal{P}$ and $X_+ = \operatorname{Ker} \mathcal{P}$. Then

$$(1.13) \qquad X = X_- \oplus X_+$$

and this decomposition reduces G. By this we mean that

$$(1.14) \qquad \operatorname{Dom} G = [\operatorname{Dom} G \cap X_-] \oplus [\operatorname{Dom} G \cap X_+],$$

G maps $\operatorname{Dom} G \cap X_-$ into X_- and $\operatorname{Dom} G \cap X_+$ into X_+. With respect to the decompositions (1.13) and (1.14), the operator G has the matrix representation

$$(1.15) \qquad G = \begin{pmatrix} G_- & 0 \\ 0 & G_+ \end{pmatrix}.$$

Here $G_\pm = G|_{X_\pm}(X_\pm \to X_\pm)$ is the restriction of G to $X_\pm$. The domain $\operatorname{Dom} G_\pm$ of $G_\pm$ is $X_\pm \cap \operatorname{Dom} G$. Thus (1.14) can be rewritten as $\operatorname{Dom} G = \operatorname{Dom} G_- \oplus \operatorname{Dom} G_+$. Following [BGK] we call the operator G *exponentially dichotomous* if the operators G_- and G_+ in (1.15) are generators of exponentially decaying strongly continuous left and right semigroups, respectively. In that case the projection $\mathcal{P}$, which turns out to be unique (see Proposition 2.1 of [BGK]) is called the *separating projection* for G. We can now state our result concerning exponentially dichotomous operators.

THEOREM 1.3. *Let $A(t)$ be a bounded and measurable $n \times n$ complex matrix valued function defined for $-\infty < t < \infty$. The operator T defined in $L_2^n(-\infty, \infty)$ by the expression*

$$(Tf)(t) = f'(t) - A(t)f(t) \qquad (-\infty < t < \infty)$$

with the domain

$\operatorname{Dom} T = \{ f \in L_2^n(-\infty, \infty): f$ is locally absolutely continuous and $f' \in L_2^n(-\infty, \infty) \}$

is exponentially dichotomous if and only if the ordinary differential equation

$$(1.16) \qquad \frac{dx}{dt} = A(t)x(t) \qquad (-\infty < t < \infty)$$

admits a dichotomy. Moreover, if P is a dichotomy for (1.16) then the separating projection $\mathcal{P}$ for T in $L_2^n(-\infty, \infty)$ given by

$$(1.17) \qquad (\mathcal{P}f)(t) = U(t)PU(t)^{-1}f(t) \qquad (f \in L_2^n(-\infty, \infty),\ -\infty < t < \infty),$$

where $U(\cdot)$ is the evolution operator of (1.16). Conversely, the dichotomy P may be obtained from the separating projection $\mathcal{P}$ via the equality

$$(1.18) \qquad Pz = \lim_{n \to 0} h^{-1} \int_0^h \mathcal{P}(\alpha z)(t)dt \qquad (z \in \mathbb{C}^n),$$

where $\alpha(t) \in L_2(-\infty, \infty)$ is continuous at 0 with $\alpha(0) = 1$ and otherwise arbitrary, and $\alpha z \in L_2^n(-\infty, \infty)$ is given by $(\alpha z)(t) = \alpha(t)z$.

Theorems 1.1–1.3 above hold also in the case when the space $L_2^n(-\infty, \infty)$ is replaced by $L_p^n(-\infty, \infty)$, $1 < p < \infty$.

This paper is divided into seven sections. The first is the introduction. The next section contains some preliminary definitions and results. Section 3 is devoted to the proof of Theorem 1.1. Section 4 is of preparatory nature and contains some results about the relations between the full line and half line cases. In Section 5 we describe the Fredholm properties of differential operators defined on the positive semiaxis. Theorems 1.2 and 1.3 are proved in Sections 6 and 7 respectively.

Similar results for discrete systems appear in [BG]. The case of differential operators with boundary conditions is analyzed in [BGK1]. The last section of the present paper was added under the influence of the latter paper.

2. PRELIMINARIES

In this section we make some preliminary remarks and fix some notation. Note first that the operator T defined in the introduction can be written as $T = D - A$ where $D, A : \operatorname{Dom} T \to L_2^n(-\infty, \infty)$ are given by

$$(Df)(t) = f'(t) \quad \text{and} \quad (Af)(t) = A(t)f(t) \qquad (-\infty < t < \infty, \ f \in \operatorname{Dom} T).$$

Since D is closed and A is bounded then T is closed. A similar remark applies to differential operators defined on a half line.

Consider now the differential equation

$$(2.1) \qquad\qquad \frac{dx}{dt} = A(t)x(t) \qquad (-\infty < t < \infty)$$

where $A(\cdot)$ is an $n \times n$ complex matrix valued function, whose entries are bounded and measurable functions on $(-\infty, \infty)$. We denote

$$\|A\|_\infty = \operatorname*{ess\,sup}_{-\infty < t < \infty} \|A(t)\|,$$

where $\|A(t)\|$ is the largest singular value of $A(t)$. It is well known, see for example [DK] that there exists a unique complex $n \times n$ matrix valued function $U(t)$ $(t \in \mathbb{R})$, which is absolutely continuous on compact interval and such that $U(0) = I_n$ and

$$\frac{dU(t)}{dt} = A(t)U(t) \qquad (-\infty < t < \infty).$$

The function $U(t)$ is called the *evolution operator* of the system (2.1). The evolution operator has the following property: for each $t_0 \in \mathbb{R}$, $U(t_0)$ is invertible, and for each $x_0 \in \mathbb{C}^n$, the function

$$x(t) = U(t)U(t_0)^{-1}x_0 \qquad (-\infty < t < \infty)$$

is the unique solution of (2.1) satisfying $x(t_0) = x_0$.

We will use the following estimates which follow from equalities (1.5) and (1.11) in §1 of Chapter III of [DK]

$$(2.2) \qquad \|U(t)\| \le e^{t\|A\|_\infty} \qquad (-\infty < t < \infty),$$

and

$$(2.3) \qquad \|U(t) - I_n\| \le e^{t\|A\|_\infty} - 1 \qquad (-\infty < t < \infty).$$

Note that for each $s \in \mathbb{R}$, $U(t_1 + s)U(s)^{-1}$ is the evolution operator of the differential equation

$$\frac{dx}{dt_1} = A(t_1 + s)x(t_1) \qquad (-\infty < t_1 < \infty).$$

Applying (2.2) and (2.3) to this differential equation and letting $t_1 = t - s$, we obtain

$$(2.4) \qquad \|U(t)U(s)^{-1}\| \le e^{|t-s|\|A\|_\infty} \qquad (-\infty < t, s < \infty),$$

and

$$(2.5) \qquad \|U(t)U(s)^{-1} - I_n\| \le e^{|t-s|\|A\|_\infty} - 1 \qquad (-\infty < t, s < \infty).$$

Note also that $U(t)U(s)^{-1} = \left(U(s)U(t)^{-1}\right)^{-1}$, hence (2.4) leads to

$$(2.6) \qquad \|U(s)U(t)^{-1}y\| \ge e^{-|t-s|\|A\|_\infty}\|y\| \qquad (-\infty < t, s < \infty,\ y \in \mathbb{C}^n).$$

Similar definitions and inequalities may be applied to differential equations defined on the positive or negative semiaxes of the real line.

3. INVERTIBILITY OF DIFFERENTIAL OPERATORS ON THE REAL LINE

In this section we prove Theorem 1.1. We begin with a proposition which gives a criterion for existence of a dichotomy. This proposition is essentially Lemma 3.2 in [DK; p. 165]. For future references we have added to it explicit estimates for the constants of dichotomy.

PROPOSITION 3.1. *Let*

$$(3.1) \qquad \frac{dx}{dt} = A(t)x(t) \qquad (-\infty < t < \infty)$$

be an ordinary differential equation where $A(\cdot)$ is an $n \times n$ complex matrix valued function which is bounded and measurable. Assume that there exist two subspaces N_+ and N_- of $\mathbb{C}^n$ such that $N_+ + N_- = \mathbb{C}^n$ and the following inequalities hold

$$(3.2) \qquad \|U(t)y\| \le e^{-\alpha(t-s)+\beta}\|U(s)y\| \qquad (t \ge s,\ y \in N_+),$$

and

$$(3.3) \qquad \|U(s)z\| \le e^{-\alpha(t-s)+\beta}\|U(t)z\| \qquad (t \ge s,\ z \in N_-),$$

where α and β are positive numbers, and $U(\cdot)$ is the evolution operator of (3.1). Then $N_+ \cap N_- = \{0\}$ and the projection P in $\mathbb{C}^n$ onto N_+ and along N_- is a dichotomy for (3.1), with constants a_1 and M_1 given by

$$(3.4) \qquad a_1 = e^{-\alpha} \qquad \text{and} \qquad M_1 = e^{\beta}(e^{(\beta+1)\|A\|_\infty/\alpha} + e).$$

PROOF. The proof is based upon the following two inequalities which are consequences of (3.2)–(3.3) and the boundedness of A:

$$(3.5) \qquad \|U(s)y\| \le C\|U(s)(y+z)\| \qquad (-\infty < s < \infty,\ y \in N_+,\ z \in N_-),$$

and

$$(3.6) \qquad \|U(s)z\| \le C\|U(s)(y+z)\| \qquad (-\infty < s < \infty,\ y \in N_+,\ z \in N_-),$$

where

$$(3.7) \qquad C = e^{(\beta+1)\|A\|_\infty/\alpha} + e.$$

Since the proofs of (3.5) and (3.6) are entirely symmetric, we only prove (3.5). Let $s \in \mathbb{R}$, $y \in N_+$ and $z \in N_-$ be arbitrary. By applying (3.2) and (3.3) with $t = s + (\beta + 1)/\alpha$, we obtain

$$\|U(t)y\| \le e^{-1}\|U(s)y\| \qquad \text{and} \qquad \|U(s)z\| \le e^{-1}\|U(t)z\|.$$

Thus,

$$\|U(t)(y+z)\| \ge \|U(t)z\| - \|U(t)y\| \ge e\|U(s)z\| - e^{-1}\|U(s)y\|,$$

which leads to

$$(3.8) \qquad \|U(t)(y+z)\| \ge e\big[\|U(s)y\| - \|U(s)(y+z)\|\big] - e^{-1}\|U(s)y\|.$$

On the other hand, since A is bounded, the left hand side of this inequality may be estimated using (2.4) to get

$$\|U(t)(y+z)\| \le \|U(t)U(s)^{-1}\|\|U(s)(y+z)\| \le e^{(t-s)\|A\|_\infty}\|U(s)(y+z)\|.$$

Combining this with (3.8), and recalling that $t - s = (\beta + 1)/\alpha$, we obtain

$$(e^{(\beta+1)\|A\|_\infty/\alpha} + e)\|U(s)(y+z)\| \ge (e - e^{-1})\|U(s)y\| \ge \|U(s)y\|.$$

Thus, (3.5) holds.

We now show that (3.5)–(3.6) and the condition $N_+ + N_- = \mathbb{C}^n$ imply the existence of the dichotomy. Let us first remark that if $y \in N_+ \cap N_-$, then applying (3.5) with $z = -y$ and $s = 0$ we obtain $\|y\| = 0$. Hence, $N_+ \cap N_- = \{0\}$. Consequently $N_+ \oplus N_- = \mathbb{C}^n$ and we may define the projection P in $\mathbb{C}^n$ onto N_+ and along N_-.

We now show that P is a dichotomy for (3.1) with constants given by (3.4). We begin by proving the dichotomy inequality (1.2). Let $x \in \mathbb{C}^n$ and t and s be real numbers with $t \geq s$. Applying (3.2) with $y = PU(s)^{-1}x \in N_+$ we obtain

$$(3.9) \qquad \|U(t)PU(s)^{-1}x\| \leq e^{-\alpha(t-s)}e^{\beta}\|U(s)PU(s)^{-1}x\|.$$

On the other hand, the inequality (3.5) with the same y and $z = (I_n - P)U(s)^{-1}x \in N_-$ implies that

$$\|U(s)PU(s)^{-1}x\| \leq C\|U(s)[PU(s)^{-1}x + (I_n - P)U(s)^{-1}x]\| = C\|x\|.$$

Combining this inequality with (3.9) we obtain

$$\|U(t)PU(s)^{-1}x\| \leq e^{-\alpha(t-s)}e^{\beta}C\|x\|.$$

Since $x \in \mathbb{C}^n$ is arbitrary and $a_1 = e^{-\alpha}$ and $M_1 = e^{\beta}C$ by the definitions (3.4) and (3.7), this inequality implies $\|U(t)PU(s)^{-1}\| \leq M_1 a_1^{(t-s)}$ $(t \geq s)$. Thus, (1.2) holds. The proof of (1.3) is entirely symmetric and is therefore omitted. Inequalities (1.2)–(1.3) show that P is a dichotomy for (3.1). $\square$

We now turn to the proof of Theorem 1.1.

PROOF OF THEOREM 1.1. Assume first that T is invertible. We define two subspaces N_+ and N_- of $\mathbb{C}^n$ in the following way

$$(3.10) \qquad N_+ = \left\{ y \in \mathbb{C}^n \colon \int_0^{\infty} \|U(t)y\|^2 dt < \infty \right\}$$

and

$$(3.11) \qquad N_- = \left\{ y \in \mathbb{C}^n \colon \int_{-\infty}^0 \|U(t)y\|^2 dt < \infty \right\},$$

where $U(\cdot)$ is the evolution operator of (1.4). We will show that N_+ and N_- satisfy the requirements of Proposition 3.1. We now divide the proof into four parts.

Part a). Here we prove the following inequalities

$$(3.12) \qquad \|U(t)y\| \leq e^{-\alpha(t-s)+\beta}\|U(s)y\| \qquad (t \geq s,\ y \in N_+)$$

and

$$(3.13) \qquad \|U(s)z\| \leq e^{-\alpha(t-s)+\beta}\|U(t)z\| \qquad (t \geq s;\ z \in N_-)$$

where α and β are given by

$$(3.14) \qquad \alpha = \frac{1}{2}\left(\|T^{-1}\|(1 + \|A\|_{\infty})\right)^{-2} \qquad \text{and} \qquad \beta = \|T^{-1}\| + 2\|A\|_{\infty}.$$

Since the proofs of (3.12) and (3.13) are entirely symmetric, we only prove (3.12). We will use the following inequality which will be proved in Part b below:

$$(3.15) \qquad \int_\tau^\infty \|U(\sigma)y\|^2 d\sigma \le (2\alpha)^{-1}\|U(\tau)y\|^2 \qquad (-\infty < \tau < \infty,\ y \in N_+),$$

where α is given by (3.14).

We now prove (3.12). Let $s \in \mathbb{R}$ and $y \in N_+$ be arbitrary. Consider the function φ defined via

$$(3.16) \qquad \varphi(t) = e^{2\alpha t} \int_t^\infty \|U(\sigma)y\|^2 d\sigma \qquad (-\infty < t < \infty),$$

where α is given by (3.14). Note that φ is well defined because $y \in N_+$ implies $\int_0^\infty \|U(\sigma)y\|^2 d\sigma < \infty$, whence $\int_t^\infty \|U(\sigma)y\|^2 d\sigma < \infty$ $(-\infty < t < \infty)$. Note also that φ is nonincreasing. In fact,

$$\varphi'(t) = e^{2\alpha t}\left[2\alpha \int_t^\infty \|U(\sigma)y\|^2 d\sigma - \|U(t)y\|^2\right],$$

and hence, (3.15) with $\tau = t$ implies that $\varphi'(t) \le 0$ $(-\infty < t < \infty)$.

Now assume further that $t \ge s$. Since φ is nonincreasing we have $\varphi(t) \le \varphi(s)$ and hence

$$\int_t^\infty \|U(\sigma)y\|^2 d\sigma \le e^{2\alpha(s-t)} \int_s^\infty \|U(\sigma)y\|^2 d\sigma \qquad (t \ge s).$$

We estimate the integral in the right hand side of this inequality using (3.15) with $\tau = s$. This leads to

$$(3.17) \qquad \int_t^\infty \|U(\sigma)y\|^2 d\sigma \le e^{2\alpha(s-t)}(2\alpha)^{-1}\|U(s)y\|^2 \qquad (t \ge s).$$

To estimate the left hand side of this inequality, note that

$$\|U(\sigma)y\| = \left\|U(\sigma)U(t)^{-1}U(t)y\right\| \ge e^{-|\sigma-t|\|A\|_\infty}\|U(t)y\|,$$

where we used (2.6). This leads to

$$\int_t^\infty \|U(\sigma)y\|^2 d\sigma \ge \int_t^\infty e^{-2|\sigma-t|\|A\|_\infty} d\sigma \|U(t)y\|^2 = \frac{\|U(t)y\|}{2\|A\|_\infty} \ge e^{-2\|A\|_\infty}\|U(t)y\|^2.$$

Combining this with (3.17), we obtain

$$e^{-2\|A\|_\infty}\|U(t)y\|^2 \le e^{2\alpha(s-t)}(2\alpha)^{-1}\|U(s)y\|^2 \qquad (t \ge s).$$

Taking into account that $(2\alpha)^{-1} \le e^{2(\|T^{-1}\|+\|A\|_\infty)}$, we obtain

$$\|U(t)y\| \le e^{-\alpha(t-s)}e^{\|T^{-1}\|+2\|A\|_\infty}\|U(s)y\| \qquad (t \ge s).$$

Since $y \in N_+$ and $s \in \mathbb{R}$ are arbitrary, this implies (3.12).

Part b). Here we prove the inequality (3.15) which was used in Part a).

Let $\tau \in \mathbb{R}$ and $y \in N_+$ be arbitrary. We apply the inequality

$$(3.18) \qquad \|f\|^2 \le \|T^{-1}\|^2 \|Tf\|^2$$

to the function f defined on $(-\infty, \infty)$ by

$$f(t) = \begin{cases} U(t)y & (\tau \le t) \\ (t - \tau + 1)U(\tau)y & (\tau - 1 < t < \tau) \\ 0 & (t \le \tau - 1). \end{cases}$$

We first show that $f \in \mathrm{Dom}\, T$. Note that $y \in N_+$ implies $\int_0^\infty \|U(t)y\|^2 dt < \infty$, and hence, $\int_\tau^\infty \|U(t)y\|^2 dt < \infty$. Thus $f \in L_2^n(-\infty, \infty)$. Moreover, it is clear that f is locally absolutely continuous and that f' is given (almost everywhere) by

$$(3.19) \qquad f'(t) = \begin{cases} A(t)U(t)y & (\tau \le t) \\ U(\tau)y & (\tau - 1 < t < \tau) \\ 0 & (t \le \tau - 1). \end{cases}$$

Since A is bounded, $f' \in L_2^n(-\infty, \infty)$ and hence, $f \in \mathrm{Dom}\, T$. In order to apply (3.18), we estimate $\|f\|^2$ from below and $\|Tf\|^2$ from above. Note first that the definition of f leads immediately to

$$(3.20) \qquad \int_\tau^\infty \|U(\sigma)y\|^2 d\sigma \le \|f\|^2.$$

Furthermore, by (3.19) we have

$$(Tf)(t) = f'(t) - A(t)f(t) = \begin{cases} 0 & (\tau \le t) \\ \left[I_n - (t - \tau + 1)A(t)\right]U(\tau)y & (\tau - 1 < t < \tau) \\ 0 & (t \le \tau - 1). \end{cases}$$

Thus,

$$\|Tf\|^2 \le \int_{\tau-1}^\tau \|I_n - (t - \tau + 1)A(t)\|^2 \|U(\tau)y\|^2 dt \le (1 + \|A\|_\infty)^2 \|U(\tau)y\|^2.$$

Combining this inequality with (3.18) and (3.20) we obtain

$$\int_\tau^\infty \|U(\sigma)y\|^2 d\sigma \le \|T^{-1}\|^2 (1 + \|A\|_\infty)^2 \|U(\tau)y\|^2.$$

In view of the definition of α in (3.14) this implies (3.15).

Part c). Here we complete the proof in the case when T is invertible by proving the existence of a dichotomy for (1.4). We achieve this by showing that the subspaces N_+ and N_- defined above satisfy the requirements of Proposition 3.1.

We first show that

$$(3.21) \qquad\qquad N_+ + N_- = \mathbb{C}^n.$$

Let $z \in \mathbb{C}^n$ be arbitrary. Define $g \in L_2^n(-\infty, \infty)$ via

$$(3.22) \qquad\qquad g(s) = \begin{cases} U(s)z & s \in [0,1], \\ 0 & s \notin [0,1], \end{cases}$$

and let

$$f = T^{-1}g.$$

It is clear from the construction of g that

$$f'(t) - A(t)f(t) = (Tf)(t) = g(t) = 0 \qquad (t > 1 \text{ or } t < 0; \text{ ta.e.}).$$

Hence we have

$$(3.23) \qquad\qquad f(t) = U(t)U(1)^{-1}f(1) \qquad (t \geq 1),$$

and

$$f(t) = U(t)f(0) \qquad (t \leq 0).$$

The last equality combined with $f \in L_2^n(-\infty, \infty)$ leads to $\int_{-\infty}^0 \|U(t)f(0)\|^2 dt < \infty$. Hence,

$$(3.24) \qquad\qquad f(0) \in N_-.$$

Similarly, (3.23) implies $\int_1^\infty \|U(t)U(1)^{-1}f(1)\|^2 dt < \infty$, whence $\int_0^\infty \|U(t)U(1)^{-1}f(1)\|^2 dt < \infty$, and therefore

$$(3.25) \qquad\qquad U(1)^{-1}f(1) \in N_+.$$

On the other hand it follows from

$$f'(t) - A(t)f(t) = (Tf)(t) = g(t) \qquad (-\infty < t < \infty),$$

that

$$f(t) = U(t)f(0) + \int_0^t U(t)U(s)^{-1}g(s)ds \qquad (-\infty < t < \infty).$$

Setting $t = 1$ in this inequality and multiplying by $U(1)^{-1}$ on the left, we obtain

$$U(1)^{-1}f(1) - f(0) = \int_0^1 U(s)^{-1}g(s)ds.$$

In view of the definition (3.22) of g, this leads to $U(1)^{-1}f(1) - f(0) = z$. Taking into account (3.24) and (3.25) we obtain $z \in N_+ + N_-$. Since $z \in \mathbb{C}^n$ is arbitrary, this proves (3.21).

It follows from (3.21) and (3.12)–(3.13) that N_+ and N_- satisfy the requirements of Proposition 3.1. Hence, (1.4) admits a dichotomy with constants a_1 and M_1 given by (3.4) where α and β are defined in (3.14). Let a and M be defined by (1.6). Since $\beta + 1 < 2(1 + \|T^{-1}\|)(1 + \|A\|_\infty)$ we have $(\beta + 1)\|A\|_\infty/\alpha < 4\big[(1 + \|T^{-1}\|)(1+\|A\|_\infty)\big]^4$. Consequently $M_1 < e^{\|T^{-1}\|+2\|A\|_\infty}\big(e^{4[(1+\|T^{-1}\|)(1+\|A\|_\infty)]^4} + e\big) < e^{5[(1+\|T^{-1}\|)(1+\|A\|_\infty)]^4} = M$. On the other hand, it is clear that $a_1 < a < 1$. Consequently a and M are also a pair of constants for the dichotomy. This completes the proof in the case when T is invertible.

Part d). We now prove the converse part of the theorem. Thus we assume that (1.4) admits a dichotomy P, and we show that T is invertible with T^{-1} given by (1.5), and that (1.7) holds.

We first prove that

$$(3.26) \qquad\qquad\qquad \operatorname{Ker} T = \{0\}.$$

Let $f \in \operatorname{Ker} T$. Then $f'(t) = A(t)f(t)$ $(-\infty < t < \infty$, a.e.$)$ and hence

$$(3.27) \qquad\qquad f(t) = U(t)f(0) \qquad (-\infty < t < \infty).$$

Note that the dichotomy inequality (1.2) with $t = 0$ lead to

$$\|Pf(0)\| = \|U(0)PU(s)^{-1}U(s)f(0)\| \le Ma^{|s|}\|U(s)f(0)\| \qquad (s \le 0),$$

while (1.3) with $s = 0$ implies

$$\|(I_n - P)f(0)\| = \|U(0)(I_n - P)U(t)^{-1}U(t)f(0)\| \le Ma^t\|U(t)f(0)\| \qquad (t \ge 0).$$

Hence, $\|f(s)\| = \|U(s)f(0)\| \ge M^{-1}a^{-|s|}\|Pf(0)\|$ $(s \le 0)$, and $\|f(t)\| = \|U(t)f(0)\| \ge M^{-1}a^{-t}\|(I_n - P)f(0)\|$ $(t \ge 0)$. Since $f \in L_2^n(-\infty, \infty)$ and $0 < a < 1$, these inequalities lead to $(I_n - P)f(0) = 0$ and $Pf(0) = 0$, whence to $f(0) = 0$. By (3.27) we obtain $f = 0$. Since f is an arbitrary function in $\operatorname{Ker} T$, this proves (3.26).

We now define an operator Γ in $L_2^n(-\infty, \infty)$ via

$$(3.28) \qquad (\Gamma f)(t) = \int_{-\infty}^{t} U(t)PU(s)^{-1}f(s)\,ds - \int_{t}^{\infty} U(t)(I_n - P)U(s)^{-1}f(s)\,ds$$

for $f \in L_2^n(-\infty, \infty)$, $-\infty < t < \infty$. The dichotomy inequalities (1.2)–(1.3) show that the right hand side is well defined, and bounded from above by a convolution type integral

$$(\Gamma f)(t) \le \int_{-\infty}^{\infty} Ma^{|t-s|}\|f(s)\|\,ds.$$

By Young's inequality, it follows that

$$\|\Gamma f\|_{L_2^n(-\infty,\infty)} \le \left(\int_{-\infty}^{\infty} Ma^{|s|}\,ds\right)\|f\|_{L_2^n(-\infty,\infty)}.$$

Hence, Γ is a bounded operator in $L_2^n(-\infty,\infty)$ with

$$(3.29) \qquad \|\Gamma\| \le \int_{-\infty}^{\infty} M a^{|s|} ds = 2M/\log a^{-1}.$$

Let $f \in L_2^n(-\infty,\infty)$. Note that $(\Gamma f)(t)$ can be rewritten as

$$(\Gamma f)(t) = U(t)\left[\int_{-\infty}^{t} PU(s)^{-1}f(s)ds - \int_{t}^{\infty}(I_n - P)U(s)^{-1}f(s)ds\right].$$

Thus, Γf being the product of two locally absolutely continuous functions, is itself locally absolutely continuous. Furthermore, the last equality also shows that

$$\begin{aligned}(\Gamma f)'(t) &= A(t)(\Gamma f)(t) + U(t)\left[PU(t)^{-1}f(t) + (I_n - P)U(t)^{-1}f(t)\right]\\ &= A(t)(\Gamma f)(t) + f(t) \qquad (-\infty < t < \infty).\end{aligned}$$

Since $f, \Gamma f \in L_2^n(-\infty,\infty)$ and $A(\cdot)$ is bounded, $(\Gamma f)' \in L_2^n(-\infty,\infty)$. Thus, $\Gamma f \in \operatorname{Dom} T$. Finally, by the last equality again we have

$$(T\Gamma f)(t) = (\Gamma f)'(t) - A(t)(\Gamma f)(t) = f(t) \qquad (-\infty < t < \infty).$$

Thus, $T\Gamma f = f$. Since f is an arbitrary element of $L_2^n(-\infty,\infty)$, this shows that Γ is a right inverse for T. However T is injective by (3.21) and therefore $\Gamma = T^{-1}$. Equalities (1.5) and (1.7) follow from (3.28) and (3.29). $\quad\square$

4. RELATIONS BETWEEN OPERATORS ON THE FULL LINE AND HALF LINE

In this section we prove two propositions which will be used in Sections 5 and 6. The first deals with the relations between the Fredholm properties of differential operators on the full line and half line.

LEMMA 4.1. *Let $A(t)$ be a bounded and measurable $n \times n$ complex matrix valued function defined for $(-\infty < t < \infty)$. The operator T defined in $L_2^n(-\infty,\infty)$ by the expression*

$$(Tf)(t) = f'(t) - A(t)f(t) \qquad (-\infty < t < \infty)$$

with the domain

$$\operatorname{Dom} T = \left\{f \in L_2^n(-\infty,\infty): f \text{ is locally absolutely continuous and } f' \in L_2^n(-\infty,\infty)\right\}$$

is Fredholm if and only if the operators G and G_1 in $L_2^n(0,\infty)$ defined via

$$(4.1) \qquad (Gf)(t) = f'(t) - A(t)f(t), \quad (G_1 f)(t) = f'(t) + A(-t)f(t) \qquad (t \ge 0),$$

with the domains

$$\operatorname{Dom} G = \operatorname{Dom} G_1 = \left\{f \in L_2^n(0,\infty): f \text{ is locally absolutely}\right.$$
$$\left.\text{continuous and } f' \in L_2^n(0,\infty)\right\}$$

are Fredholm. Moreover, if T is Fredholm then

$$(4.2) \qquad \text{index}\, T = \text{index}\, G + \text{index}\, G_1 - n.$$

In the proof we will use the following lemma which may be found in [GK].

LEMMA 4.2. *Let L be a closed operator in a Hilbert space H, and $\mathcal{D}$ a subspace of finite codimension in $\text{Dom}\, L$ such that $L|_{\mathcal{D}}$ is closed. Then L is Fredholm if and only if $L|_{\mathcal{D}}$ is Fredholm, and in this case*

$$(4.3) \qquad \text{index}\, L = \text{index}\, L|_{\mathcal{D}} + \dim((\text{Dom}\, L)/\mathcal{D}).$$

PROOF OF LEMMA 4.1. Let $V \colon L_2^n(0,\infty) \oplus L_2^n(0,\infty) \to L_2^n(-\infty,\infty)$ be the unitary equivalence given by

$$(4.4) \qquad (V(f \oplus g))(t) = \begin{cases} f(t) & (t > 0) \\ g(-t) & (t < 0). \end{cases}$$

We consider the following operator in $L_2^n(-\infty,\infty)$ given by

$$(4.5) \qquad L = V(G \oplus (-G_1))V^{-1},$$

where $G \oplus (-G_1)$ is the operator acting in $L_2^n(0,\infty) \oplus L_2^n(0,\infty)$ via $(G \oplus (-G_1))(f \oplus g) = (Gf) \oplus (-G_1 g)$ with the domain $\text{Dom}(G \oplus (-G_1)) = \text{Dom}\, G \oplus \text{Dom}\, G_1$. The domain of L is given by $\text{Dom}\, L = V(\text{Dom}\, G \oplus \text{Dom}\, G_1)$. In view of the definition 4.4 of V, and the definitions of $\text{Dom}\, G$ and $\text{Dom}\, G_1$ we can give the following description of $\text{Dom}\, L$

$$\text{Dom}\, L = \{ f \in L_2^n(-\infty,\infty) \colon f \text{ is absolutely continuous on each}$$
$$\text{finite interval not containing } 0,\ f'|_{(0,\infty)} \in L_2^n(0,\infty)$$
$$\text{and } f'|_{(-\infty,0)} \in L_2^n(-\infty,0)\}.$$

Consequently $\text{Dom}\, L \supset \text{Dom}\, T$, and it is easy to see that

$$(4.6) \qquad \dim(\text{Dom}\, L/\text{Dom}\, T) = n.$$

We now show that $T = L|_{\text{Dom}\, T}$. Let $h \in \text{Dom}\, T$. Define f and g in $L_2^n(0,\infty)$ via

$$(4.7) \qquad f(t) = h(t) \quad \text{and} \quad g(t) = h(-t) \qquad (t \geq 0).$$

Then $h = V(f \oplus g)$, $f,g \in \text{Dom}\, G = \text{Dom}\, G_1$ and

$$Lh = V(G \oplus (-G_1))V^{-1}h = V(G \oplus (-G_1))(f \oplus g) = V(Gf \oplus (-G_1 g)).$$

Taking into account the definition (4.1) of G and G_1 and the construction (4.7) of f and g we obtain

$$(Lh)(t) = (Gf)(t) = f'(t) - A(t)f(t) = h'(t) - A(t)h(t) = (Th)(t) \qquad (t > 0)$$

and

$$(Lh)(t) = (-G_1 g)(-t) = \big(g'(-t) + A(t)g(-t)\big)$$
$$= -\big(-h'(t) + A(t)h(t)\big) = (Th)(t) \qquad (t < 0).$$

Thus $Lh = Th$. Since $h \in \mathrm{Dom}\, T$ is arbitrary, this leads to $T = L|_{\mathrm{Dom}\, T}$.

We now apply Lemma 4.2 with $\mathcal{D} = \mathrm{Dom}\, T$. It follows that T is Fredholm if and only if L is Fredholm, and in this case

$$(4.8) \qquad \mathrm{index}\, T = \mathrm{index}\, L - \dim(\mathrm{Dom}\, L / \mathrm{Dom}\, T) = \mathrm{index}\, L - n,$$

where we made use of (4.6). Finally, it is clear from the definition (4.5) that L is Fredholm if and only if G and G_1 are Fredholm, and in this case

$$(4.9) \qquad \mathrm{index}\, L = \mathrm{index}\, G + \mathrm{index}\, G_1.$$

Thus, T is Fredholm if and only if G and G_1 are Fredholm, and (4.2) follows from (4.8) and (4.9). $\square$

The next result connects the dichotomy of linear differential equations on the full line and on the half line.

LEMMA 4.3. *Let $A(t)$ be a bounded and measurable $n \times n$ complex matrix valued function defined on $(-\infty, \infty)$. A projection P in $\mathbf{C}^n$ is a dichotomy for the differential equation on the full line*

$$(4.10) \qquad \frac{dx}{dt} = A(t)x(t) \qquad (-\infty < t < \infty)$$

if and only if the following two differential equations on the half line

$$(4.11) \qquad \frac{dx}{dt} = A(t)x(t) \qquad (t \geq 0)$$

and

$$(4.12) \qquad \frac{dx}{dt} = -A(-t)x(t) \qquad (t \geq 0)$$

admit the dichotomies P and $I_n - P$ respectively.

PROOF. Let $U(t)$ $(-\infty < t < \infty)$ be the evolution operator of (4.10). Then $U(t)$ and $U(-t)$ $(t \geq 0)$ are respectively the evolution operators of (4.11) and (4.12).

Assume that P and $I_n - P$ are dichotomies for (4.11) and (4.12) respectively, with the pairs of constants (a_1, M_1) and (a_2, M_2). Set $a = \max(a_1, a_2)$ and $M = \max(M_1, M_2)$. Then (a, M) is a pair of constants for the dichotomies P and $I_n - P$ of (4.11) and (4.12). The dichotomy inequalities for P are

$$(4.13) \quad \|U(t)PU(s)^{-1}\| \leq Ma^{t-s}, \quad \|U(s)(I_n - P)U(t)^{-1}\| \leq Ma^{t-s} \qquad (t \geq s \geq 0).$$

Moreover, since $U(-t)$ is the evolution operator of (4.12), the dichotomy inequalities for the dichotomy $I_n - P$ of (4.12) are

$$(4.14) \quad \|U(-t_1)(I_n - P)U(-s_1)^{-1}\| \leq Ma^{t_1 - s_1}, \qquad \|U(-s_1)PU(-t_1)^{-1}\| \leq Ma^{t_1 - s_1}$$

for $t_1 \geq s_1 \geq 0$. We now show that

$$(4.15) \qquad \|U(t)PU(s)^{-1}\| \leq \max(M, M^2)a^{t-s} \qquad (t \geq s).$$

If $s \geq 0$, this follows from the left hand side inequality in (4.13). Similarly, if $t \leq 0$ then (4.15) follows from the right hand side inequality in (4.14) with $t_1 = -s$ and $s_1 = -t$. Assume that $t > 0 > s$. By the left hand side inequality in (4.13) we obtain $\|U(t)P\| \leq Ma^t$, while the right hand side equality in (4.14) with $t_1 = -s$ and $s_1 = 0$ implies $\|PU(s)^{-1}\| \leq Ma^{-s}$. Combining these two inequalities we obtain

$$\|U(t)PU(s)^{-1}\| \leq \|U(t)P\|\|PU(s)^{-1}\| \leq Ma^t Ma^{-s} = M^2 a^{t-s}.$$

Thus (4.15) holds.

By considering the right hand side inequality in (4.13) and the left hand side inequality in (4.14), one obtains in a similar way

$$(4.16) \qquad \|U(s)(I_n - P)U(t)^{-1}\| \leq Ma^{t-s} \qquad (t \geq s).$$

It follows from (4.15) and (4.16) that P is a dichotomy for (4.10).

Conversely, assume that P is a dichotomy for (4.10) with pair of constants a and M. Then (4.15) and (4.16) hold, and consequently (4.13) and (4.14) hold as well. By (4.13), P is a dichotomy for (4.11), and by (4.14) $I_n - P$ is a dichotomy for (4.12). $\square$

Finally, we will use the following two simple results which we state here for convenience.

LEMMA 4.4. *Let $A(t)$ be a bounded and measurable $n \times n$ complex matrix valued function such that the differential equation*

$$(4.17) \qquad \frac{dx}{dt} = A(t)x(t) \qquad (t \geq 0)$$

admits the dichotomy P. For $x \in \mathbb{C}^n$, the function f on $[0, \infty)$ defined by

$$(4.18) \qquad f(t) = U(t)x \qquad (t \geq 0),$$

where $U(\cdot)$ is the evolution operator of (4.17), belongs to $L_2^n(0, \infty)$ if and only if

$$x \in \operatorname{Im} P.$$

PROOF. If $x \in \operatorname{Im} P$ then the dichotomy inequality (1.2) leads to

$$\|U(t)x\| = \|U(t)PU(0)^{-1}x\| \leq Ma^t\|x\| \qquad (t \geq 0).$$

Thus, $f \in L_2^n(0, \infty)$.

Conversely, let $x \in \mathbb{C}^n$ and assume that the function f defined by (4.18) belongs to $L_2^n(0, \infty)$. Define functions g and h via

$$g(t) = U(t)Px \quad \text{and} \quad h(t) = U(t)(I_n - P)x \qquad (x \geq 0).$$

By the first part of the proof $g \in L_2^n(0, \infty)$. Thus, $h = f - g \in L_2^n(0, \infty)$. Note that the dichotomy inequality (1.3) with $s = 0$ leads to

$$\|U(0)(I_n - P)U(t)^{-1}\| \leq Ma^t \qquad (t \geq 0).$$

Thus,

$$\|(I_n - P)x\| = \|U(0)(I_n - P)U(t)^{-1}U(t)(I_n - P)x\| \leq Ma^t\|U(t)(I_n - P)x\| \qquad (t \geq 0).$$

Hence, $\|h(t)\| \geq M^{-1}a^{-t}\|(I_n - P)x\|$ $(t \geq 0)$. Since $h \in L_2^n(0, \infty)$ we have $(I_n - P)x = 0$, whence $x \in \operatorname{Im} P$. $\square$

LEMMA 4.5. *Let $A(t)$ be a bounded and measurable $n \times n$ complex matrix valued function such that the differential equation*

$$(4.19) \qquad \frac{dx}{dt} = A(t)x(t) \qquad (t \geq 0)$$

admits the dichotomy P. Then the differential equation

$$(4.20) \qquad \frac{dx}{dt} = -A(t)^*x(t) \qquad (t \geq 0)$$

admits the dichotomy $I - P^*$.

PROOF. Let $U(t)$ be the evolution operator of (4.19). Since $dU(t)/dt = A(t)U(t)$, then

$$\begin{aligned}
dU(t)^{*-1}/dt &= -U(t)^{*-1}\big(dU(t)^*/dt\big)U(t)^{*-1} \\
&= -U(t)^{*-1}U(t)^*A(t)^*U(t)^{*-1} = -A(t)^*U(t)^{*-1} \qquad (t \geq 0).
\end{aligned}$$

Thus, $U(t)^{*-1}$ is the evolution operator of (4.20).

Now, by taking adjoints in the dichotomy inequality (1.3) we obtain

$$\|U(t)^{*-1}(I - P^*)[U(s)^{*-1}]^{-1}\| \leq Ma^{t-s} \qquad (t \geq s),$$

while (1.2) implies similarly

$$\|U(s)^{*-1}(I - (I - P^*))[U(t)^{*-1}]^{-1}\| \leq Ma^{t-s} \qquad (t \geq s).$$

Since $U(t)^{*-1}$ is the evolution operator of (4.20), these two inequalities mean that $I - P^*$ is a dichotomy for (4.20). $\square$

5. FREDHOLM PROPERTIES OF DIFFERENTIAL OPERATORS ON A HALF LINE

The following result describes the connection between the Fredholm properties of differential operators on the half line and dichotomy.

THEOREM 5.1. *Let $A(t)$ be a bounded and measurable $n \times n$ complex matrix valued function defined on $[0, \infty)$. The operator G in $L_2^n(0, \infty)$ defined by*

$$(Gf)(t) = f'(t) - A(t)f(t) \qquad (t \geq 0),$$

with the domain

$$\operatorname{Dom} G = \left\{ f \in L_2^n(0, \infty) : f \text{ is locally absolutely continuous and } f' \in L_2^n(0, \infty) \right\}$$

is Fredholm if and only if the ordinary differential equation

$$(5.1) \qquad \frac{dx}{dt} = A(t)x(t) \qquad (t \geq 0)$$

admits a dichotomy. Moreover, if G is Fredholm and P is a dichotomy for (5.1) then

$$(5.2) \qquad \operatorname{Im} G = L_2^n(0, \infty),$$

and

$$(5.3) \qquad \operatorname{Ker} G = \left\{ f : f(t) = U(t)x \ (t \geq 0), \ x \in \operatorname{Im} P \right\},$$

where $U(t)$ is the evolution operator of (5.1). In particular,

$$(5.4) \qquad \operatorname{index} G = \operatorname{Rank} P.$$

For analogous results on differential operators in the function spaces $L_1^n(0, \infty)$, $C([0, \infty))$ and the space M of locally integrable functions $f : [0, \infty) \to \mathbb{C}^n$ with norm $\|f\|_M = \sup_{t \geq 0} \int_t^{t+1} \|f(s)\| ds$, see Lecture 3 in [C].

In the proof of Theorem 1.1 we will use the following result.

LEMMA 5.2. *Let $A(t)$ and G be as in Theorem 5.1. Then G has a dense range, namely*

$$(5.5) \qquad \overline{\operatorname{Im} G} = L_2^n(0, \infty).$$

PROOF. We first prove (5.5). Let $g : [0, \infty) \to \mathbb{C}^n$ be a continuous function with compact support. Define f via

$$f(t) = - \int_t^\infty U(t)U(s)^{-1}g(s)ds \qquad (t \geq 0).$$

Note that $f(t) = 0$ for $t > \max \operatorname{supp} g$. Thus, $f \in L_2^n(0, \infty)$. Furthermore f is locally absolutely continuous and $f'(t) = A(t)f(t) + g(t)$ $(t \geq 0)$. Thus $f \in \operatorname{Dom} G$ and $Gf = g$. Since the set of continuous function $g : [0, \infty) \to \mathbb{C}^n$ with compact support is dense in $L_2^n(0, \infty)$, this proves (5.5). $\square$

Before we present the proof of Theorem 5.1 we make some remarks about the time independent case. Let B be a complex $n \times n$ matrix with no purely imaginary or zero eigenvalues. Consider the operator G in $L_2^n(0, \infty)$ given by

$$(Gf)(t) = f'(t) - Bf(t) \qquad (t \geq 0)$$

with $\operatorname{Dom} G = \{f \in L_2^n(0,\infty): f$ is locally absolutely continuous and $f' \in L_2^n(0,\infty)\}$. It is easily checked that the following integral operator Γ is a right inverse for G

$$(\Gamma g)(t) = \int_0^t e^{(t-s)B} Q g(s)\,ds - \int_t^\infty e^{(t-s)B}(I_n - Q)g(s)\,ds \qquad (t \geq 0,\ g \in L_2^n(0,\infty)),$$

where Q is the Riesz projection of B corresponding to the left half plane of $\mathbb{C}$. Thus $\operatorname{Im} G = L_2^n(0,\infty)$. Moreover $f \in \operatorname{Ker} G$ if and only if $f(t) = e^{tB}x$ for some x in $\mathbb{C}^n$ and $f \in L_2^n(0,\infty)$. The last condition is clearly equivalent to $x \in \operatorname{Im} Q$. Thus,

$$\operatorname{Ker} G = \{f: f(t) = e^{tB}x \ (t \geq 0),\ x \in \operatorname{Im} Q\}.$$

Since G is onto we obtain

$$(5.6) \qquad\qquad\qquad \operatorname{index} G = \operatorname{Rank} Q.$$

We now turn to the proof of Theorem 5.1.

PROOF OF THEOREM 5.1. Assume first that G is Fredholm. It is clear that a function f belongs to $\operatorname{Ker} G$ if and only if $f(t) = U(t)x \ (t \geq 0)$ where $x = f(0) \in \mathbb{C}^n$ and $f \in L_2^n(0,\infty)$. This may be restated as

$$(5.7) \qquad\qquad\qquad \operatorname{Ker} G = \{f: f(t) = U(t)x \ (t \geq 0),\ x \in N\}$$

where N is the subspace of $\mathbb{C}^n$ defined by

$$(5.8) \qquad\qquad\qquad N = \left\{x \in \mathbb{C}^n: \int_0^\infty \|U(t)x\|^2\,dt < \infty\right\}.$$

Let R be an arbitrary projection in $\mathbb{C}^n$ with

$$(5.9) \qquad\qquad\qquad\qquad \operatorname{Im} R = N.$$

Note first that (5.5) and (5.7)–(5.9) lead to

$$(5.10) \qquad\qquad\qquad\qquad \operatorname{index} G = \operatorname{Rank} R.$$

We now extend $A(\cdot)$ to the negative half line via

$$A(-t) = I_n - 2R \qquad (t > 0),$$

and we consider the operator G_1 defined in $L_2^n(0,\infty)$ by

$$(G_1 f)(t) = f'(t) + A(-t)f(t) \qquad (t \geq 0),$$

with $\operatorname{Dom} G_1 = \operatorname{Dom} G$. Note that G_1 is time independent. We apply the remarks preceding the proof with $B = -A(-t) = 2R - I_n$. Thus, G_1 is Fredholm. Moreover, since $Q = I_n - R$ is the Riesz projection of B corresponding to the left half plane we have by (5.6)

$$(5.11) \qquad\qquad\qquad\qquad \operatorname{index} G_1 = \operatorname{Rank}(I_n - R).$$

Consider now the differential operator T on the full line defined by

$$(Tf)(t) = f'(t) - A(t)f(t) \qquad (-\infty < t < \infty),$$

with

$$\mathrm{Dom}\, T = \{f \in L_2^n(-\infty, \infty) : f \text{ is locally absolutely continuous and } f' \in L_2^n(-\infty, \infty)\},$$

and $A(t)$ as above $(-\infty < t < \infty)$. Since G and G_1 are Fredholm, we can apply Lemma 4.1 to T. It follows that T is Fredholm with

$$(5.12) \qquad \mathrm{index}\, T = \mathrm{index}\, G + \mathrm{index}\, G_1 - n = 0$$

where we used (5.10) and (5.11).

Let us now prove that

$$(5.13) \qquad \mathrm{Ker}\, T = \{0\}.$$

Let $f \in \mathrm{Ker}\, T$. Then the restriction of f to the positive half line belongs to $\mathrm{Ker}\, G$. It follows from (5.7) and (5.9) that

$$(5.14) \qquad f(t) = U(t)f(0) \qquad (t \geq 0)$$

and

$$(5.15) \qquad f(0) \in N = \mathrm{Im}\, R.$$

Moreover, since $A(t) = I_n - 2R$ $(t < 0)$, we obtain

$$f'(t) - (I_n - 2R)f(t) = f'(t) - A(t)f(t) = (Tf)(t) = 0 \qquad (t < 0).$$

Thus,

$$(5.16) \qquad f(t) = e^{(I_n - 2R)t} f(0) \qquad (t < 0).$$

Since $f \in L_n^2(-\infty, \infty)$ this immediately implies $Rf(0) = 0$. In view of (5.15) this leads to $f(0) = 0$, and hence (5.14) and (5.16) show that $f = 0$. Thus (5.13) holds.

It follows from (5.12)–(5.13) and the Fredholm property of T that T is invertible. By Theorem 1.1 the ordinary differential equation on the full line

$$\frac{dx}{dt} = A(t)x(t) \qquad (-\infty < t < \infty)$$

admits a dichotomy. Hence, (5.1) admits a dichotomy.

Conversely, assume that the differential equation (5.1) admits a dichotomy P. We extend $A(\cdot)$ to the negative half line via

$$A(-t) = I - 2P \qquad (t > 0).$$

Then the following time independent differential equation

$$\frac{dx}{dt} = -A(-t)x(t) = (P - (I_n - P))x(t) \qquad (t \geq 0),$$

admits the dichotomy $I_n - P$. Since (5.1) admits the dichotomy P, it follows from Lemma 4.3 that the differential equation

$$(5.17) \qquad \frac{dx}{dt} = A(t)x(t) \qquad (-\infty < t < \infty)$$

admits the dichotomy P.

Consider the operator T defined by

$$(Tf)(t) = f'(t) - A(t)f(t) \qquad (-\infty < t < \infty)$$

with $\mathrm{Dom}\, T = \{f \in L_2^n(-\infty, \infty): f \text{ is locally absolutely continuous and } f' \in L_2^n(-\infty, \infty)\}$. Since (5.17) admits a dichotomy, it follows from Theorem 1.1 that T is invertible. Hence, Lemma 4.1 shows that G is Fredholm. In particular, $\mathrm{Im}\, G$ is closed and therefore, Lemma 5.2 implies (5.2). Note also that $f \in \mathrm{Ker}\, G$ if and only if $f(t) = u(t)x$ $(-\infty < t < \infty)$ for some $x \in \mathbb{C}^n$, and $f \in L_2^n(0, \infty)$. By Lemma 4.4 the last condition is equivalent to $x \in \mathrm{Im}\, P$. Thus, (5.3) holds. Finally, (5.4) is a consequence of (5.2) and (5.3). $\square$

6. FREDHOLM PROPERTIES OF DIFFERENTIAL OPERATORS ON A FULL LINE

In this section we prove Theorem 1.2. We will use the following result which describes the adjoint of T.

LEMMA 6.1. *Let $A(t)$ be a bounded and measurable $n \times n$ complex matrix valued function defined for $-\infty < t < \infty$. The operator T in $L_2^n(-\infty, \infty)$ defined by the formula*

$$(6.1) \qquad (Tf)(t) = f'(t) - A(t)f(t) \qquad (-\infty < t < \infty)$$

with domain given by

$$\mathrm{Dom}\, T = \{f \in L_2^n(-\infty, \infty): f \text{ is locally absolutely continuous and } f' \in L_2^n(-\infty, \infty)\},$$

has an adjoint T^* given by the formula

$$(6.3) \qquad (T^*g)(t) = -g'(t) - A(t)^*g(t) \qquad (-\infty < t < \infty)$$

with domain given by

$$(6.4) \qquad \mathrm{Dom}\, T^* = \mathrm{Dom}\, T.$$

In the proof we will use the well known fact that if a function $g \in L_2^n(-\infty, \infty)$ admits a derivative $h \in L_2^n(-\infty, \infty)$ in the distributional sense, then g is locally absolutely continuous and $g'(t) = h(t)$ a.e.

PROOF. The proof is based on the following equality

(6.5)
$$\langle Tf, g \rangle = \int_{-\infty}^{\infty} g^*(t)(f'(t) - A(t)f(t))\,dt = \int_{-\infty}^{\infty} g^*(t)f'(t)\,dt - \int_{-\infty}^{\infty} (A(t)^* g(t))^* f(t)\,dt,$$

for $f \in \operatorname{Dom} T$ and $g \in L_2^n(-\infty, \infty)$. Here, $\langle \cdot, \cdot \rangle$ denotes the inner product in $L_2^n(-\infty, \infty)$.

Assume that $g \in \operatorname{Dom} T^*$. Then

$$\langle Tf, g \rangle = \langle f, T^* g \rangle = \int_{-\infty}^{\infty} ((T^* g)(t))^* f(t)\,dt \qquad (f \in \operatorname{Dom} T).$$

By (6.5) this leads to

(6.6)
$$\int_{-\infty}^{\infty} g^*(t)f'(t)\,dt = \int_{-\infty}^{\infty} [(T^* g)(t) + A(t)^* g(t)]^* f(t)\,dt \qquad (f \in \operatorname{Dom} T).$$

This holds, in particular for every C^∞ function f with compact support. Hence, this equality shows that $g(t)$ admits the derivative $-((T^* g)(t) + A(t)^* g(t))$ in the distributional sense. Moreover, $-((T^* g)(t) + A(t)^* g(t)) \in L_2^n(-\infty, \infty)$. Hence, $g \in \operatorname{Dom} T$ and

$$g'(t) = -(T^* g)(t) - A(t)^* g(t) \qquad (-\infty < t < \infty, \text{ a.e.}).$$

Since $g \in \operatorname{Dom} T^*$ is arbitrary, this implies (6.3) and

$$\operatorname{Dom} T^* \subset \operatorname{Dom} T.$$

In order to conclude the proof we have to show that

(6.7)
$$\operatorname{Dom} T^* \supset \operatorname{Dom} T.$$

Let $g \in \operatorname{Dom} T$. As is well known, we have

$$\int_{-\infty}^{\infty} g(t)^* f'(t)\,dt = -\int_{-\infty}^{\infty} g'(t)^* f(t)\,dt \qquad (f \in \operatorname{Dom} T).$$

Hence, (6.5) leads to

$$\langle Tf, g \rangle = -\int_{-\infty}^{\infty} [g'(t) + A(t)^* g(t)]^* f(t)\,dt = -\langle f, g' + A^* g \rangle \qquad (f \in \operatorname{Dom} T),$$

where $(A^* g)(t) = A(t)^* g(t)$. This shows that $g \in \operatorname{Dom} T^*$ and hence, that (6.7) holds.
$\square$

We now prove Theorem 1.2.

PROOF OF THEOREM 1.2. We define two operators G and G_1 in $L_2^n(0, \infty)$ via $(Gf)(t) = f'(t) - A(t)f(t)$ and $(G_1 f)(t) = f'(t) + A(-t)f(t)$ $(t \geq 0)$, with $\operatorname{Dom} G = \operatorname{Dom} G_1 = \{f \in L_2^n(0, \infty) \colon f \text{ is locally absolutely continuous and } f' \in L_2^n(0, \infty)\}$. By Lemma 4.1, T is Fredholm if and only if G and G_1 are Fredholm, and in this case

(6.8)
$$\operatorname{index} T = \operatorname{index} G + \operatorname{index} G_1 - n.$$

By Theorem 5.1 the operator G (respectively G_1) is Fredholm if and only if the differential equation (1.8) (respectively (1.9)) admits a dichotomy. Hence, T is Fredholm if and only if the equations (1.8) and (1.9) admit dichotomies. This proves the first part of the theorem.

Assume now that (1.8) and (1.9) admit respectively, the dichotomies P and P_1. By Theorem 5.1, we have

$$\text{index}\, G = \text{Rank}\, P \qquad \text{and} \qquad \text{index}\, G_1 = \text{Rank}\, P_1.$$

Combining this with (6.8) we obtain (1.12).

We now prove (1.10). For a function f, it is clear that $f \in \text{Ker}\, T$ if and only if $f(t) = U(t)x$ $(-\infty < t < \infty)$ for some $x \in \mathbb{C}^n$ and $f \in L_2^n(-\infty, \infty)$. For $x \in \mathbb{C}^n$ define a function f_x via $f_x(t) = U(t)x$ $(-\infty < t < \infty)$, and define two functions g and h on $[0, \infty)$ via

$$g(t)U(t)x \quad \text{and} \quad h(t) = U(-t)x \qquad (t \geq 0).$$

Then $f_x \in L_2^n(-\infty, \infty)$ if and only if $g, h \in L_2^n(0, \infty)$. Since P is a dichotomy for (1.8), Lemma 4.4 shows that $g \in L_2^n(0, \infty)$ is equivalent to the condition $x \in \text{Im}\, P$. Furthermore, note that $U(-t)$ $(t \geq 0)$ is the evolution operator of (1.9). Hence, applying Lemma 4.4 again shows that $h \in L_2^n(0, \infty)$ if and only if $x \in \text{Im}\, P_1$. This shows that for $x \in \mathbb{C}^n$, the function $f_x : (-\infty, \infty) \to \mathbb{C}^n : t \to U(t)x$ belongs to $L_2^n(-\infty, \infty)$ if and only if $x \in \text{Im}\, P \cap \text{Im}\, P_1$. Since each $f \in \text{Ker}\, T$ is of the type $f(t) = U(t)x = f_x(t)$ for some $x \in \mathbb{C}^n$, this proves (1.10).

Finally, we prove (1.11). Since T is Fredholm, T^* is Fredholm too, and we have

$$(6.9) \qquad\qquad \text{Im}\, T = (\text{Ker}\, T^*)^{\perp}.$$

By Lemma 6.1, T^* is given by

$$(T^*f)(t) = -f'(t) - A(t)^* f(t) \qquad (-\infty < t < \infty)$$

with $\text{Dom}\, T^* = \text{Dom}\, T$. We may thus apply the first part of the proof to the operator $-T^*$ given by

$$(-T^*f)(t) = f'(t) + A(t)^* f(t) \qquad (-\infty < t < \infty).$$

The corresponding differential equation on the full line is

$$\frac{dx}{dt} = -A^*(t)x(t) \qquad (-\infty < t < \infty)$$

with evolution operator $U(t)^{*-1}$. Moreover, by Lemma 4.5, $I - P^*$ and $I - P_1^*$ are dichotomies of the differential equations on the half line

$$\frac{dx}{dt} = -A(t)^* x(t) \qquad (t \geq 0),$$

and

$$\frac{dx}{dt} = A(-t)^* x(t) \qquad (t \geq 0),$$

respectively. Applying (1.10) to $-T^*$ we thus obtain

$$(6.10) \qquad \mathrm{Ker}(-T^*) = \{f \colon f(t) = U(t)^{*-1}x, \ x \in \mathrm{Im}(I - P^*) \cap \mathrm{Im}(I - P_1^*)\}.$$

Note that

$$\mathrm{Im}(I - P^*) \cap \mathrm{Im}(I - P_1^*) = \mathrm{Ker}\, P^* \cap \mathrm{Ker}\, P_1^* = (\mathrm{Im}\, P)^\perp \cap (\mathrm{Im}\, P_1)^\perp = (\mathrm{Im}\, P + \mathrm{Im}\, P_1)^\perp.$$

Thus, (6.10) leads to

$$\mathrm{Ker}\, T^* = \{f \colon f(t) = U(t)^{*-1}x, \ x \in (\mathrm{Im}\, P + \mathrm{Im}\, P_1)^\perp\}.$$

This inequality and (6.9) imply (1.11). $\quad\square$

7. EXPONENTIALLY DICHOTOMOUS OPERATORS

In this section we prove Theorem 1.3.

PROOF OF THEOREM 1.3. Assume that T is exponentially dichotomous. It follows from a result in Section 1 of [BGK] that T is invertible. Hence, Theorem 1.1 implies that (1.16) admits a dichotomy.

Conversely, assume that (1.16) admits a dichotomy P. Define $\mathcal{P}$ via (1.17). We will show that T is exponentially dichotomous with separating projection $\mathcal{P}$. Note first that (1.2) with $s = t$ implies

$$(7.1) \qquad\qquad \|U(t)PU(t)^{-1}\| \leq M \qquad (-\infty < t < \infty).$$

Hence, $\mathcal{P}$ is a bounded operator in $L_2^n(-\infty, \infty)$. It is clear from its definition, and the fact that P is a projection, that $\mathcal{P}$ is a projection in $L_2^n(-\infty, \infty)$.

We now show that $\mathcal{P}$ commutes with T. Let $f \in \mathrm{Dom}\, T$. Then $\mathcal{P}f \in L_2^n(-\infty, \infty)$. Moreover, since $f(t)$, $U(t)$ and $U(t)^{-1}$ are locally absolutely continuous, so is $(\mathcal{P}f)(t) = U(t)PU(t)^{-1}f(t)$. Furthermore, we have

$$\begin{aligned}(\mathcal{P}f)'(t) = {}& A(t)U(t)PU(t)^{-1}f(t) - U(t)PU(t)^{-1}A(t)U(t)U(t)^{-1}f(t)\\ & + U(t)PU(t)^{-1}f'(t) \qquad (-\infty < t < \infty).\end{aligned}$$

It follows from $f, f' \in L_2^n(-\infty, \infty)$, inequality (7.1) and the boundedness of $A(t)$ that $(\mathcal{P}f)' \in L_2^n(-\infty, \infty)$. Thus, $\mathcal{P}f \in \mathrm{Dom}\, T$. Finally, the above equality also shows that

$$\begin{aligned}(T(\mathcal{P}f))(t) &= (\mathcal{P}f)'(t) - A(t)(\mathcal{P}f)(t)\\ &= -U(t)PU(t)^{-1}A(t)f(t) + U(t)PU(t)^{-1}f'(t)\\ &= U(t)PU(t)^{-1}\big(f'(t) - A(t)f(t)\big) = (\mathcal{P}(Tf))(t) \qquad (-\infty < t < \infty).\end{aligned}$$

Hence $\mathcal{P}$ and T commute.

To show that T is an exponentially dichotomous operator with separating projection $\mathcal{P}$, we have to show that there exists a right (respectively left) exponentially decaying C_0-semigroup $S(\cdot)$ (respectively $R(\cdot)$) on $\operatorname{Ker}\mathcal{P}$ (respectively $\operatorname{Im}\mathcal{P}$) whose infinitesimal generator is $T|_{Ker\,\mathcal{P}}$ (respectively $T|_{\operatorname{Im}\mathcal{P}}$).

For each real number t we define an operator $L(t)$ on $L_2^n(-\infty,\infty)$ via

$$(7.2)\qquad (L(t)f)(s) = U(s)U(s+t)^{-1}f(s+t)\qquad (-\infty < s < \infty,\ f \in L_2^n(-\infty,\infty)).$$

For each nonnegative number t we set

$$(7.3)\qquad\qquad\qquad S(t) = L(t)|_{Ker\,\mathcal{P}},$$

and for each nonpositive t we set

$$R(t) = L(t)|_{\operatorname{Im}\mathcal{P}}.$$

We will show that $S(\cdot)$ and $R(\cdot)$ satisfy the above requirements. Since the proofs for $S(\cdot)$ and $R(\cdot)$ are entirely symmetric, we only give the proof for $S(\cdot)$.

We first show that $S(t)$ is a bounded operator in $\operatorname{Ker}\mathcal{P}$ $(t \geq 0)$. Note that

$$(7.4)\qquad \operatorname{Ker}\mathcal{P} = \{f \in L_2^n(-\infty,\infty): U(s)^{-1}f(s) \in \operatorname{Ker}P\ (-\infty < s < \infty)\}.$$

Let $f \in \operatorname{Ker}\mathcal{P}$ and $t \geq 0$. Then $U(t+s)^{-1}f(t+s) \in \operatorname{Ker}P$, and therefore

$$(7.5)\ \ U(s)^{-1}\big(S(t)f\big)(s) = U(s)^{-1}U(s)U(t+s)^{-1}f(t+s) \in \operatorname{Ker}P\qquad (-\infty < s < \infty),$$

and

$$U(t+s)^{-1}f(t+s) = (I_n - P)U(t+s)^{-1}f(t+s)\qquad (-\infty < s < \infty).$$

Applying the dichotomy inequality (1.3) we obtain that

$$(7.6)\qquad \big\|(S(t)f)(s)\big\| = \big\|U(s)(I_n - P)U(t+s)^{-1}f(t+s)\big\| \leq Ma^t\|f(t+s)\|.$$

This shows in particular that $S(t)f \in L_2^n(-\infty,\infty)$, combining this with (7.5) and (7.4) we obtain that $S(t)f \in \operatorname{Ker}\mathcal{P}$. Furthermore, (7.6) also implies that $\|S(t)f\| \leq Ma^t\|f\|$. Hence $S(t)$ is a bounded operator in $\operatorname{Ker}\mathcal{P}$ with

$$(7.7)\qquad\qquad\qquad \|S(t)\| \leq Ma^t\qquad (t \geq 0).$$

Note also that $S(0) = I|_{Ker\,\mathcal{P}}$. Furthermore, the identity

$$(S(t_1)(S(t_2)f))(s) = U(s)U(s+t_1)^{-1}\big((S(t_2)f)(s+t_1)\big)$$
$$= U(s)U(s+t_1)^{-1}U(s+t_1)U(s+t_1+t_2)^{-1}f(s+t_1+t_2)$$
$$= U(s)U(s+t_1+t_2)^{-1}f(s+t_1+t_2) = (S(t_1+t_2)f)(s),$$

shows that $S(t)$ $(t \geq 0)$ is a semigroup of operators on $\operatorname{Ker} \mathcal{P}$.

We now show that the semigroup $S(t)$ $(t \geq 0)$ is strongly continuous. Let $f \in \operatorname{Ker} \mathcal{P}$. Then

$$
\begin{aligned}
(7.8) \qquad (S(t)f)(s) - f(s) &= U(s)U(s+t)^{-1}f(s+t) - f(s) \\
&= (U(s)U(s+t)^{-1} - I_n)f(s+t) + f(s+t) - f(s).
\end{aligned}
$$

By the inequality (2.5) we have

$$
\begin{aligned}
\left[\int_{-\infty}^{\infty} \|(U(s)U(s+t)^{-1} - I_n)f(s+t)\|^2 ds\right]^{1/2} &\leq (e^{t\|A\|_\infty} - 1)\left[\int_{-\infty}^{\infty} \|f(s+t)\|^2 ds\right]^{1/2} \\
&= (e^{t\|A\|_\infty} - 1)\|f\|.
\end{aligned}
$$

Hence,

$$
\lim_{t\to 0+} (U(\cdot)U(\cdot + t)^{-1} - I_n)f(\cdot + t) = 0 \quad \text{in} \quad L_2^n(-\infty, \infty).
$$

Moreover, the translation operator is strongly continuous in $L_2^n(-\infty, \infty)$ and therefore

$$
\lim_{t\to 0+} [f(\cdot + t) - f(\cdot)] = 0 \quad \text{in} \quad L_2^n(-\infty, \infty).
$$

Combining these two limits and (7.8) it follows that $\lim_{t\to 0+} (S(t)f)(\cdot) - f(\cdot) = 0$ in $L_2^n(-\infty, \infty)$. Thus $\lim_{t\to 0+} S(t) = I|_{\operatorname{Ker} \mathcal{P}}$ strongly. Therefore, $S(t)$ $(t \geq 0)$ is a strongly continuous semigroups of operators on $\operatorname{Ker} \mathcal{P}$. Furthermore, the semigroup $S(t)$ $(t \geq 0)$ is exponentially decaying by (7.7).

There remains to be shown that $T|_{\operatorname{Ker} \mathcal{P}}$ is the infinitesimal generator of $S(t)$ $(t \geq 0)$. Denote by J the infinitesimal generator of $S(t)$ $(t \geq 0)$.

We first show that

$$
(7.9) \qquad\qquad J \subset T|_{\operatorname{Ker} \mathcal{P}}.
$$

Let $f \in \operatorname{Dom} J$. Then the following limit exists in $L_2^n(-\infty, \infty)$

$$
(7.10) \qquad\qquad \lim_{t\to 0+} \frac{S(t)f - f}{t} = Jf.
$$

Let φ be a C^∞ function defined on $(-\infty, \infty)$, with values in $\mathbb{C}^n$ and of compact support. By taking the product of (7.10) with φ^* on the left, we obtain that

$$
\lim_{t\to 0+} \int_{-\infty}^{\infty} \varphi^*(s)t^{-1}\big((S(t)f)(s) - f(s)\big)\,ds = \int_{-\infty}^{\infty} \varphi^*(s)\big((Jf)(s)\big)\,ds.
$$

The definition of $S(t)$ leads to

$$
\lim_{t\to 0+} \int_{-\infty}^{\infty} \varphi^*(s)t^{-1}\big(U(s)U(s+t)^{-1}f(s+t) - f(s)\big)\,ds = \int_{-\infty}^{\infty} \varphi^*(s)\big((Jf)(s)\big)\,ds.
$$

However, the first summand in the left hand side integral of this equality is

$$\int_{-\infty}^{\infty} \varphi^*(s)t^{-1}U(s)U(s+t)^{-1}f(s+t)ds = \int_{-\infty}^{\infty} \varphi^*(s-t)t^{-1}U(s-t)U(s)^{-1}f(s)ds.$$

Hence, the above limit implies

$$(7.11) \quad \lim_{t\to0+}\int_{-\infty}^{\infty} t^{-1}\left[\varphi^*(s-t)U(s-t)U(s)^{-1} - \varphi^*(s)\right]f(s)ds = \int_{-\infty}^{\infty} \varphi^*(s)((Jf)(s))ds.$$

We now evaluate the limit on the left. Since φ is smooth, we have at each point s where $U(s)$ is differentiable, and in particular s-almost everywhere,
(7.12)

$$\lim_{t\to0+} t^{-1}\left[\varphi^*(s-t)U(s-t)U(s)^{-1} - \varphi^*(s)\right] = \frac{d}{dt}\left[\varphi^*(s-t)U(s-t)U(s)^{-1}\right]\Big|_{t=0}$$
$$= -\varphi'^*(s) - \varphi^*(s)A(s).$$

In addition the following inequality holds

$$\left\| t^{-1}\left[\varphi^*(s-t)U(s-t)U(s)^{-1} - \varphi^*(s)\right]\right\|$$
$$\leq \|\varphi^*(s-t)\|\,\|t^{-1}\|\,\|U(s-t)U(s)^{-1} - I_n\| + \left\|t^{-1}\left[\varphi^*(s-t) - \varphi^*(s)\right]\right\|$$
$$\leq \|\varphi\|_\infty t^{-1}(e^{t\|A\|_\infty} - 1) + \|\varphi'\|_\infty,$$

where we used (2.5). Hence, the term

$$t^{-1}\left[\varphi^*(s-t)U(s-t)U(s)^{-1} - \varphi^*(s)\right]$$

is uniformly bounded for $-\infty < s < \infty$ and t bounded. Since φ has compact support and $f \in L_2^n(-\infty, \infty)$ is locally integrable, it follows that we may apply Lebesgue's dominated convergence theorem to (7.11). By (7.12) we obtain

$$-\int_{-\infty}^{\infty}\left(\varphi'^*(s) + \varphi^*(s)A(s)\right)f(s)ds = \int_{-\infty}^{\infty} \varphi^*(s)((Jf)(s))\,ds.$$

Hence,

$$-\int_{-\infty}^{\infty} \varphi'^*(s)f(s)ds = \int_{-\infty}^{\infty} \varphi^*(s)\big(A(s)f(s) + (Jf)(s)\big)\,ds.$$

This means that $Af + Jf \in L_2^n(-\infty, \infty)$ is the distributional derivative of f, where A is the operator given by $(Af)(s) = A(s)f(s)$ $(-\infty < s < \infty)$. As is well known this implies that f is locally absolutely continuous and $f' = Af + Jf$. Thus, $f \in \mathrm{Dom}\,T$ and $Jf = f' - Af = Tf$. Thus, $f \in \mathrm{Dom}\,T \cap \mathrm{Ker}\,\mathcal{P} = \mathrm{Dom}(T|_{Ker\,\mathcal{P}})$ and $Jf = T|_{Ker\,\mathcal{P}}f$. Hence, (7.9) holds.

We now prove the converse inclusion

$$(7.13) \qquad\qquad J \supset T|_{Ker\,\mathcal{P}}.$$

Let $f \in \mathrm{Dom}(T|_{Ker\,\mathcal{P}})$. Then $f \in \mathrm{Dom}\,T \cap \mathrm{Ker}\,\mathcal{P}$. We prove that the following limit exists in $L_2^n(-\infty, \infty)$

$$(7.14) \qquad \lim_{t \to 0+} t^{-1}\big[U(s)U(s+t)^{-1}f(s+t) - f(s)\big] = f'(s) - A(s)f(s).$$

Note first that

$$\int_{-\infty}^{\infty} \big\|t^{-1}\big(U(s)U(s+t)^{-1} - I_n\big)\big(f(s+t) - f(s)\big)\big\|^2 ds$$

$$\leq \big[t^{-1}(e^{t\|A\|_\infty} - 1)\big]^2 \int_{-\infty}^{\infty} \|f(s+t) - f(s)\|^2 ds,$$

where we used (2.5). Since $t^{-1}(e^{t\|A\|_\infty} - 1)$ is bounded for t bounded and $\lim_{t \to 0+} \int_{-\infty}^{\infty} \|f(s+t) - f(s)\|^2 ds = 0$, it follows that we have the following limit in $L_2^n(-\infty, \infty)$

$$(7.15) \qquad \lim_{t \to 0+} t^{-1}\big[U(s)U(s+t)^{-1} - I_n\big]\big(f(s+t) - f(s)\big) = 0.$$

In addition, since f is locally absolutely continuous with $f, f' \in L_2^n(-\infty, \infty)$, it is well known that we have the following limit in $L_2^n(-\infty, \infty)$

$$(7.16) \qquad \lim_{t \to 0+} t^{-1}\big(f(s+t) - f(s)\big) = f'(s).$$

Finally, we prove that the following limit exists in $L_2^n(-\infty, \infty)$

$$(7.17) \qquad \lim_{t \to 0+} t^{-1}\big(U(s)U(s+t)^{-1} - I_n\big)f(s) = -A(s)f(s).$$

To do so consider the integral

$$(7.18) \qquad \int_{-\infty}^{\infty} \big\|t^{-1}\big(U(s)U(s+t)^{-1} - I_n\big) + A(s)\big\|^2 \|f(s)\|^2 ds.$$

Note that for each s at which $U(s)$ is differentiable, in particular s-almost everywhere, we have
(7.19)

$$\lim_{t \to 0+} t^{-1}\big(U(s)U(s+t)^{-1} - I_n\big) + A(s) = \frac{d}{dt}\big(U(s)U(s+t)^{-1}\big)|_{t=0} + A(s)$$

$$= -U(s)U(s)^{-1}A(s)U(s)U(s)^{-1} + A(s) = 0.$$

In addition, inequality (2.5) implies that

$$\big\|t^{-1}\big(U(s)U(s+t)^{-1} - I_n\big) + A(s)\big\| \leq t^{-1}(e^{t\|A\|_\infty} - 1) + \|A\|_\infty \qquad (-\infty < s < \infty).$$

Since $t^{-1}(e^{t\|A\|_\infty} - 1)$ is bounded for t bounded, and $\|f(s)\|^2$ is integrable, the last equality shows that we may apply Lebesgue's dominated convergence theorem to the integral (7.18). By (7.19) this leads to

$$\lim_{t \to 0+} \int_{-\infty}^{\infty} \big\|t^{-1}\big(U(s)U(s+t)^{-1} - I_n\big) + A(s)\big\|^2 \|f(s)\|^2 ds = 0.$$

This implies the following limit in $L_2^n(-\infty, \infty)$

$$\lim_{t \to 0+} \left[t^{-1}\left(U(s)U(s+t)^{-1} - I_n\right) + A(s) \right] f(s) = 0.$$

Hence, (7.17) holds. Adding (7.15), (7.16) and (7.17) we obtain (7.14). Now (7.14) means that the following limit holds

$$\lim_{t \to 0+} t^{-1}\left(S(t)f - f\right) = f' - Af.$$

Hence, $f \in \operatorname{Dom} J$ and $Jf = Tf$. Thus $Jf = T|_{Ker\,\mathcal{P}}f$, and therefore (7.13) holds.

Combining (7.9) and (7.13) we obtain that $T|_{Ker\,\mathcal{P}}$ is the infinitesimal generator of $S(t)$ $(t \geq 0)$. This completes the proof of Theorem 3 except for the formula (1.18). To prove (1.18) let $z \in \mathbb{C}^n$ be arbitrary. Note that $\mathcal{P}(\alpha z)$ is given by $(\mathcal{P}(\alpha z))(t) = \alpha(t)U(t)PU(t)^{-1}z$. Since $\lim_{t \to 0} U(t) = I_n$ and $\lim_{t \to 0} \alpha(t) = 1$, then $\lim_{t \to 0}(\mathcal{P}(\alpha z))(t) = Pz$. This implies (1.18). $\square$

8. REFERENCES

[BGK]   H. Bart, I. Gohberg and M. A. Kaashoek, Wiener-Hopf factorization, inverse Fourier transform and exponentially dichotomous operators. *J. Functional Analysis* **68** (1986) 1–42.

[BG]   A. Ben-Artzi and I. Gohberg, Band matrices and dichotomies, *Operator Theory: Advances and Applications* **50** (1990) 137–170.

[BGK1]   A. Ben-Artzi, I. Gohberg, and M. A. Kaashoek, Invertibility and dichotomy of differential operators on a half line, submitted.

[CS]   C. Chicone and R. C. Swanson, Spectral theory for linearizations of dynamical systems, *J. Differential Equations* **40** (1981) 155–167.

[C]   W. A. Coppel, Dichotomies in stability theory, *Lecture Notes in Mathematics* **629**, Springer-Verlag, Berlin, 1978.

[DK]   Ju. L. Daleckii and M. G. Krein, Stability of solutions of differential equations in Banach space, *Transl. Math. Monographs* **43**, Amer. Math. Soc., Providence, Rhode Island, 1974.

[DS]   N. Dunford and J. T. Schwartz, *Linear Operators*, Vol. 1, Interscience, New York, 1981.

[GGK]   I. Gohberg, S. Goldberg, and M. A. Kaashoek, *Classes of Linear Operators*, Vol. 1, Birkhäuser Verlag, 1990.

[GK]   I. Gohberg and M. G. Krein, The basic propositions on defect numbers, root numbers and indices of linear operators, *Amer. Math. Soc. Transl.* **13**(2) (1960) 185–264.

[GKvS]   I. Gohberg, M. A. Kaashoek and F. van Schagen, Non-compact integral operators with semi-separable kernels and their discrete analogous: inversion and Fredholm properties, *Integral Equations and Operator Theory* **7** (1984) 642–703.

[HP]   E. Hille and R. S. Phillips, *Functional Analysis and Semigroups*, Amer. Math. Soc., Providence, Rhode Island, 1957.

[MS]   J. L. Massera and J. J. Schäffer, *Linear Differential Equations and Function Spaces*, Academic Press, New York, 1966. ·

[P]   A. Pazy, Semigroups of linear operators and applications to partial differential equations, *Applied Mathematical Sciences* 44, Springer-Verlag, New York, 1983.

[SaS1]   R. J. Sacker and G. R. Sell, Existence of dichotomies and invariant splittings for linear differential system I, *J. Differential Equations* 15 (1974) 429–458.

[SaS2]   R. J. Sacker and G. R. Sell, Existence of dichotomies and invariant splittings for linear differential system II, *J. Differential Equations* 22 (1976) 478–496.

[SaS3]   R. J. Sacker and G. R. Sell, Existence of dichotomies and invariant splittings for linear differential system III, *J. Differential Equations* 22 (1976) 497–522.

[Y]   K. Yosida, *Functional Analysis*, 6th ed., Springer Verlag, New York, 1980.

A. Ben-Artzi
Department of Mathematics
Indiana University
Swain East 222
Bloomington, IN 47405, U.S.A.

I. Gohberg
School of Mathematical Sciences
The Raymond and Beverly Sackler Faculty of Exact Sciences
Tel-Aviv University
Tel-Aviv, Ramat-Aviv 69978, Israel

120

Operator Theory:
Advances and Applications, Vol. 56
© 1992 Birkhäuser Verlag Basel

INERTIA THEOREMS FOR BLOCK WEIGHTED SHIFTS AND APPLICATIONS

Asher Ben-Artzi and **Israel Gohberg**

In this paper, we study connections between Fredholm properties of block weighted shifts in Hilbert spaces, Stein inequalities for nonstationary systems, and dichotomies for such systems.

1. INTRODUCTION

The Fredholm properties of block weighted shifts in Hilbert spaces are related to dichotomies and Stein inequalities for such shifts. Theorem 1.1 below describes these relations in one special case and may serve as a sample.

THEOREM 1.1. *Let* $G = (\delta_{i,j+1} A_j)_{i,j=0}^{\infty}$ *be a block weighted shift in* ℓ_r^2, *where* $\{A_n\}_{n=0}^{\infty}$ *is a bounded sequence of invertible* $r \times r$ *matrices. Then the following conditions are equivalent:*

I) *The operator* $I - G$ *is a Fredholm operator in* ℓ_r^2.

II) *There exists a bounded sequence* $\{X_n\}_{n=0}^{\infty}$ *of self-adjoint matrices such that*

$$(1.1) \qquad X_n - A_n^* X_{n+1} A_n \geq \varepsilon I_r \qquad (n = 0, 1, \ldots)$$

for some positive number ε.

III) *There exists a projection* P *in* $\mathbb{C}^r$, *and two positive numbers* a *and* M, *with* $a < 1$, *such that*

$$(1.2) \qquad \|U_i P U_j^{-1}\| \leq M a^{i-j} \qquad \text{and} \qquad \|U_j (I_r - P) U_i^{-1}\| \leq M a^{i-j}$$

for $j = 0, 1, \ldots; i = j, j+1, \ldots$. *Here* $U_0 = I_r$ *and* $U_n = A_{n-1} \cdots A_0$ $(n = 1, 2, \ldots)$. *Furthermore, if* P *is a projection satisfying* (1.2) *then*

$$(1.3) \qquad \operatorname{Im}(I - G) = \left\{ (U_n^{*-1} v)_{n=0}^{\infty} : v \perp \operatorname{Im} P \right\}^{\perp},$$

$$(1.4) \qquad \operatorname{Ker}(I - G) = \{0\},$$

and

$$(1.5) \qquad \operatorname{index}(I - G) = \operatorname{Rank} P - r.$$

Finally, if II) *and* III) *hold then there is an integer* N *such that*

$$(1.6) \qquad In(X_n) = (\operatorname{Rank} P, 0, r - \operatorname{Rank} P) \qquad (n = N, N+1, \ldots).$$

In the above statement and in the sequel we use the following notation.

For a finite self-adjoint matrix X, we define the *inertia of* X to be the triple of integers $In(X) = (\nu_+, \nu_0, \nu_-)$, where $\nu_0 = \dim \operatorname{Ker} X$ and ν_+ (respectively ν_-) is the number of positive (respectively negative) eigenvalues of X, counting multiplicities. We also say that a sequence $(X_n)_{n=0}^{\infty}$ of self-adjoint matrices is of constant inertia if $In(X_n) = In(X_{n+1})$ $(n = 0, 1, \ldots)$. We denote by ℓ_r^2 the Hilbert space of all square summable sequences $(x_n)_{n=0}^{\infty}$ with entries $x_n \in \mathbb{C}^r$. We also denote by I the identity operator, the space being clear from the context. We occasionally denote $I_r = (\delta_{ij})_{i,j=1}^r$.

Theorem 2.11 in the next section generalizes Theorem 1.1 above, except for (1.3), to the case when the matrices A_n are not necessarily invertible. The proof of (1.3) and the complete proof of Theorem 1.1 appear at the end of Section 3. In Section 3 we also give preliminary results about left and two sided systems. These results are then used in Section 4 to derive some Fredholm properties of two sided block weighted shifts. Section 5 contains an asymptotic inertia theorem.

2. ONE SIDED BLOCK WEIGHTED SHIFTS

In this section we give some preliminary definitions and results about dichotomy, nonstationary inertia theorems, and Fredholm operators with band characteristic. Theorem 2.11 at the end of this section generalizes most of Theorem 1.1 to the case when the matrix weights A_n are not necessarily invertible.

Let $\{A_n\}_{n=0}^{\infty}$ be a bounded sequence of complex $r \times r$ matrices. A bounded sequence of projections $\{P_n\}_{n=0}^{\infty}$ in $\mathbb{C}^r$, having constant rank, is called a dichotomy for the system

$$(2.1) \qquad x_{n+1} = A_n x_n \qquad (n = 0, 1, \ldots),$$

if the following commutation relations hold

$$(2.2) \qquad P_{n+1} A_n = A_n P_n \qquad (n = 0, 1, \ldots),$$

and if there exist two positive numbers a and M, with $a < 1$, such that the following inequalities hold

$$(2.3) \qquad \|A_{n+k-1} \cdots A_n P_n x\| \leq M a^k \|P_n x\|,$$

and

$$(2.4) \qquad \|A_{n+k-1} \cdots A_n (I_r - P_n) x\| \geq \frac{1}{M a^k} \|(I_r - P_n) x\|,$$

for $x \in \mathbb{C}^r$, $n = 0, 1, \ldots$ and $k = 1, 2, \ldots$. A pair of numbers a and M as above is called a *pair of constants* for the dichotomy. We also call the number $\operatorname{Rank} P_n$, which is independent of n, the *rank of the dichotomy*. This notion appears in [BG1], [BGK], [Bo], [CS] and [GKvS]. In the latter paper, the notion of dichotomy is connected with semi-separable operators.

Assume now that the matrices A_n $(n = 0, 1, \ldots)$ are invertible, and denote $U_0 = I_r$ and $U_n = A_{n-i} \cdots A_0$ $(n = 1, 2, \ldots)$. In this case, it follows from (2.2) that the initial projection $P = P_0$ determines the dichotomy $\{P_n\}_{n=0}^{\infty}$ completely. In fact, (2.2) leads to

$$(2.5) \qquad\qquad P_n = U_n P U_n^{-1} \qquad (n = 0, 1, \ldots).$$

Moreover, starting from a projection P, the sequence of projections $\{P_n\}_{n=0}^{\infty}$ defined by (2.5) is a dichotomy for (2.1), if and only if there exists two positive numbers a and M, with $a < 1$, such that

$$(2.6) \qquad \|U_i P U_j^{-1}\| \le M a^{i-j} \qquad \text{and} \qquad \|U_j (I_r - P) U_i^{-1}\| \le M a^{i-j}$$

for $j = 0, 1, \ldots; i = j, j+1, \ldots$. We also call the single projection $P = P_0$ the dichotomy.

We will use the following characterization of $\operatorname{Im} P_n$.

LEMMA 2.1. *If $\{P_n\}_{n=0}^{\infty}$ is a dichotomy for the system $x_{n+1} = A_n x_n$ $(n = 0, 1, \ldots)$, then*

$$(2.7) \qquad \operatorname{Im} P_n = \left\{ x \in \mathbb{C}^r \colon \lim_{k \to \infty} \|A_{n+k} \cdots A_n x\| = 0 \right\},$$

and

$$(2.8) \qquad \operatorname{Im} P_n = \left\{ x \in \mathbb{C}^r \colon \sum_{k=0}^{\infty} \|A_{n+k} \cdots A_n x\|^2 < \infty \right\},$$

for $n = 0, 1, \ldots$.

PROOF. Equality (2.7) is precisely the Corollary 6.2 of [BG1]. In addition, (2.7) implies that

$$(2.9) \qquad \operatorname{Im} P_n \supseteq \left\{ x \in \mathbb{C}^r \colon \sum_{k=0}^{\infty} \|A_{n+k} \cdots A_n x\|^2 < \infty \right\}.$$

Finally, assume $x \in \operatorname{Im} P_n$. Then, by (2.3)

$$\|A_{n+k} \cdots A_n x\| = \|A_{n+k} \cdots A_n P_n x\| \le M a^{k+1} \|x\|.$$

Since $0 < a < 1$, $\sum_{k=0}^{\infty} \|A_{n+k} \cdots A_n x\|^2 < \infty$. Hence

$$\operatorname{Im} P_n \subseteq \left\{ x \in \mathbb{C}^r \colon \sum_{k=0}^{\infty} \|A_{n+k} \cdots A_n x\|^2 < \infty \right\}.$$

This inclusion and (2.9) prove (2.8). □

REMARK. The above lemma shows that all the dichotomies of the system $x_{n+1} = A_n x_n$ $(n = 0, 1, \ldots)$ have the same rank. See [BGK] for a description of the set of all dichotomies of a discrete system.

We also use systems indexed in different ways. We say that $\{P_n\}_{n=N}^{\infty}$ is a dichotomy for the system $x_{n+1} = A_n x_n$ $(n = N, N + 1, \ldots)$, where N is an integer, if $\{P_{n+N}\}_{n=0}^{\infty}$ is a dichotomy for the system $x_{n+1} = A_{n+N} x_n$ $(n = 0, 1, \ldots)$. As a rule, theorems and proofs about systems of the form $x_{n+1} = A_n x_n$ $(n = 0, 1, \ldots)$ apply equally well to systems of the form $x_{n+1} = A_n X_n$ $(n = N, N + 1, \ldots)$. Thus, we will use such shift in indices without further explanations.

We say that the system $x_{n+1} = A_n x_n$ $(n = 0, 1, \ldots)$ admits an *asymptotic dichotomy* if for some positive integer N, the system $x_{n+1} = A_n x_n$ $(n = N, N + 1, \ldots)$ admits a dichotomy.

The following result is an immediate consequence of Lemma 2.1.

COROLLARY 2.2. *Assume that the systems* $x_{n+1} = A_n x_n$ $(n = N_1, N_1 + 1, \ldots)$ *and* $x_{n+1} = A_n x_n$ $(n = N_2, N_2 + 1, \ldots)$ *admit the dichotomies* $\{P_{n,1}\}_{n=N_1}^{\infty}$ *and* $\{P_{n,2}\}_{n=N_2}^{\infty}$ *respectively. Then*

$$\operatorname{Im} P_{n,1} = \operatorname{Im} P_{n,2} \qquad (n \geq \max\{N_1, N_2\}).$$

In particularly, the ranks of $\{P_{n,1}\}_{n=N_1}^{\infty}$ *and* $\{P_{n,2}\}_{n=N_2}^{\infty}$ *are equal.*

As a consequence of this result, we may speak about the *rank of an asymptotic dichotomy* of a system $x_{n+1} = A_n x_n$ $(n = 0, 1, \ldots)$, being defined as the rank of an arbitrary dichotomy of $x_{n+1} = A_n x_n$ $(n = N, N + 1, \ldots)$, where $N \geq 0$ is also arbitrary.

We will also use the following simple result in order to extend dichotomies to larger systems.

LEMMA 2.3. *Let* $\{A_n\}_{n=0}^{\infty}$ *be a sequence of invertible matrices. If the system*

$$(2.10) \qquad x_{n+1} = A_n x_n \qquad (n = 0, 1, \ldots)$$

admits an asymptotic dichotomy, then it admits a dichotomy.

PROOF. Assume that (2.10) admits an asymptotic dichotomy. Then, there exists a positive integer N such that the system

$$(2.11) \qquad x_{n+1} = A_n x_n \qquad (n = N, N + 1, \ldots)$$

admits a dichotomy $\{P_n\}_{n=N}^{\infty}$. Thus $\{P_n\}_{n=N}^{\infty}$ is a bounded sequence of projection having constant rank, such that

$$(2.13) \qquad P_{n+1} A_n = A_n P_n \qquad (n = N, N + 1, \ldots),$$

and the following inequalities hold

$$(2.14) \qquad \|A_{n+k-1} \cdots A_n P_n x\| \leq M a^k \|P_n x\|,$$

$$(2.15) \qquad \|A_{n+k-1}\cdots A_n(I_r - P_n)x\| \geq \frac{1}{Ma^k}\|(I_r - P_n)x\|,$$

for $n = N, N+1,\ldots$; $k = 1,2,\ldots$; $x \in \mathbb{C}^r$. Here a and M are two positive numbers with $a < 1$. We can assume without loss of generality that $M \geq 1$, and hence that (2.14)–(2.15) hold also for $k = 0$.

Define

$$P_n = A_n^{-1}\cdots A_{N-1}^{-1} P_N A_{N-1}\cdots A_n \qquad (n = 0, 1, \ldots, N-1).$$

It follows from this definition and (2.13) that

$$P_{n+1}A_n = A_n P_n \qquad (n = 0, 1, \ldots).$$

Furthermore, it is clear that $\{P_n\}_{n=N-1}^{\infty}$ is a bounded sequence of projections, having a constant rank.

Now denote

$$M_1 = Ma^{-N}\left[\max_{0 \leq n < N}(\|A_n\|, \|A_n^{-1}\|)\right]^N.$$

Then (2.14)–(2.15) lead to

$$\|A_{n+k-1}\cdots A_n P_n x\| \leq M_1 a^k \|P_n x\|$$

and

$$\|A_{n+k-1}\cdots A_n(I_r - P_n)x\| \geq \frac{1}{M_1 a^k}\|(I_r - P_n)x\|$$

for $n = 0, 1, \ldots$; $k = 1, 2, \ldots$ and $x \in \mathbb{C}^r$. Hence, $\{P_n\}_{n=0}^{\infty}$ is a dichotomy for (2.10). $\quad\square$

There are various ways to show that a system admits a dichotomy. We will use the following result which follows from Theorem 4.2, Proposition 6.1, and Theorem 6.4 of [BG1].

THEOREM 2.4. *Let $\{A_n\}_{n=0}^{\infty}$ be a bounded sequence of $r \times r$ matrices. Then the system*

$$(2.16) \qquad x_{n+1} = A_n x_n \qquad (n = 0, 1, \ldots),$$

admits a dichotomy if and only if there exists a bounded sequence of self-adjoint matrices $\{X_n\}_{n=0}^{\infty}$, having constant inertia, and such that

$$(2.17) \qquad X_n - A_n^* X_{n+1} A_n \geq \varepsilon I_r \qquad (n = 0, 1, \ldots),$$

for some positive number ε. Moreover, if (2.17) holds for some bounded sequence $\{X_n\}_{n=0}^{\infty}$ of self-adjoint matrices, having constant inertia given by

$$(2.18) \qquad In(X_n) = (\nu_+, \nu_0, \nu_-) \qquad (n = 0, 1, \ldots),$$

then $\nu_0 = 0$ and ν_+ is equal to the rank of a dichotomy of the system (2.16).

A system of inequalities as in (2.18) is called a nonstationary system of Stein inequalities. We will use some results about such inequalities. We begin with a single inequality. Here and in the sequel, we denote for an $r \times r$ matrix A, $V(\lambda, A) = \mathrm{Ker}\big((\lambda I_r - A)^r\big)$ $(\lambda \in \mathbb{C})$. Thus $V(\lambda, A)$ is the image of the Riesz projection of A corresponding to λ.

LEMMA 2.5. *Let A, X and Y be complex $r \times r$ matrices, with X and Y self-adjoint, and denote $In(X) = (\nu_+, \nu_0, \nu_-)$, and $In(Y) = (\mu_+, \mu_0, \mu_-)$. Then, the following statements hold.*

a) *If $X - A^*YA \geq 0$ then $\nu_+ + \nu_0 \geq \mu_+ + \mu_0$.*

b) *If $X - A^*YA > 0$ then $\nu_+ \geq \mu_+ + \mu_0$.*

c) *If $X - A^*YA \geq \varepsilon I_r$ for some real number ε, then, for each nonpositive number α,*

$$(2.19) \qquad \sum_{t \geq \varepsilon + \alpha \|A\|^2} \dim V(t, X) \geq \sum_{t \geq \alpha} \dim V(t, Y).$$

PROOF. First note that a) follows from c) with $\alpha = \varepsilon = 0$. Similarly c) implies also b). In fact, if $X - A^*YA > 0$ then $X - A^*YA \geq \varepsilon I_r$ for some $\varepsilon > 0$. Apply c) with $\alpha = 0$. Then (2.18) leads to

$$\nu_+ \geq \sum_{t \geq \varepsilon} \dim V(t, X) \geq \sum_{t \geq 0} \dim V(t, Y) = \mu_+ + \mu_0.$$

We now prove c). Denote,

$$N = \bigoplus_{t \geq \alpha} V(t, Y),$$

and

$$L = \{x \in \mathbb{C}^r : Ax \in N\}.$$

Since A is a square matrix,

$$(2.20) \qquad \dim L \geq \dim N = \sum_{t \geq \alpha} \dim V(t, Y).$$

Assume that $u \in L$. Then $X - A^*YA \geq \varepsilon I_r$ leads to

$$(2.21) \qquad u^*Xu \geq u^*A^*YAu + \varepsilon\|u\|^2.$$

However, $Au \in N$ and hence,

$$(2.22) \qquad u^*A^*YAu = (Au)^*YAu \geq \alpha\|Au\|^2.$$

On the other hand, $\|Au\|^2 \leq \|A\|^2 \|u\|^2$, and since $\alpha \leq 0$ we obtain $\alpha \|Au\|^2 \geq \alpha \|A\|^2 \|u\|^2$. By (2.22) this leads to $u^* A^* Y A u \geq \alpha \|A\|^2 \|u\|^2$. Inserting this in (2.21) we obtain

$$u^* X u \geq (\varepsilon + \alpha \|A\|^2) \|u\|^2.$$

This holds for each $u \in L$, and consequently $L \cap \left(\bigoplus_{t < \varepsilon + \alpha \|A\|^2} V(t, X) \right) = \{0\}$. Hence,

$$\dim L \leq \sum_{t \geq \varepsilon + \alpha \|A\|^2} \dim V(t, X).$$

Combining this inequality with (2.20) we obtain (2.19). $\quad\square$

We will apply the previous lemma in the following way.

COROLLARY 2.6. *Let* $\{A_n\}_{n=N}^{\infty}$ *and* $\{X_n\}_{n=N}^{\infty}$ *be two sequences of* $r \times r$ *matrices with* X_n *self-adjoint* $(n = N, N+1, \ldots)$. *Assume that*

$$(2.23) \qquad X_n - A_n^* X_{n+1} A_n > 0 \qquad (n = N, N+1, \ldots).$$

Then there exists an integer N_1, *with* $N_1 \geq N$, *such that*

$$(2.24) \qquad In(X_n) = (\nu_+, 0, \nu_-) \qquad (n = N_1, N_1+1, \ldots),$$

where ν_+ *and* ν_- *are independent of* n. *Similarly, if* $\{A_n\}_{n=-\infty}^{-N}$ *and* $\{X_n\}_{n=-\infty}^{-N}$ *are two sequences of matrices, with* X_n *self-adjoint* $(n = -N+1, -N, \ldots)$, *and the inequalities*

$$(2.25) \qquad X_n - A_n^* X_{n+1} A_n > 0 \qquad (n = -N, -N-1, \ldots),$$

hold, then there exists an integer N_1, *with* $N_1 \leq -N$, *such that*

$$(2.26) \qquad In(X_n) = (\mu_+, 0, \mu_-) \qquad (n = N_1, N_1-1, \ldots)$$

where μ_+ *and* μ_- *are independent of* n.

PROOF. Since the proofs of the first and the second parts of the lemma are similar, we prove only the first part.

Denote the inertia of X_n by

$$In(X_n) = \big(\nu_+(X_n), \nu_0(X_n), \nu_-(X_n)\big).$$

By Lemma 2.5 we have

$$(2.27) \qquad \nu_+(X_n) \geq \nu_+(X_{n+1}) + \nu_0(X_{n+1}) \qquad (n = N, N+1, \ldots).$$

Thus,

$$\nu_+(X_N) \geq \nu_+(X_n) + \sum_{i=N+1}^{n} \nu_0(X_i) \qquad (n = N, N+1, \ldots).$$

Since $0 \leq \nu_+(X_n) \leq r$, it follows that $\nu_0(X_n)$ can be different from 0 for at most a finite number of indices n. Let n_1 be an integer, with $n_1 \geq N$, such that $\nu_0(X_n) = 0$ $(n = n_1, n_1 + 1, \ldots)$. Note also that (2.27) implies that the sequence $\{\nu_+(X_n)\}_{n=N}^{\infty}$ is monotone. Consequently, there exists an integer N_1, with $N_1 \geq n_1$, such that $\nu_+(X_n)$ $(n = N_1, N_1 + 1, \ldots)$ is constant, say ν_+. If $n = N_1, N_1 + 1, \ldots$, we have $In(X_n) = (\nu_+, 0, \nu_-)$ where $\nu_- = r - \nu_+$. $\square$

The dichotomy is also connected with Fredholm properties of block weighted shifts. We will use a special case of a result in [BG2] where the more general case of band matrices is considered. In the statement we will use the notion of *Fredholm operator with $(1,0)$ band characteristic* which we describe first. Let $K = (M_j \delta_{ij} + N_j \delta_{i,j+1})_{ij=0}^{\infty}$ be a bounded operator in ℓ_r^2, where $\{M_n\}_{n=0}^{\infty}$ and $\{N_n\}_{n=0}^{\infty}$ are two sequences of $r \times r$ matrices. We say that K is a Fredholm operator with $(1,0)$ band characteristic if there exist two bounded sequences $\{M_n\}_{n=-\infty}^{-1}$ and $\{N_n\}_{n=-\infty}^{-1}$ of $r \times r$ matrices such that $H = (M_j \delta_{ij} + N_j \delta_{i,j+1})_{ij=-\infty}^{\infty}$ defines an invertible operator in $\ell_r^2(\mathbb{Z})$. Here and in the sequel, we denote by $\ell_r^2(\mathbb{Z})$ the Hilbert space of all two sided sequences $x = (x_n)_{n=-\infty}^{\infty}$, with $x_n \in \mathbb{C}^r$ $(n = 0, \pm 1, \ldots)$, such that the norm $\|x\| = (\sum_{n=-\infty}^{\infty} \|x_n\|^2)^{1/2}$ is finite. The connection between this notion and dichotomy is stated in the next theorem. In the proof we will use results from [BG2].

THEOREM 2.7. *The operator $I-G$, where $G = (\delta_{i,j+1} A_j)_{ij=0}^{\infty}$ and $\{A_n\}_{n=0}^{\infty}$ is a bounded sequence of $r \times r$ matrices, defines a Fredholm operator in ℓ_r^2 with $(1,0)$ band characteristic if and only if the system*

$$(2.28) \qquad x_{n+1} = A_n x_n \qquad (n = 0, 1, \ldots)$$

admits a dichotomy. Furthermore, if (2.28) admits a dichotomy of rank p, then

$$\mathrm{index}(I - G) = p - r.$$

In order to apply the results of [BG2], we need the following lemma.

LEMMA 2.8. *Let $\{P_n\}_{n=0}^{\infty}$ be a dichotomy of the system*

$$(2.29) \qquad x_{n+1} = A_n x_n \qquad (n = 0, 1, \ldots),$$

with constants a and M. Then the mappings

$$(2.30) \qquad \left(I|_{\mathrm{Im}\, P_{n+1}}, A_n|_{Ker\, P_n}\right): \mathrm{Im}\, P_{n+1} \oplus \mathrm{Ker}\, P_n \to \mathbb{C}^r \qquad (n = 0, 1, \ldots)$$

are invertible and

$$(2.31) \qquad \sup_n \left\|\left(I|_{\mathrm{Im}\, P_{n+1}}, A_n|_{Ker\, P_n}\right)^{-1}\right\| \leq M a \sup_n(\|P_n\|, \|I - P_n\|).$$

In the last equality the norm of

$$\left(I|_{\mathrm{Im}\, P_{n+1}}, A_n|_{Ker\, P_n}\right)^{-1}: \mathbb{C}^r \to \mathrm{Im}\, P_{n+1} \oplus \mathrm{Ker}\, P_n$$

is computed relative to the usual norm on $\mathbf{C}^r$, and $\|u \oplus v\| = (\|u\|^2 + \|v\|^2)^{1/2}$ for $u \oplus v \in \operatorname{Im} P_{n+1} \oplus \operatorname{Ker} P_n$.

PROOF. The commutation relations (2.2) imply that $A_n|_{Ker\,P_n} : \operatorname{Ker} P_n \to \operatorname{Ker} P_{n+1}$ $(n = 0, 1, \ldots)$. Furthermore, the dichotomy inequalities (2.4) with $k = 1$ lead to $\|A_n x\| \geq \frac{1}{Ma}\|x\|$ $(x \in \operatorname{Ker} P_n; n = 0, 1, \ldots)$. Since $\operatorname{Rank} P_n = \operatorname{Rank} P_{n+1}$, this implies that $A_n|_{Ker\,P_n} : \operatorname{Ker} P_n \to \operatorname{Ker} P_{n+1}$ is invertible and

$$\sup_n \left\| \left(A_n|_{Ker\,P_n} \right)^{-1} \right\| \leq Ma.$$

Hence, the diagonal map

$$D_n = \begin{pmatrix} I|_{\operatorname{Im} P_{n+1}} & 0 \\ 0 & A_n|_{Ker\,P_n} \end{pmatrix} : \operatorname{Im} P_{n+1} \oplus \operatorname{Ker} P_n \to \operatorname{Im} P_{n+1} \oplus \operatorname{Ker} P_{n+1}$$

is invertible, and

$$(2.32) \qquad \sup_n \|D_n^{-1}\| \leq Ma.$$

On the other hand, the boundedness of the sequence $\{P_n\}_{n=0}^{\infty}$ shows that the natural mapping

$$\Sigma_n : \operatorname{Im} P_{n+1} \oplus \operatorname{Ker} P_{n+1} \to \mathbf{C}^r : u \oplus v \to u + v$$

is invertible and

$$(2.33) \qquad \sup_n \|\Sigma_n^{-1}\| \leq \sup_n (\|P_n\|, \|I - P_n\|).$$

This follows from $\Sigma_n^{-1} x = P_{n+1} x \oplus (I_r - P_{n+1}) x$.

Finally, it is clear that

$$(2.34) \qquad \left(I|_{\operatorname{Im} P_{n+1}}, A_n|_{Ker\,P_n} \right) = \Sigma_n D_n \qquad (n = 0, 1, \ldots).$$

Hence, (2.31) follows from (2.32)–(2.33). $\square$

REMARK. In the terminology of Theorem 6.7 of [BG2], the above lemma shows that all the dichotomies of (2.30) are normal.

PROOF OF THEOREM 2.7. Note first that $\{P_n\}_{n=0}^{\infty}$ is a dichotomy for (2.28) if and only if it is a dichotomy for

$$(2.35) \qquad x_{n+1} = -A_n x_n \qquad (n = 0, 1, \ldots).$$

We will use Theorem 6.7 of [BG2] and the corresponding terminology. Note that the system (2.35) is the second companion system of $I - G$ as defined in [BG2]. Moreover, the special structure of G leads to $\operatorname{Ker}(I + G) = \{0\}$. Thus, the equivalence of conditions I) and III) in Theorem 6.7 of [BG2] implies that $I + G$ is a Fredholm operator with $(1, 0)$ band characteristic if and only if the system (2.35), or equivalently (2.28), admits a normal dichotomy. In view of Lemma 2.8 and the remark following its proof, this implies the first part of the theorem.

We now prove the second part of the theorem. Assume that (2.28), and hence (2.35), admit a dichotomy $\{P_n\}_{n=0}^\infty$ of rank p. By Lemma 2.8 above, the inequalities (2.31) hold. We now apply Theorem 6.6 of [BG2] to the operator

$$
G_1 = \begin{pmatrix} -A_0 & I & 0 & \cdots \\ 0 & -A_1 & I & \cdots \\ . & . & . & \cdots \\ . & . & . & \cdots \end{pmatrix}.
$$

It follows that index $G_1 = p$. This implies immediately that index$(I - G) = p - r$. $\square$

In order to apply the above theorem to general Fredholm block weighted shifts we need the following lemma.

LEMMA 2.9. *Let $G = (\delta_{i,j+1}A_j)_{ij=0}^\infty$ be an operator in ℓ_r^2, where $\{A_n\}_{n=0}^\infty$ is a bounded sequence of $r \times r$ matrices. For each nonnegative integer n denote the operator $G_n = (\delta_{i,j+1}A_{j+n})_{ij=0}^\infty$ in ℓ_r^2. If $I - G$ is Fredholm, then there exists an integer N such that, for each $n \geq N$, the operator $I - G_n$ is a Fredholm operator with $(1,0)$ band characteristic.*

In the proof we will use the following result, which is a special case of Theorem 6.1 in [BG2], and gives another characterization of Fredholm operators with band characteristic.

LEMMA 2.10. *Let $K = (M_j\delta_{ij} + N_j\delta_{i,j+1})_{ij=0}^\infty$ be a bounded operator in ℓ_r^2, where $\{M_n\}_{n=0}^\infty$ and $\{N_n\}_{n=0}^\infty$ are two bounded sequences of $r \times r$ matrices. Then K is a Fredholm operator with $(1,0)$ band characteristic if and only if the following two conditions hold.*

a) *K is Fredholm and $\mathrm{Ker}\, K = \{0\}$.*

b) *If $0 \neq (x_i)_{i=0}^\infty \in (\mathrm{Im}\, K)^\perp$, where $x_i \in \mathbb{C}^r$ $(i = 0, 1, \ldots)$, then $x_0 \neq 0$.*

PROOF OF LEMMA 2.9. Let $u_1, \ldots, u_s$ be a basis for $\mathrm{Ker}(I - G^*)$. Here $u_j = (u_{ij})_{i=0}^\infty$ is a vector in ℓ_r^2 $(j = 1, \ldots, s)$, where $u_{ij} \in \mathbb{C}^r$ $(i = 0, 1, \ldots)$. Since $u_1, \ldots, u_s$ are linearly independent, there exists a positive integer N such that the following vectors in $(\mathbb{C}^r)^N$ are linearly independent:

$$
u_1^N = (u_{i1})_{i=0}^{N-1}, u_2^N = (u_{i2})_{i=0}^{N-1}, \ldots, u_s^N = (u_{is})_{i=0}^{N-1}.
$$

We claim that N satisfies the requirement of the lemma. Let $n \geq N$. We first prove the following statement:

(2.36) $\qquad$ If $x = (x_i)_{i=0}^\infty \in \big(\mathrm{Im}(I - G_n)\big)^\perp$, and $x_0 = 0$, then $x = 0$.

In fact, assume $x = (x_i)_{i=0}^\infty \in \big(\mathrm{Im}(I - G_n)\big)^\perp$ and $x_0 = 0$. Then $x \in \mathrm{Ker}(I - G_n^*) = \mathrm{Ker}(\delta_{ij}I_r - \delta_{i+1,j}A_i^*)_{ij=0}^\infty$. Define the vector $y = (y_i)_{i=0}^\infty \in \ell_r^2$ via $y_0 = \cdots = y_{n-1} = 0$ and $y_{n+k} = x_k$ $(k = 0, 1, \ldots)$. It follows from $x \in \mathrm{Ker}(I - G_n^*)$, $x_0 = 0$, and the special structure of G_n^*, that $y \in \mathrm{Ker}(I - G^*)$. Therefore $y = \sum_{j=1}^s \alpha_j u_j$ ($\alpha_j \in \mathbb{C}; j = 1, \ldots, s$). By taking the N first coordinates, it follows that $0 = (y_i)_{i=0}^{N-1} = \sum_{j=1}^s \alpha_j u_j^N$. Since $u_1^N, \ldots, u_s^N$ are linearly independent, $\alpha_1 = \alpha_2 = \cdots = \alpha_s = 0$. Consequently, $y = 0$ and hence, $x = 0$. This proves (2.36). Finally, it is clear that $I - G_n$ is Fredholm and that $\mathrm{Ker}(I - G_n) = \{0\}$. Thus, by Lemma 2.10, $I - G_n$ is a Fredholm operator with $(1, 0)$ band characteristic. $\square$

Theorem 2.11 below generalizes most of Theorem 1.1 of the introduction to the case when the matrices A_n are not necessarily invertible.

THEOREM 2.11. *Let* $G = (\delta_{i,j+1} A_j)_{ij=0}^{\infty}$ *be a block weighted shift in* ℓ_r^2, *where* $\{A_n\}_{n=0}^{\infty}$ *is a bounded sequence of* $r \times r$ *matrices. Then the following conditions are equivalent:*

I) *The operator* $I - G$ *is Fredholm.*

II) *There exists a bounded sequence* $\{X_n\}_{n=0}^{\infty}$ *of self-adjoint matrices such that*

$$(2.37) \qquad X_n - A_n^* X_{n+1} A_n \geq \varepsilon I_r \qquad (n = 0, 1, \ldots)$$

for some positive number ε.

III) *The system*

$$(2.38) \qquad x_{n+1} = A_n x_n \qquad (n = 0, 1, \ldots)$$

admits an asymptotic dichotomy.

Furthermore, if (2.38) *admits an asymptotic dichotomy of rank* p, *and* $\{X_n\}_{n=0}^{\infty}$ *is a bounded sequence of self-adjoint matrices satisfying* (2.37), *then*

$$(2.39) \qquad \dim \mathrm{Ker}(I - G) = 0, \qquad \dim \mathrm{Coker}(I - G) = r - p,$$

$$(2.40) \qquad \mathrm{index}(I - G) = p - r,$$

and there is an integer N *such that*

$$(2.41) \qquad In(X_n) = (p, 0, r - p) \qquad (n = N, N + 1, \ldots).$$

PROOF. Note first that if (2.38) admits an asymptotic dichotomy, then the rank of the asymptotic dichotomy is uniquely determined by Corollary 2.2. We now divide the proof into three parts.

Part a). We show that I$\rightarrow$III. Assume that $I - G$ is Fredholm. Denote $G_n = (\delta_{i,j+1} A_{j+n})_{ij=0}^{\infty}$. By Lemma 2.9, there is an integer N such that $I - G_N$ is a Fredholm operator with $(1, 0)$ band characteristic. By Theorem 2.7, the system (2.38) admits an asymptotic dichotomy and (2.40) holds. Thus, I$\rightarrow$III. Moreover, the triangular structure of $I - G$ implies $\mathrm{Ker}(I - G) = \{0\}$. Combining this with (2.40) we obtain (2.39).

Part b). We prove the implication II→III. Assume that $\{X_n\}_{n=0}^{\infty}$ is a bounded sequence of self-adjoint matrices such that (2.37) holds. By Corollary 2.6 there is an integer N such that

$$(2.42) \qquad \operatorname{Im}(X_n) = (\nu_+, 0, \nu_-) \qquad (n = N, N+1, \ldots)$$

for some constant values of ν_+ and ν_-. Hence, we can apply Theorem 2.4 to the inequalities

$$X_n - A_n^* X_{n+1} A_n \geq \varepsilon I_r \qquad (n = N, N+1, \ldots).$$

Thus, the system $x_{n+1} = A_n x_n$ $(n = N, N+1, \ldots)$ admits a dichotomy of rank ν_+, and therefore, (2.38) admits an asymptotic dichotomy of rank $p = \nu_+$. Thus I→III. Moreover, (2.41) follows from $p = \nu_+$ and (2.42).

Part c). Here we prove the implications III→I and III→II. Assume that for some positive integer N, the system $x_{n+1} = A_n x_n$ $(n = N, N+1, \ldots)$ dichotomy. By Lemmas 2.7 and 2.9, the operator $I - G_N$, where $G_N = (\delta_{i,j+1} A_{j+N})_{ij=N}^{\infty}$ is Fredholm. Hence, $I - G$ is Fredholm. Thus III→I. Furthermore, Theorem 2.4 shows that there exists a bounded sequence of self-adjoint matrices $\{X_n\}_{n=N}^{\infty}$ such that

$$X_n - A_n^* X_{n+1} A_n \geq \varepsilon I_r \qquad (n = N, N+1, \ldots)$$

for some positive number ε. Define

$$X_i = A_i^* X_{i+1} A_i + \varepsilon I_r \quad \text{recursively for} \quad i = N-1, \ldots, 0.$$

Then the sequence $\{X_n\}_{n=0}^{\infty}$ satisfies (2.37). Thus, III→II.

This completes the proof of the equivalence of conditions I, II and III. Finally note that (2.39)–(2.40) and (2.41) were proved in Parts a) and b) respectively. □

3. DICHOTOMIES FOR LEFT SYSTEMS AND TWO SIDED SYSTEMS

In this section we consider systems that are indexed in one of the following ways

$$(3.1) \qquad x_{n+1} = A_n x_n \qquad (n = 0, \pm 1, \ldots),$$

$$(3.2) \qquad x_{n+1} = A_n x_n \qquad (n = N, N+1, \ldots),$$

or

$$(3.3) \qquad x_{n+1} = A_n x_n \qquad (n = -N, -N-1, \ldots),$$

where N is an integer and A_n are $r \times r$ complex matrices. We will occasionally refer to systems of type (3.1), (3.2) or (3.3) as *two sided systems, right systems* or *left systems* respectively.

The definition of dichotomy given in (2.2)–(2.4) of Section 2 can be modified in an obvious way to yield the definition of dichotomy for systems of the form (3.1), (3.2) or (3.3). For the latter system, we wish to specify that a dichotomy for (3.3) is a bounded sequence of projections $\{P_n\}_{n=-\infty}^{-N+1}$, having constant rank and satisfying (2.2)–(2.4) for $n = -N, -N - 1, \ldots$ and $k = 1, 2, \ldots, -n - N + 1$.

Most of the definitions and results about dichotomies of right systems carry over to left or two sided systems without essential changes. For example, we speak of the rank of a dichotomy. The motion of asymptotic dichotomy also extends in an obvious way to two sided or left systems. To be more precise, the asymptotic dichotomy as defined in Section 2 will sometimes be referred to as an *asymptotic dichotomy in the positive direction*. We will also say that a system of the form (3.1) or (3.3) has an *asymptotic dichotomy in the negative direction* if for some integer N', the system $x_{n+1} = A_n x_n$ $(n = -N', -N' - 1, \ldots)$ admits a dichotomy.

In the case when all the matrices A_n are invertible, the dichotomy is determined by a single projection via the commutativity conditions (2.2), and the analogue of condition (2.6) holds for two sided or left systems. As for the case of right systems, we say that P is a dichotomy for the left system $x_{n+1} = A_n x_n$ $(n = -1, -2, \ldots)$, where A_n are invertible, if the sequence $\{P_n\}_{n=-\infty}^{0}$ given by $P_n = A_{n+1}^{-1} \cdots A_{-1}^{-1} P A_{-1} \cdots A_{n+1}$ $(n = 0, -1, -2, \ldots)$ is a dichotomy for the system. Similar terminology applies to two sided systems.

There are two methods for passing from right systems to left systems and vice versa. First assume that all the matrices A_n in the system

$$(3.4) \qquad\qquad x_{n+1} = A_n x_n \qquad (n = 0, 1, \ldots)$$

are invertible, and consider the system

$$(3.5) \qquad\qquad x_{n+1} = A_{-n-1}^{-1} x_n \qquad (n = -1, -2, \ldots).$$

It follows easily from the definition (2.2)–(2.4) that a sequence of projections $\{P_n\}_{n=0}^{\infty}$ is a dichotomy for (3.4) and only if $\{I_r - P_{-n}\}_{n=-\infty}^{0}$ is a dichotomy for (3.5). Applying this transformation and taking into account Lemma 2.1 of Section 2, we obtain the following result.

LEMMA 3.1. *If* $\{P_n\}_{n=-\infty}^{0}$ *is a dichotomy for the left system*

$$(3.6) \qquad\qquad x_{n+1} = A_n x_n \qquad (n = -1, -2, \ldots),$$

where A_n are invertible $r \times r$ matrices, then

$$
\begin{aligned}
\operatorname{Ker} P_n &= \left\{ x \in \mathbb{C}^r : \lim_{k \to \infty} \|A_{n-k}^{-1} \cdots A_{n-1}^{-1} x\| = 0 \right\} \\
&= \left\{ x \in \mathbb{C}^r : \sum_{k=0}^{\infty} \|A_{n-k}^{-1} \cdots A_{n-1}^{-1} x\|^2 < \infty \right\}.
\end{aligned}
\qquad (3.7)
$$

In a similar way, the following analogue of Lemma 2.3 of Section 2 also holds.

LEMMA 3.2. *Let* $\{A_n\}_{n=-\infty}^{-1}$ *be a sequence of invertible matrices. If the left system*

$$(3.8) \qquad\qquad x_{n+1} = A_n x_n \qquad (n = -1, -2, \ldots)$$

admits an asymptotic dichotomy in the negative direction, then it admits a dichotomy.

Another method to relate left and right systems is via duality. We say that the systems

$$(3.9) \qquad x_{n+1} = A_n x_n \qquad (n = N, N+1, \ldots)$$

and

$$(3.10) \qquad x_{n+1} = A^*_{-n} x_n \qquad (n = -N, -N-1, \ldots)$$

are *dual* to each other. For the two sided case, the systems

$$(3.11) \qquad x_{n+1} = A_n x_n \qquad (n = 0, \pm 1, \ldots)$$

and

$$(3.12) \qquad x_{n+1} = A^*_{-n} x_n \qquad (n = 0, \pm 1, \ldots)$$

are also called dual to each other. The relation between the dichotomies of dual systems is given by the following result.

PROPOSITION 3.3. *The system* (3.9) *admits the dichotomy* $\{P_n\}_{n=N}^{\infty}$ *if and only if the dual system* (3.10) *admits the dichotomy* $\{P^*_{-n+1}\}_{n=-\infty}^{-N+1}$. *Similarly, the system* (3.11) *admits the dichotomy* $\{P_n\}_{n=-\infty}^{\infty}$ *if and only if* (3.12) *admits the dichotomy* $\{P^*_{-n+1}\}_{n=-\infty}^{\infty}$.

PROOF. We will only prove that if $\{P_n\}_{n=N}^{\infty}$ is a dichotomy for (3.9) then $\{P^*_{-n+1}\}_{n=-\infty}^{-N+1}$ is a dichotomy for (3.10), the other implications being proved in a similar way. We will also assume $N = 0$ for simplicity. Thus, we assume that $\{P_n\}_{n=0}^{\infty}$ is a dichotomy for

$$(3.13) \qquad x_{n+1} = A_n x_n \qquad (n = 0, 1, \ldots),$$

and prove that $\{P^*_{-n+1}\}_{n=-\infty}^{1}$ is a dichotomy for

$$(3.14) \qquad x_{n+1} = A^*_{-n} x_n \qquad (n = 0, -1, \ldots).$$

Denote

$$(3.15) \qquad B_n = A^*_{-n} \qquad (n = 0, -1, \ldots),$$

and

$$(3.16) \qquad Q_n = P^*_{-n+1} \qquad (n = 1, 0, -1, \ldots).$$

It follows from the commutation relation (2.2) that $P^*_n A^*_n = A^*_n P^*_{n+1}$ $(n = 0, 1, \ldots)$. Hence, $P^*_{-n} A^*_{-n} = A^*_{-n} P^*_{-n+1}$ $(n = 0, -1, \ldots)$, and therefore

$$(3.17) \qquad Q_{n+1} B_n = B_n Q_n \qquad (n = 0, -1, \ldots).$$

In addition, the commutation relation (2.2) also lead to

$$P_{n+k}A_{n+k-1}\cdots A_n = A_{n+k-1}\cdots A_n P_n \qquad (n = 0,1,\ldots; k = 1,2,\ldots).$$

Taking into account the dichotomy inequalities (2.3), and denoting $L = \sup_{n=0,1,\ldots}(\|P_n\|, \|I_r - P_n\|)$, we obtain

$$\|P_{n+k}A_{n+k-1}\cdots A_n x\| \le Ma^k\|P_n x\| \le MLa^k\|x\|$$

for $x \in \mathbb{C}^r$, $n = 0,1,\ldots$, and $k = 1,2,\ldots$. This inequality means that $\|P_{n+k}A_{n+k-1}\cdots A_n\| \le MLa^k$, and hence leads to

$$\|A_n^*\cdots A_{n+k-1}^* P_{n+k}^*\| \le MLa^k \qquad (n = 0,1,\ldots; k = 1,2,\ldots).$$

Setting $m = -n - k + 1$, we obtain

$$\|A_{-(m+k-1)}^*\cdots A_{-m}^* P_{-m+1}^*\| \le MLa^k \qquad (m = 0,-1,\ldots; k = 1,2,\ldots,-m+1)$$

and hence,

$$\|B_{m+k-1}\cdots B_m Q_m\| \le MLa^k \qquad (m = 0,-1,\ldots; k = 1,2,\ldots,-m+1).$$

In particular, since Q_m is a projection, we obtain

$$(3.18) \qquad \|B_{m+k-1}\cdots B_m Q_m x\| \le MLa^k\|Q_m x\|,$$

for $x \in \mathbb{C}^r$, $m = 0,-1,\ldots$, and $k = 1,2,\ldots,-m+1$.

To obtain the second set of inequalities let $n = 0,1,\ldots$ and $k = 1,2,\ldots$ be arbitrary, and denote

$$(3.19) \qquad A = A_{n+k-1}\cdots A_n(I - P_n).$$

Let $(\lambda_i)_{i=1}^r$ be the eigenvalues of A^*A, counting multiplicities, ordered so that

$$\lambda_1 \le \lambda_2 \le \cdots \le \lambda_r.$$

Note that $\operatorname{Im} P_n \subset \operatorname{Ker} A^*A$. Since $\dim \operatorname{Im} P_n = p$ this leads to

$$(3.20) \qquad \lambda_1 = \cdots = \lambda_p = 0,$$

where p is the rank of the dichotomy of (3.9). In addition, it follows from the dichotomy inequality (2.4) that

$$\langle A^*Au, u\rangle = \|Au\|^2 \ge (Ma^k)^{-2}\|u\|^2 \qquad (u \in \operatorname{Ker} P_n).$$

Since $\dim \operatorname{Ker} P_n = r - p$, it follows from this inequality and (3.20), that

$$(3.21) \qquad \frac{1}{M^2 a^{2k}} \le \lambda_{p+1} \le \lambda_{p+2} \le \cdots \le \lambda_r.$$

Now recall that $\{\lambda_i\}_{i=1}^r$ are also the eigenvalues of AA^*. In addition, the commutation relation (2.2) show that

$$(3.22) \qquad A = (I_r - P_{n+k})A_{n+k-1}\cdots A_n.$$

Therefore,

$$(3.23) \qquad (\operatorname{Ker} P_{n+k})^\perp = \operatorname{Im} P_{n+k}^* \subseteq \operatorname{Ker} A^* = \operatorname{Ker} AA^*.$$

Note that $\dim(\operatorname{Ker} P_{n+k})^\perp = p$. Hence, (3.21) shows that $AA^*|_{\operatorname{Ker} P_{n+k}} \geq (Ma^k)^{-2}I_r|_{\operatorname{Ker} P_{n+k}}$, and hence

$$(3.24) \qquad \|A^*y\| \geq (Ma^k)^{-1}\|y\| \qquad (y \in \operatorname{Ker} P_{n+k}).$$

Now let $x \in \mathbb{C}^r$ be arbitrary, and let $y \in \operatorname{Ker} P_{n+k}$ and $z \in (\operatorname{Ker} P_{n+k})^\perp = \operatorname{Im} P_{n+k}^*$ be such that $x = y + z$. Then $z \in \operatorname{Im} P_{n+k}^*$ implies

$$(3.25) \qquad \|(I_r - P_{n+k}^*)x\| = \|(I_r - P_{n+k}^*)y\| \leq L\|y\|,$$

where L is as above. In addition, $y \in \operatorname{Ker} P_{n+k}$, $z \in (\operatorname{Ker} P_{n+k})^\perp$ and (3.23)–(3.24) lead to

$$\|A^*x\| = \|A^*y\| \geq (Ma^k)^{-1}\|y\|.$$

Combining this inequality with (3.25), we obtain

$$\|A^*x\| \geq (LMa^k)^{-1}\|(I_r - P_{n+k}^*)x\| \qquad (x \in \mathbb{C}^r).$$

Taking into account (3.22) this implies

$$\|A_n^*\cdots A_{n+k-1}^*(I - P_{n+k}^*)x\| \geq (LMa^k)^{-1}\|(I - P_{n+k}^*)x\|$$

for $x \in \mathbb{C}^r$. Setting $m = -n - k + 1$, this leads to

$$\|A_{-(m+k-1)}^*\cdots A_{-m}^*(I - P_{-m+1}^*)x\| \geq (LMa^k)^{-1}\|(I - P_{-m+1}^*)x\|.$$

Recalling the definitions (3.15) and (3.16), this shows that

$$(3.26) \qquad \|B_{m+k-1}\cdots B_m(I - Q_m)x\| \geq (LMa^k)^{-1}\|(I - Q_m)x\|,$$

for $x \in \mathbb{C}^r$, $m = 0, -1, \ldots$; $k = 1, 2, \ldots, -m + 1$.

Combining (3.17), (3.18) and (3.26), we obtain that $\{Q_m\}_{m=-\infty}^1$ is a dichotomy for the system

$$x_{n+1} = B_n x_n \qquad (n = 0, -1, \ldots).$$

By (3.15) and (3.16), it follows that $\{P_{-m+1}^*\}_{m=-\infty}^1$ is a dichotomy for the dual system

$$x_{n+1} = A_{-n}^* x_n \qquad (n = 0, -1, \ldots). \qquad \square$$

COROLLARY 3.4. *All the (asymptotic) dichotomies of the left system*

$$(3.27) \qquad x_{n+1} = A_n x_n \qquad (n = 0, -1, \ldots)$$

have the same rank. The two sided system

$$(3.28) \qquad x_{n+1} = A_n x_n \qquad (n = 0, \pm 1, \ldots)$$

admits at most one dichotomy.

PROOF. The first part follows from Corollary 2.2 and Proposition 3.3. For the second part, note that the subspaces $\mathrm{Im}(P_n)$ $(n = 0, \pm 1, \ldots)$ are fixed for any dichotomy of (3.28). Using duality and Proposition 3.3 we see that the subspaces $\mathrm{Im}(P^*_{-n+1}) = \left(\mathrm{Ker}(P_{-n+1})\right)^{-1}$ are also uniquely determined. Thus the dichotomy is unique. $\square$

COROLLARY 3.5. *Let $\{A_n\}_{n=0}^{\infty}$ be a sequence of invertible $r \times r$ matrices. Then, a projection P is a dichotomy for the system*

$$(3.29) \qquad x_{n+1} = A_n x_n \qquad (n = 0, 1, \ldots)$$

if and only if P^* is a dichotomy for

$$(3.30) \qquad x_{n+1} = A^*_{-n-1} x_n \qquad (n = -1, -2, \ldots).$$

PROOF. Note first that P is a dichotomy for (3.29) if and only if the sequence $\{P_n\}_{n=0}^{\infty}$ defined by $P_0 = P$ and $P_n = A_{n-1} \cdots A_0 P A_0^{-1} \cdots A_{n-1}^{-1}$ $(n = 0, 1, \ldots)$, is a dichotomy for (3.29). By Proposition 3.3, $\{P_n\}_{n=0}^{\infty}$ is a dichotomy for (3.29) if and only if $\{P^*_{-n+1}\}_{n=-\infty}^{1}$ is a dichotomy for

$$(3.31) \qquad x_{n+1} = A^*_{-n} x_n \qquad (n = 0, -1, \ldots).$$

Setting $j = n - 1$, it is easily seen that $\{P^*_{-n+1}\}_{n=-\infty}^{1}$ is a dichotomy for (3.31) if and only if $\{P^*_{-j}\}_{j=-\infty}^{0}$ is a dichotomy for

$$x_{j+1} = A^*_{-j-1} x_j \qquad (j = -1, -2, \ldots).$$

Since the matrices A^*_{-j-1} are invertible, the last condition is equivalent to $P^* = P_0^*$ being a dichotomy for (3.30). $\square$

The following result shows how to extend dichotomies from one sided systems to two sided systems.

LEMMA 3.6. *If $\{P_n\}_{n=N}^{\infty}$ is a dichotomy for the system*

$$(3.32) \qquad x_{n+1} = A_n x_n \qquad (n = N, N+1, \ldots)$$

then there exists a bounded sequence of matrices $\{A_n\}_{n=-\infty}^{N-1}$ and a bounded sequence of projections $\{P_n\}_{n=-\infty}^{N-1}$ such that $\{P_n\}_{n=-\infty}^{\infty}$ is a dichotomy for

$$(3.33) \qquad x_{n+1} = A_n x_n \qquad (n = 0, \pm 1, \ldots).$$

Similarly, if $\{P_n\}_{n=-\infty}^{N+1}$ is a dichotomy for

(3.34) $$x_{n+1} = A_n x_n \qquad (n = N, N-1, \ldots),$$

then there exists a bounded sequence of matrices $\{A_n\}_{n=N+1}^{\infty}$ and a bounded sequence of projections $\{P_n\}_{n=N+2}^{\infty}$ such that $\{P_n\}_{n=-\infty}^{\infty}$ is a dichotomy for

(3.35) $$x_{n+1} = A_n x_n \qquad (n = 0, \pm 1, \ldots).$$

PROOF. The first part of the lemma follows from Proposition 6.1 in [BG1]. Assume that $\{P_n\}_{n=-\infty}^{N+1}$ is a dichotomy for (3.34). Then, by Proposition 3.3, the sequence $\{P_{-n+1}^*\}_{n=-N}^{\infty}$ is a dichotomy for

$$x_{n+1} = A_{-n}^* x_n \qquad (n = -N, -N+1, \ldots).$$

By the first part of the lemma, there is a bounded sequence of matrices, which we denote by $\{A_{-n}^*\}_{n=-\infty}^{-N-1}$, and a bounded sequence of projections, which we denote by $\{P_{-n+1}^*\}_{n=-\infty}^{-N-1}$, such that $\{P_{-n+1}^*\}_{n=-\infty}^{\infty}$ is a dichotomy for the system

$$x_{n+1} = A_{-n}^* x_n \qquad (n = 0, \pm 1, \ldots).$$

By Proposition 3.3 again, it follows that $\{P_n\}_{n=-\infty}^{\infty}$ is a dichotomy for (3.35). $\square$

Theorem 2.4 in Section 2 states a connection between dichotomies and non-stationary Stein equations. This result can be generalized to two sided and left systems. The two sided version is obtained by replacing $n = 0, 1, \ldots$ with $n = 0, \pm 1, \ldots$ in Theorem 2.4, and follows from Theorems 4.2 and 5.3 of [BG1]. Here we prove the corresponding result for left systems.

THEOREM 3.7. *Let $\{A_n\}_{n=-\infty}^{-1}$ be a sequence of $r \times r$ matrices. Then the system*

(3.36) $$x_{n+1} = A_n x_n \qquad (n = -1, -2, \ldots)$$

admits a dichotomy if and only if there exists a bounded sequence $\{X_n\}_{n=-\infty}^{0}$ of self-adjoint matrices, having constant inertia, such that

(3.37) $$X_n - A_n^* X_{n+1} A_n \geq \varepsilon I_r \qquad (n = -1, -2, \ldots)$$

for some positive number ε. Furthermore, if (3.37) holds for some bounded sequence of self-adjoint matrices, having constant inertia given by

(3.38) $$In(X_n) = (\mu_+, \mu_0, \mu_-) \qquad (n = 0, -1, \ldots).$$

Then $\mu_0 = 0$ and μ_+ is equal to the rank of every dichotomy of the system (3.36).

PROOF. If the system (3.36) admits a dichotomy, then by Lemma 3.6, there exist matrices $A_0, A_1, \ldots$ such that the system $x_{n+1} = A_n x_n$ $(n = 0, \pm 1, \ldots)$ admits a dichotomy. By Theorem 4.2 of [BG1], there exists a bounded sequence $\{X_n\}_{n=-\infty}^{\infty}$

of self-adjoint matrices, having constant inertia and such that $X_n - A_n^* X_{n+1} A_n \geq \varepsilon I_r$ $(n = 0, \pm 1, \ldots)$ for some $\varepsilon > 0$. Hence, (3.37) holds.

Conversely, assume that (3.37)–(3.38) hold. It follows from Lemma 2.5 that $\mu_+ \geq \mu_+ + \mu_0$. Hence, $\mu_0 = 0$. In particular X_0 is invertible. Let A be an $r \times r$ matrix such that $X_0 - A^* X_0 A > 0$. For example let $A = \frac{1}{2}Q + 2(I - Q)$ where Q is the spectral projection of X_0 corresponding to the positive eigenvalues. Then $X_0 - A^* X_0 A \geq \varepsilon_1 I$ for some $\varepsilon_1 > 0$. Define $X_n = X_0$ $(n = 1, 2, \ldots)$ and $A_n = A$ $(n = 0, 1, \ldots)$. Then $X_n - A_n^* X_{n+1} A_n \geq \varepsilon_1 I_1$ $(n = 0, 1, \ldots)$. Combining this with (3.37), we obtain

$$(3.39) \qquad X_n - A_n^* X_{n+1} A_n \geq \min(\varepsilon, \varepsilon_1) I \qquad (n = 0, \pm 1, \ldots).$$

Note also $\{X_n\}_{n=-\infty}^{\infty}$ is a bounded sequence of self-adjoint matrices with constant inertia $(\mu_+, 0, \mu_-)$. By Theorem 5.3 of [BG1], the two sided system $x_{n+1} = A_n x_n$ $(n = 0, \pm 1, \ldots)$ admits a dichotomy $\{P_n\}_{n=-\infty}^{\infty}$ of rank μ_+. Then $\{P_n\}_{n=-\infty}^{0}$ is a dichotomy for (3.36) of rank μ_+. Finally, all the dichotomies of (3.36) have the same rank by Corollary 3.4. $\square$

We can now complete the proof of Theorem 1.1.

PROOF OF THEOREM 1.1. Note first that condition III of Theorem 1.1 is equivalent to P being a dichotomy for the system

$$(3.40) \qquad x_{n+1} = A_n x_n \qquad (n = 0, 1, \ldots).$$

Hence, all of Theorem 1.1, except for the equality (1.3) is a consequence of Theorem 2.11 and Lemma 2.3.

We now prove (1.3). Assume that P satisfies (1.2). Then P is a dichotomy for (3.40) and $I - G$ is Fredholm by Theorem 2.11. Let $x = (x_n)_{n=0}^{\infty}$ be an arbitrary vector in ℓ_r^2. It follows from the special two diagonal structure of $I - G^*$, that $x \in \left(\mathrm{Im}(I - G)\right)^{\perp} = \mathrm{Ker}(I - G^*)$ if and only if $x_n - A_n^* x_{n+1} = 0$ $(n = 0, 1, \ldots)$. This is equivalent to

$$x_{n+1} = A_n^{*-1} x_n \qquad (n = 0, 1, \ldots),$$

and hence to

$$x_n = A_{n-1}^{*-1} \cdots A_0^{*-1} x_0 = (A_{n-1} \cdots A_0)^{*-1} x_0 = U_n^{*-1} x_0.$$

Thus, denoting $v = x_0$, we obtain

$$(3.41) \qquad \left(\mathrm{Im}(I - G)\right)^{\perp} = \left\{ (U_n^{*-1} v)_{n=0}^{\infty} : v \in \mathbb{C}^r, \sum_{n=0}^{\infty} \|U_n^{*-1} v\|^2 < \infty \right\}.$$

Consider now the left system

$$(3.42) \qquad x_{n+1} = A_{-n-1}^* x_n \qquad (n = -1, -2, \ldots).$$

By Corollary 3.5, P^* is a dichotomy for (3.42). Thus, it follows from Lemma 3.1 that

$$\mathrm{Ker}\, P^* = \left\{ v \in \mathbb{C}^r : \sum_{k}^{\infty} \|A_{k=0}^{*-1} \cdots A_0^{*-1} v\|^2 < \infty \right\}.$$

Since $U_k^{*-1} = A_{k-1}^{*-1} \cdots A_0^{*-1}$ $(k = 0, 1, \ldots)$, it follows that

$$\operatorname{Ker} P^* = \left\{ v \in \mathbb{C}^r \colon \sum_{k=0}^{\infty} \|U_k^{*-1} v\|^2 < \infty \right\}.$$

This equality and (3.41) lead to

$$\left(\operatorname{Im}(I - G) \right)^{\perp} = \left\{ (U_n^{*-1} v)_{n=0}^{\infty} \colon v \in \operatorname{Ker} P^* \right\}.$$

However, $\operatorname{Ker} P^* = (\operatorname{Im} P)^{\perp}$, and hence,

$$\left(\operatorname{Im}(I - G) \right)^{\perp} = \left\{ (U_n^{*-1} v)_{n=0}^{\infty} \colon v \in (\operatorname{Im} P)^{\perp} \right\}.$$

Finally, $I - G$ is Fredholm and therefore $I - G$ is a closed subspace of ℓ_r^2. Thus, the last equality implies (1.3). $\square$

4. TWO SIDED BLOCK WEIGHTED SHIFTS

Here we complete the results of Section 2 to the two sided case. The first result deals with a block weighted shift $S = (\delta_{i,j+1} A_j)_{ij=0}^{\infty}$ where the $r \times r$ matrices A_n are invertible, and gives necessary and sufficient conditions for $I - S$ to be a Fredholm operator in $\ell_r^2(\mathbb{Z})$. Note that the Stein inequalities (4.1) below are not required to hold for $n = -1$. Hence the system of Stein inequalities (4.1) is a disjoint union of two independent systems of inequalities: one for $n = 0, 1, \ldots$ and the other for $n = -2, -3, \ldots$. We will show in Lemma 4.3 at the end of this section that requiring inequality (4.1) for $n = -1$ is equivalent to adding the following additional condition on

$$S \colon \operatorname{Ker}(I - S) = \{0\}.$$

THEOREM 4.1. *Let $S = (\delta_{i,j+1} A_j)_{ij=-\infty}^{\infty}$ be a block weighted shift in $\ell_r^2(\mathbb{Z})$, where $(A_n)_{n=-\infty}^{\infty}$ is a bounded sequence of invertible $r \times r$ matrices. Then, the following conditions are equivalent.*

I) *The operator $I - S$ is a Fredholm operator in $\ell_r^2(\mathbb{Z})$.*

II) *There exists a bounded sequence $\{X_n\}_{n=-\infty}^{\infty}$ of self-adjoint $r \times r$ matrices, such that*

$$(4.1) \qquad X_n - A_n^* X_{n+1} A_n \geq \varepsilon I_r \qquad (n = 0, 1, \pm 2, \pm 3, \ldots)$$

for some positive number ε.

III) *The right system*

$$(4.2) \qquad x_{n+1} = A_n x_n \qquad (n = 0, 1, \ldots),$$

and the left system

$$(4.3) \qquad x_{n+1} = A_n x_n \qquad (n = -1, -2, \ldots),$$

admit dichotomies.

Moreover, assume that P (respectively Q) is a dichotomy of (4.2) (respectively (4.3)), and that $\{X_n\}_{n=-\infty}^{\infty}$ is a bounded sequence of self-adjoint matrices satisfying (4.1). Then the following equalities hold

$$(4.4) \qquad \mathrm{Ker}(I - S) = \{(U_n v)_{n=-\infty}^{\infty} : v \in \mathrm{Im}\, P \cap \mathrm{Ker}\, Q\},$$

$$(4.5) \qquad \mathrm{Im}(I - S) = \{(U_n^{*-1} v)_{n=-\infty}^{\infty} : v \perp (\mathrm{Im}\, P + \mathrm{Ker}\, Q)\}^{\perp},$$

where $U_n = A_{n-1} \cdots A_0$, $U_{-n} = A_{-n}^{-1} \cdots A_{-1}^{-1}$ $(n = 1, 2, \ldots)$ and $U_0 = I_r$, and

$$(4.6) \qquad \mathrm{index}(I - S) = \mathrm{Rank}\, P - \mathrm{Rank}\, Q.$$

Furthermore, there exists an integer N such that

$$(4.7) \qquad In(X_n) = (\mathrm{Rank}\, P, 0, r - \mathrm{Rank}\, P) \qquad (n = N, N+1, \ldots),$$

$$(4.8) \qquad In(X_n) = (\mathrm{Rank}\, Q, 0, r - \mathrm{Rank}\, Q) \qquad (n = -N, -N-1, \ldots).$$

The following theorem considers the case when the matrices A_n are not necessarily invertible.

THEOREM 4.2. *Let $S = (\delta_{i,j+1} A_j)_{ij=-\infty}^{\infty}$ be a block weighted shift in $\ell_r^2(\mathbf{Z})$, where A_n is a bounded sequence of $r \times r$ matrices. Then the following conditions are equivalent.*

I) *The operator $I - S$ is Fredholm.*

II) *There exists a bounded sequence $\{X_n\}_{n=-\infty}^{\infty}$ of self-adjoint matrices, and a positive integer n_0, such that*

$$(4.9) \qquad X_n - A_n^* X_{n+1} A_n \geq \epsilon I_r \qquad (n = n_0, \pm(n_0 + 1), \pm(n_0 + 2), \ldots)$$

for some positive number ϵ.

III) *The system*

$$(4.10) \qquad x_{n+1} = A_n x_n \qquad (n = 0, \pm 1, \ldots)$$

admits asymptotic dichotomies in the positive and negative directions.

Moreover, if (4.10) admits asymptotic dichotomies of ranks p and q in the positive and negative directions respectively, then

$$(4.11) \qquad \mathrm{index}(I - S) = p - q.$$

Finally, assume that $\{X_n\}_{n=-\infty}^{\infty}$ is a bounded sequence of self-adjoint matrices satisfying (4.9). Then there exists an integer N such that

$$(4.12) \qquad In(X_n) = (p, 0, r - p) \qquad (n = N, N+1, \ldots),$$

and

$$(4.13) \qquad In(X_n) = (q, 0, r - q) \qquad (n = -N, -N - 1, \ldots).$$

We now turn to the proofs of these statements.

PROOF OF THEOREM 4.2. Note first that if the system (4.10) has asymptotic dichotomies, then, by Corollary 3.4, the ranks of these dichotomies are uniquely determined. Moreover, the equivalence of conditions II and III, as well as the equalities (4.12)–(4.13) follow from Corollary 2.6, and from Theorems 2.4 and 3.7 with obvious change of indices.

We now prove the equivalence of conditions I and III and equality (4.11). We denote by $\ell_r^{2\prime}$ the Hilbert space of all sequences $x = (x_n)_{n=-\infty}^0$, with $x_n \in \mathbb{C}^r$ $(n = 0, 1, \ldots)$, with norm $\|x\| = \left(\sum_{n=-\infty}^0 \|x_n\|^2 \right)^{1/2}$. Note that the operator

$$S_1 = S - (\delta_{i,0}\delta_{j,-1}A_{-1})_{ij=-\infty}^{\infty}$$

is a finite rank perturbation of S. Furthermore, S_1 is unitarily equivalent to the direct sum of the operators

$$(4.14) \qquad G_1 = \begin{pmatrix} 0 & 0 & 0 & \cdots \\ A_0 & 0 & 0 & \cdots \\ 0 & A_1 & 0 & \cdots \\ \cdot & \cdot & \cdot & \cdots \end{pmatrix} \quad \text{and} \quad G_2 = \begin{pmatrix} \cdots & \cdot & \cdot & \cdot \\ \cdots & 0 & 0 & 0 \\ \cdots & A_{-3} & 0 & 0 \\ \cdots & 0 & A_{-2} & 0 \end{pmatrix}$$

on ℓ_r^2 and $\ell_r^{2\prime}$ respectively. Denote the unitary transformation $V \colon (x_n)_{n=0}^{\infty} \to (x_{-n})_{n=-\infty}^0$ from ℓ_r^2 onto $\ell_r^{2\prime}$. Then VG_2V^{-1} is the following operator on ℓ_r^2

$$VG_2V^{-1} = \begin{pmatrix} 0 & A_{-2} & 0 & \cdots \\ 0 & 0 & A_{-3} & \cdots \\ 0 & 0 & 0 & \cdots \\ \cdot & \cdot & \cdot & \cdots \end{pmatrix}.$$

Thus, $I - S$ is a Fredholm operator in $\ell_r^2(\mathbb{Z})$ if and only if $I - G_1$ and $I - VG_2V^{-1}$ are Fredholm operators in ℓ_r^2, and in this case

$$\text{index}(I - S) = \text{index}(I - G_1) + \text{index}(I - VG_2V^{-1}).$$

Furthermore, $I - VG_2V^{-1}$ is Fredholm if and only if $I - (VG_2V^{-1})^*$ is Fredholm, and $\text{index}(I - VG_2V^{-1}) = -\text{index}(I - (VG_2V^{-1})^*)$. Denote $G_3 = (VG_2V^{-1})^*$. Then, the operator $I - S$ is a Fredholm operator in $\ell_r^2(\mathbb{Z})$ if and only if $I - G_1$ and $I - G_3$ are Fredholm operators in ℓ_r^2, and in this case

$$(4.15) \qquad \text{index}(I - S) = \text{index}(I - G_1) - \text{index}(I - G_3).$$

Note that G_1 and G_3 are block weighted shifts in ℓ_r^2 given by (4.14) and

$$(4.16) \qquad G_3 = \begin{pmatrix} 0 & 0 & 0 & \cdots \\ A_{-2}^* & 0 & 0 & \cdots \\ 0 & A_{-3}^* & 0 & \cdots \\ \cdot & \cdot & \cdot & \cdots \end{pmatrix}.$$

We apply Theorem 2.11 to G_1 and G_3. Combining this with the above, it follows that $I - S$ is a Fredholm operator if and only if the systems

$$(4.17) \qquad x_{n+1} = A_n x_n \qquad (n = 0, 1, \ldots)$$

and

$$(4.18) \qquad x_{n+1} = A^*_{-n} x_n \qquad (n = 0, 1, \ldots)$$

admit asymptotic dichotomies. Moreover, if p and q denote the ranks of the asymptotic dichotomies of (4.17) and (4.18), then (4.15) and the equality (2.40) in Theorem 2.11 lead to

$$(4.19) \qquad \mathrm{index}(I - S) = (p - r) - (q - r) = p - q.$$

Moreover, Proposition 3.3, with an obvious shift of indices, shows that (4.18) admits an asymptotic dichotomy of rank q if and only if the dual system

$$x_{n+1} = A_n x_n \qquad (n = 0, -1, \ldots)$$

admits an asymptotic dichotomy of rank q in the negative direction. Clearly, asymptotic dichotomy conditions for the latter system and (4.17) are equivalent to condition III. Hence, $I - S$ is a Fredholm operator if and only if condition III holds. This proves the equivalence of conditions II and III. Finally, (4.11) follows from (4.19). $\square$

PROOF OF THEOREM 4.1. By Lemmas 2.3 and 3.2 and the invertibility of A_n, conditions III in Theorems 4.1 and 4.2 are equivalent. Let us also show that conditions II in these theorems are equivalent. In fact, (4.1) obviously imply (4.9) with $n_0 = 1$. Conversely, assume that (4.9) holds for some n_0. Define $X_n = \epsilon I_r + A^*_n X_{n+1} A_n$ inductively for $n = n_0 - 1, n_0 - 2, \ldots, 0$, and $X_{n+1} = A^{*-1}_n (X_n - \epsilon I_r) A^{-1}_n$ inductively for $n = -n_0 - 1, -n_0, \ldots, -2$. Then (4.1) holds. These equivalences, and Theorem 4.2 immediately prove the equivalence of conditions I, II and III. In addition, Theorem 4.2 also imply equalities (4.6), (4.7) and (4.8).

There remains to prove (4.4) and (4.5). We first consider (4.4). Let v be an arbitrary vector in $\mathrm{Im}\, P \cap \mathrm{Ker}\, Q$. Let a and M be a common pair of constants for the dichotomies P and Q. Then we have $\|U_n v\| \leq M a^n \|v\|$ $(n = 0, 1, \ldots)$, and $\|U_n v\| \leq M a^{|n|} \|v\|$ $(n = -1, -2, \ldots)$. Hence, $(U_n v)_{n=-\infty}^{\infty} \in \ell_r^2(\mathbf{Z})$. In addition, it is clear that $-A_{n-1} U_{n-1} v + U_n v = 0$ $(n = 0, \pm 1, \ldots)$. Thus, $(U_n v)_{n=-\infty}^{\infty} \in \mathrm{Ker}(I - S)$. Conversely, let $z = (z_n)_{n=-\infty}^{\infty} \in \mathrm{Ker}(I - S)$. Then $-A_{n-1} z_{n-1} + z_n = 0$ $(n = 0, \pm 1, \ldots)$. Therefore $z_n = U_n z_0$ $(n = 0, \pm 1, \ldots)$. Since $z = (U_n z_0)_{n=-\infty}^{\infty} \in \ell_r^2(\mathbf{Z})$, we have

$$\lim_{n \to +\infty} \|U_n z_0\| = \lim_{n \to -\infty} \|U_n z_0\| = 0.$$

By Lemmas 2.1 and 3.1 this implies $z_0 \in \mathrm{Im}\, P \cap \mathrm{Ker}\, Q$. Thus, (4.4) holds.

We now prove (4.5). Let W be the unitary transformation in $\ell_r^2(\mathbf{Z})$ defined by $W\big((y_n)_{n=-\infty}^{\infty}\big) = (y_{-n})_{n=-\infty}^{\infty}$. Note that $W = W^{-1} = W^*$. We have $S^* = (\delta_{j,i+1} A^*_i)_{ij=-\infty}^{\infty}$, and hence

$$W S^* W = (\delta_{-j,-i+1} A^*_{-i})_{ij=-\infty}^{\infty} = (\delta_{i-1,j} A^*_{-i})_{ij=-\infty}^{\infty} = (\delta_{i,j+1} A^*_{-j-1})_{ij=-\infty}^{\infty}.$$

Denote $B_n = A^*_{-n-1}$ $(n = 0, \pm 1, \ldots)$. Then

$$WS^*W = (\delta_{i,j+1} B_j)_{ij=-\infty}^{\infty}.$$

We now apply (4.4) to the operator WS^*W. It follows that

$$(4.20) \qquad \mathrm{Ker}(I - WS^*W) = \{(U'_n v)_{n=-\infty}^{\infty} : v \in \mathrm{Im}\, P' \cap \mathrm{Ker}\, Q'\}.$$

Here $U'_n = B_{n-1} \cdots B_0$ $(n = 1, 2, \ldots)$, $U'_0 = I_r$ and $U' = B_n^{-1} \cdots B_{-1}^{-1}$ $(n = -1, -2, \ldots)$, and P' and Q' are the dichotomies of the systems $x_{n+1} = B_n x_n$ $(n = 0, 1, \ldots)$ and $x_{n+1} = B_n x_n$ $(n = -1, -2, \ldots)$ respectively. Note that P' and Q' exist because $I - WS^*W = W(I - S)^*W$ is Fredholm.

Let us first note that $U'_0 = I_r = U_0$ and
$$(4.21)$$
$$U'_n = B_{n-1} \cdots B_0 = A^*_{-(n-1)-1} \cdots A^*_{-0-1} = A^*_{-n} \cdots A^*_{-1} = (A_{-1} \cdots A_{-n})^* = U_{-n}^{*-1}$$

for $n = 1, 2, \ldots$, while for $n = -1, -2, \ldots$ we have similarly

$$(4.22) \qquad U'_n = B_n^{-1} \cdots B_{-1}^{-1} = A^{*-1}_{-n-1} \cdots A_0^{*-1} = (A_{-n-1} \cdots A_0)^{*-1} = U_{-n}^{*-1}.$$

In addition, P' is a dichotomy of the system

$$(4.23) \qquad\qquad x_{n+1} = A^*_{-n-1} x_n \qquad (n = 0, 1, \ldots),$$

and Q' is a dichotomy of the system

$$(4.24) \qquad\qquad x_{n+1} = A^*_{-n-1} x_n \qquad (n = -1, -2, \ldots).$$

On the other hand, Corollary 3.5 and the dichotomy properties of P and Q, show that Q^* is a dichotomy for (4.23) while P^* is a dichotomy for (4.24). Thus Lemmas 2.1 and 3.1 lead to

$$\mathrm{Im}\, P' = \mathrm{Im}\, Q^* = (\mathrm{Ker}\, Q)^{\perp},$$

and

$$\mathrm{Ker}\, Q' = \mathrm{Ker}\, P^* = (\mathrm{Im}\, P)^{\perp}.$$

These inequalities and (4.20)–(4.22) imply that

$$\mathrm{Ker}(I - WS^*W) = \{(U_{-n}^{*-1} v)_{n=-\infty}^{\infty} : v \in (\mathrm{Ker}\, Q)^{\perp} \cap (\mathrm{Im}\, P)^{\perp}\}.$$

However $I - WS^*W = W(I - S^*)W$, and hence $\mathrm{Ker}(I - S^*) = W(\mathrm{Ker}(I - WS^*W))$. By the preceding equality, this leads to

$$\mathrm{Ker}(I - S^*) = \{(U_n^{*-1} v)_{n=-\infty}^{\infty} : v \in (\mathrm{Ker}\, Q)^{\perp} \cap (\mathrm{Im}\, P)^{\perp}\}.$$

In addition, it is clear that $(\mathrm{Ker}\, Q)^{\perp} \cap (\mathrm{Im}\, P)^{\perp} = (\mathrm{Ker}\, Q + \mathrm{Im}\, P)^{\perp}$. Consequently,

$$\mathrm{Ker}(I - S^*) = \{(U_n^{*-1} v)_{n=-\infty}^{\infty} : v \perp (\mathrm{Im}\, P + \mathrm{Ker}\, Q)\}.$$

Finally, $I - S$ is Fredholm and therefore $\mathrm{Im}(I - S) = \left(\mathrm{Ker}(I - S^*)\right)^{\perp}$. This and the preceding equality prove (4.5). $\square$

We now consider the case when equalities (4.1) hold for all integers n.

LEMMA 4.3.   *Let* $S = (\delta_{i,j+1}A_j)_{ij=-\infty}^{\infty}$ *be a block weighted shift in* $\ell_r^2(\mathbb{Z})$, *where* $\{A_n\}_{n=-\infty}^{\infty}$ *is a bounded sequence of invertible* $r \times r$ *matrices. There exists a bounded sequence* $\{X_n\}_{n=-\infty}^{\infty}$ *of self-adjoint* $r \times r$ *matrices such that*

$$(4.25) \qquad X_n - A_n^* X_{n+1} A_n \geq \varepsilon I_r \qquad (n = 0, \pm 1, \ldots)$$

for some positive number ε, *if and only if* $I - S$ *is a Fredholm operator in* $\ell_r^2(\mathbb{Z})$, *and*

$$(4.26) \qquad \mathrm{Ker}(I - S) = \{0\}.$$

PROOF.   Assume first that (4.25) holds for some sequence $\{X_n\}_{n=-\infty}^{\infty}$ as in the statement. By Theorem 4.1, the operator $I - S$ is Fredholm. We now prove (4.26). Define $X = (\delta_{ij} X_j)_{n=-\infty}^{\infty}$. Then X is a bounded self-adjoint operator in $\ell_r^2(\mathbb{Z})$ and (4.25) leads to

$$(4.27) \qquad X - S^* X S \geq \varepsilon I.$$

Assume that $u \in \mathrm{Ker}(I - S)$. Then $Su = u$, and hence, taking into account (4.27) we obtain

$$0 = (Xu, u) - (XSu, Su) = ((X - S^* X S)u, u) \geq \varepsilon \|u\|^2.$$

Consequently $u = 0$, and therefore (4.26) holds.

Conversely, assume that $I - S$ is a Fredholm operator and that (4.26) holds. By Theorem 4.1, the right system

$$(4.28) \qquad x_{n+1} = A_n x_n \qquad (n = 0, 1, \ldots)$$

admits a dichotomy P, and the left system

$$(4.29) \qquad x_{n+1} = A_n x_n \qquad (n = -1, -2, \ldots)$$

admits a dichotomy Q. Moreover, there exists a bounded sequence of self-adjoint matrices $\{W_n\}_{n=-\infty}^{\infty}$ such that

$$(4.30 \qquad W_n - A_n^* W_{n+1} A_n \geq \varepsilon I_r \qquad (n = 0, 1, \pm 2, \ldots).$$

Furthermore, equality (4.4) and condition (4.26) imply that

$$(4.31) \qquad \mathrm{Im}\, P \cap \mathrm{Ker}\, Q = \{0\}.$$

It is clear that the sequences of projections $\{P_n\}_{n=0}^{\infty} = \{U_n P U_n^{-1}\}_{n=0}^{\infty}$ and $\{Q_n\}_{n=-\infty}^{0} = \{U_n Q U_n^{-1}\}_{n=-\infty}^{0}$ are dichotomies for the systems (4.28) and (4.29). This follows from the equivalence of the two descriptions of the dichotomy given at the beginning of Sections 2 and 3.

We now define two sequences of nonnegative $r \times r$ matrices $\{Y_n\}_{n=0}^{\infty}$ and $\{Z_n\}_{n=-\infty}^{0}$ via

$$Y_n = (I_r - P_n)^*(I_r - P_n) + \sum_{k=0}^{n-1} A_{n-1}^{*-1} \cdots A_k^{*-1}(I_r - P_k)^*(I_r - P_k)A_k^{-1} \cdots A_{n-1}^{-1}$$

for $n = 0, 1, \ldots$, and

$$Z_n = Q_n^* Q_n + \sum_{k=n}^{-1} A_n^* \cdots A_k^* Q_{k+1}^* Q_{k+1} A_k \cdots A_n,$$

for $n = 0, -1, \ldots$.

Let a and M be a common pair of constants for the dichotomies $\{P_n\}_{n=0}^{\infty}$ and $\{Q_n\}_{n=-\infty}^{0}$ of (4.28) and (4.29). Denote also $L = \sup_{n=0,1,\ldots}\{\|P_n\|, \|I-P_n\|, \|Q_n\|, \|I-Q_n\|\}$. It follows from the commutation relations (2.2) that

$$A_{n-1} \cdots A_k(I_r - P_k) = (In - P_n)A_{n-1} \cdots A_k \qquad (n = 0, 1, \ldots; k = 0, \ldots, n-1).$$

Thus, (2.4) implies $\|(I_r - P_n)A_{n-1} \cdots A_k x\| \geq (Ma^{n-k})^{-1}\|(I_r - P_k)x\|(x \in \mathbf{C}^r)$. Setting $y = A_{n-1} \cdots A_k x$, we obtain $Ma^{n-k}\|(I_r - P_n)y\| \geq \|(I_r - P_k)A_k^{-1} \cdots A_{n-1}^{-1}y\|(y \in \mathbf{C}^r)$. Since $\|(I_r - P_n)y\| \leq L\|y\|$, this leads to

$$\|(I_r - P_k)A_k^{-1} \cdots A_{n-1}^{-1}\| \leq Ma^{n-k}L \qquad (n = 0, 1, \ldots; k = 0, \ldots, n-1).$$

Hence, $\|Y_n\| \leq L^2 + \sum_{k=0}^{n-1} M^2 L^2 a^{2(n-k)} < L^2 + M^2 L^2/(1 - a^2)$, and therefore $\{Y_n\}_{n=0}^{\infty}$ is a bounded sequence of self-adjoint matrices. Similarly, we have

$$\|Q_{k+1} A_k \cdots A_n\| = \|A_k \cdots A_n Q_n\| \leq MLa^{k-n} \qquad (n = 0, -1, \ldots; k = n, \ldots, -1),$$

which implies

$$\|Z_n\| \leq L^2 + \sum_{k=n}^{-1} M^2 L^2 a^{2(k-n)} < L^2 + M^2 L^2/(1 - a^2).$$

Thus, $\{Z_n\}_{n=-\infty}^{0}$ is also a bounded sequence of self-adjoint matrices.

The definition of Y_n also leads to

$$A_n^* Y_{n+1} A_n = A_n^*(I_r - P_{n+1})^*(I_r - P_{n+1})A_n$$
$$+ \sum_{k=0}^{n} A_n^* A_n^{-1*} \cdots A_k^{-1*}(I_r - P_k)^*(I_r - P_k)A_k^{-1} \cdots A_n^{-1} \cdot A_n$$
$$= A_n^*(I_r - P_{n+1})^*(I_r - P_{n+1})A_n + (I_r - P_n)^*(I_r - P_n)$$
$$+ \sum_{k=0}^{n-1} A_{n-1}^{-1*} \cdots A_k^{-1*}(I_r - P_k)^*(I_r - P_k)A_k^{-1} \cdots A_{n-1}^{-1} \geq Y_n.$$

Thus,

$$(4.32) \qquad Y_n - A_n^* Y_{n+1} A_n \leq 0 \qquad (n = 0, 1, \ldots).$$

Similarly, we have for $n = -1, -2, \ldots$

$$A_n^* Z_{n+1} A_n = A_n^* Q_{n+1}^* Q_{n+1} A_n + \sum_{k=n+1}^{-1} A_n^* A_{n+1}^* \cdots A_k^* Q_{k+1}^* Q_{k+1} A_k \cdots A_{n+1} A_n$$

$$= \sum_{k=n}^{-1} A_n^* \cdots A_k^* Q_{k+1}^* Q_{k+1} A_k \cdots A_n \leq Z_n.$$

Hence,

$$(4.33) \qquad Z_n - A_n^* Z_{n+1} A_n \geq 0 \qquad (n = -1, -2, \ldots).$$

Furthermore, the above definitions and the equalities $P_0 = P$ and $Q_0 = Q$ lead to

$$(4.34) \qquad Y_0 = (I_r - P)^*(I_r - P) \quad \text{and} \quad Z_0 = Q^* Q.$$

Now note that $Y_0 + Z_0 \geq 0$. Furthermore, the equality $((Y_0 + Z_0)u, u) = 0$ implies $u = 0$. In fact, by (4.34), $((Y_0 + Z_0)u, u) = 0$ is equivalent to $(I_r - P)u = Qu = 0$. Thus, $((Y_0 + Z_0)u, u) = 0$ holds only for $u \in \operatorname{Im} P \cap \operatorname{Ker} Q$. By (4.31) we have that $((Y_0 + Z_0)u, u) = 0$ implies $u = 0$. Since $Y_0 + Z_0 \geq 0$ this leads to $(Y_0 + Z_0) > 0$. In particular, there exists a positive number α such that

$$(4.35) \qquad \alpha(Z_0 + Y_0) \geq A_0^* W_1 A_0 + \varepsilon I_r,$$

and

$$(4.36) \qquad \alpha A_{-1}^*(Z_0 + Y_0)A_{-1} \geq \varepsilon I_r - W_1,$$

where in the last equality we used the fact that A_{-1} is invertible.

We now define a sequence of self-adjoint matrices $\{X_n\}_{n=-\infty}^{\infty}$ via

$$X_n = \begin{cases} W_n - 2\alpha Y_n & (n = 1, 2, \ldots) \\ \alpha Z_0 - \alpha Y_0 & (n = 0) \\ W_n + 2\alpha Z_n & (n = -1, -2, \ldots). \end{cases}$$

Since the sequences $\{W_n\}_{n=-\infty}^{\infty}$, $\{Y_n\}_{n=0}^{\infty}$ and $\{Z_n\}_{n=-\infty}^{0}$ are bounded, the sequence $\{X_n\}_{n=-\infty}^{\infty}$ is bounded.

We now prove that (4.25) holds. We consider the cases $n = 0$, $n = -1$, $n = 1, 2, \ldots$, and $n = -2, -3, \ldots$, separately.

$\underline{n = 0}$: From the definition of X_n, we have

$$X_0 - A_0^* X_1 A_0 = \alpha Z_0 - \alpha Y_0 - A_0^* W_1 A_0 + 2\alpha A_0^* Y_1 A_0$$
$$= \alpha Z_0 + \alpha Y_0 - A_0^* W_1 A_0 - 2\alpha(Y_0 - A_0^* Y_1 A_0).$$

Since $Y_0 - A_0^* Y_1 A_0 \leq 0$ by (4.32), $\alpha > 0$, and $\alpha Z_0 + \alpha Y_0 - A_0^* W_1 A_0 \geq \varepsilon I_r$ by (4.35), this inequality leads to

$$(4.37) \qquad\qquad X_0 - A_0^* X_1 A_0 \geq \varepsilon I_r.$$

$\underline{n = -1}$: In this case we have

$$\begin{aligned}
X_{-1} - A_{-1}^* X_0 A_{-1} &= W_{-1} + 2\alpha Z_{-1} - \alpha A_{-1}^* Z_0 A_{-1} + \alpha A_{-1}^* Y_0 A_{-1} \\
&= W_{-1} + 2\alpha(Z_{-1} - A_{-1}^* Z_0 A_{-1}) + \alpha A_{-1}^*(Z_0 + Y_0)A_{-1}.
\end{aligned}$$

By (4.36) we obtain $\alpha A_{-1}^*(Z_0 + Y_0)A_{-1} + W_{-1} \geq \varepsilon I_r$, while (4.33) and $\alpha > 0$ lead to $2\alpha(Z_{-1} - A_{-1}^* Z_0 A_{-1}) \geq 0$. Hence,

$$(4.38) \qquad\qquad X_{-1} - A_{-1}^* X_0 A_{-1} \geq \varepsilon I_r.$$

$\underline{n = 1, 2, \ldots}$: The definition of X_n leads to

$$X_n - A_n^* X_{n+1} A_n = W_n - A_n^* W_{n+1} A_n - 2\alpha(Y_n - A_n^* Y_{n+1} A_n).$$

By (4.30), (4.32) and $\alpha > 0$, this implies

$$(4.39) \qquad\qquad X_n - A_n^* X_{n+1} A_n \geq \varepsilon I_r \qquad (n = 1, 2, \ldots).$$

$\underline{n = -2, -3, \ldots}$: Similarly, (4.30), (4.33) and $\alpha > 0$ imply that in this case

$$(4.40) \qquad X_n - A_n^* X_{n+1} A_n = W_n - A_n^* W_{n+1} A_n + 2\alpha(Z_n - A_n^* Z_{n+1} A_n) \geq \varepsilon I_r.$$

The inequalities (4.25) follow from (4.37)–(4.40). $\square$

5. ASYMPTOTIC INERTIA

We begin with some notation. For a finite self adjoint matrix A we denote $\nu_0(A) = \dim \mathrm{Ker}(A)$, and let $\nu_+(A)$ (respectively $\nu_-(A)$) be the number of positive (respectively negative) eigenvalues of A, counting multiplicities.

Let $X = (X_{ij})_{ij=0}^{\infty}$ be a bounded self-adjoint operator in ℓ_r^2, where X_{ij} ($i, j = 0, 1, \ldots$) are $r \times r$ complex matrices. If the following limits exist

$$\nu_{\pm} = \lim_{n \to \infty} \frac{1}{n} \nu_{\pm}((X_{ij})_{ij=0}^n) \qquad \text{and} \qquad \nu_0 = \lim_{n \to \infty} \frac{1}{n} \nu_0((X_{ij})_{ij=0}^n),$$

then we say that X has the asymptotic inertia (ν_+, ν_0, ν_-). Note that the existence of any two of the above limits imply the existence of the third and $\nu_+ + \nu_0 + \nu_- = r$.

In this section we prove the following result.

THEOREM 5.1. *Let* $G = (\delta_{i,j+1} A_j)_{ij=0}^{\infty}$ *be a bounded block weighted shift in* ℓ_r^2, *and* $X = (X_{ij})_{ij=0}^{\infty}$ *be a bounded self-adjoint operator in* ℓ_r^2. *If the inequality*

$$(5.1) \qquad\qquad X - G^* X G \geq \varepsilon I$$

holds for some positive number ε, then $I - G$ is Fredholm and X has an asymptotic inertia (ν_+, ν_0, ν_-) given by

$$(5.2) \qquad (\nu_+, \nu_0, \nu_-) = \left(r + \mathrm{index}(I - G), 0, -\mathrm{index}(I - G) \right).$$

PROOF. Let us first prove the theorem under the additional assumption that the matrices $X_{n,n}$ ($n = 0, 1, \ldots$) have constant inertia. Note that (5.1) leads to

$$(5.3) \qquad X_{n,n} - A_n^* X_{n+1,n+1} A_n \geq \varepsilon I_r \qquad (n = 0, 1, \ldots).$$

By Theorem 2.4 it follows that the system

$$(5.4) \qquad x_{n+1} = A_n x_n \qquad (n = 0, 1, \ldots)$$

admits a dichotomy $(P_n)_{n=0}^\infty$. Denote by p the rank of this dichotomy. Theorem 2.7 implies that $I - G$ is Fredholm and

$$(5.5) \qquad \mathrm{index}(I - G) = p - r.$$

Let a and M be a pair of constants for the dichotomy $\{P_n\}_{n=0}^\infty$ of (5.4), and let ℓ be a positive integer such that

$$(5.6) \qquad M^2 a^{2\ell} \|X\| \leq \varepsilon/2.$$

It follows from (5.1) that

$$(5.7) \qquad G^{*n} X G^n - G^{*(n+1)} X G^{(n+1)} \geq \varepsilon G^{*n} G^n \qquad (n = 0, 1, \ldots).$$

Adding the preceding inequalities for $n = 0, 1, \ldots, k - 1$ we obtain

$$(5.8) \qquad X - G^{*k} X G^k \geq \varepsilon(I + G^* G + \cdots + G^{*(k-1)} G^{k-1}) \qquad (k = 1, 2, \ldots).$$

Let n be an arbitrary integer with $n > \ell$. We first prove that

$$(5.9) \qquad \nu_+\left((X_{ij})_{ij=0}^n \right) \geq p(n + 1) \qquad (n = 1, 2, \ldots).$$

Denote

$$X(n) = (X_{ij})_{ij=0}^n \qquad (n = 0, 1, \ldots),$$

and let L be the following subspace of $\mathbb{C}^{r(n+1)}$

$$(5.10) \qquad L = \left\{ (u_i)_{i=0}^n : u_i \in \mathrm{Im}\, P_i, \ i = 0, \ldots, n \right\}.$$

Since $\mathrm{Rank}\, P_i = p$ ($i = 0, 1, \ldots$), then

$$(5.11) \qquad \dim L = p(n + 1).$$

We now show that $(X_{ij})_{ij=0}^n$ is positive definite on L. Let $u = (u_i)_{i=0}^n \in L$ be arbitrary. Set $u_i = 0$ for $i = n + 1, n + 2, \ldots$ and denote $u' = (u_i)_{i=0}^\infty \in \ell_r^2$. Then

$$(5.12) \qquad (X(n)u, u)_{\mathbb{C}^{(n+1)r}} = (Xu', u')_{\ell_r^2}.$$

On the other hand, it follows from (5.8) that $X - G^{*\ell}XG^\ell \geq \varepsilon I$. Hence,

$$(5.13) \qquad (Xu', u')_{\ell_r^2} \geq \varepsilon \|u'\|^2 + (XG^\ell u', G^\ell u').$$

Denote $G^\ell u' = (v_i)_{i=0}^\infty$ where $v_i \in \mathbb{C}^2$ $(i = 0, 1, \ldots)$. It follows from the special structure of G and u' that $v_i = 0$ for $i = 0, \ldots, \ell - 1$ and for $i = n + \ell + 1, n + \ell + 2, \ldots$, while

$$v_{i+\ell} = A_{i+\ell-1} \cdots A_i u_i \qquad (i = 0, \ldots, n).$$

Since $(u_i)_{i=0}^n \in L$, we have $u_i \in \operatorname{Im} P_i$ $(i = 0, \ldots, n)$. Hence, (2.4) leads to

$$\|v_{i+\ell}\| = \|A_{i+\ell-1} \cdots A_i P_i u_i\| \leq Ma^\ell \|P_i u_i\| \leq Ma^\ell \|u_i\| \qquad (i = 0, 1, \ldots).$$

This implies

$$\|G^\ell u'\| \leq Ma^\ell \|u'\|.$$

Consequently

$$|(XG^\ell u', G^\ell u')\| \leq (Ma^\ell \|u'\|)^2 \|X\| \leq \frac{\varepsilon}{2} \|u'\|^2,$$

where we used (5.6). This inequality and (5.12)–(5.13) imply that $\big(X(n)u, u\big) = (Xu', u') \geq \|u'\|^2 \varepsilon/2 = \|u\|^2 \varepsilon/2$. Hence, $X(n)$ is positive definite on L. By (5.11) this implies (5.9).

Next, we show that

$$(5.14) \qquad \nu_-\big((X_{ij})_{ij=0}^n\big) \geq (r - p)(n - \ell) \qquad (n = \ell + 1, \ell + 2, \ldots).$$

Note that A_j maps $\operatorname{Ker} P_j$ into $\operatorname{Ker} P_{j+1}$ by the commutation relations (2.2) $(j = 0, 1, \ldots)$. Furthermore, $\operatorname{Ker} A_j|_{\operatorname{Ker} P_j} = \{0\}$ by the dichotomy inequalities (2.4). Since $\operatorname{Rank} P_j = \operatorname{Rank} P_{j+1}$ it follows that the mapping

$$A_j|_{\operatorname{Ker} P_j}: \operatorname{Ker} \Gamma_j \to \operatorname{Ker} \Gamma_{j+1} \qquad (j - 0, 1, \ldots)$$

is invertible. Consequently, for each vector $u \in \operatorname{Ker} P_{j+1}$ there exists a vector $v \in \operatorname{Ker} P_j$ such that $A_j v = u$.

Let M be the subspace of $\mathbb{C}^{r(n+1)}$ defined by

$$(5.15) \qquad M = \{(u_i)_{i=0}^n : u_0 = \cdots = u_\ell = 0;\ u_i \in \operatorname{Ker} P_i\ (i = \ell + 1, \ldots, n)\}.$$

Then

$$(5.16) \qquad \dim M = (r - p)(n - \ell).$$

We now prove that $(X_{ij})_{ij=0}^n$ is negative definite on M. Let $u = (u_i)_{i=0}^n \in M$ be an arbitrary nonzero vector in M. Since $u_i \in \operatorname{Ker} P_i$ $(i = \ell + 1, \ldots, n)$, it follows from the above that there exist vectors $v_j \in \operatorname{Ker} P_j$ $(j = 0, \ldots, n - \ell - 1)$ such that

$$(5.17) \qquad u_i = A_{i-1} \cdots A_{i-\ell-1} v_{i-\ell-1} \qquad (i = \ell + 1, \ldots, n).$$

New set $u_i = 0$ for $i = n+1, n+2, \ldots$ and $v_i = 0$ for $i = n-\ell, n-\ell+1, \ldots$. Then $u' = (u_i)_{i=0}^{\infty}$ and $v' = (v_i)_{i=0}^{\infty}$ belong to ℓ_r^2, and furthermore

$$(5.18) \qquad\qquad u' = G^{\ell+1}v'.$$

In particular, $v' \neq 0$. Moreover, we have

$$(5.19) \qquad\qquad (X(n)u, u)_{\mathbb{C}^{(n+1)r}} = (Xu', u')_{\ell_r^2}.$$

The dichotomy inequalities (2.4) and $v_j \in \operatorname{Ker} P_j$ $(j = 0, 1, \ldots, n-\ell-1)$ leads to

$$\|A_{j+\ell-1} \cdots A_j v_j\| = \|A_{j+\ell-1} \cdots A_j (I_r - P_j)v_j\| \geq \frac{1}{Ma^\ell}\|v_j\|$$

for $j = 0, 1, \ldots, n-\ell-1$. Hence

$$(5.20) \qquad\qquad \|G^\ell v'\| \geq \frac{1}{Ma^\ell}\|v'\|.$$

Inequality (5.8) implies that

$$G^{*\ell+1} X G^{\ell+1} \leq X - \varepsilon G^{*\ell} G^\ell.$$

Hence,

$$(X G^{\ell+1} v', G^{\ell+1} v') \leq (Xv', v') - \varepsilon\|G^\ell v'\|^2.$$

By (5.18), and $(Xv', v') \leq \|X\|\|v\|^2$, this leads to

$$(Xu', u') \leq \|X\|\|v\|^2 - \varepsilon\|G^\ell v'\|^2.$$

Combining this with (5.20) and (5.6), we obtain

$$(Xu', u') \leq \left(\|X\| - \frac{\varepsilon}{M^2 a^{2\ell}}\right)\|v'\|^2 \leq \frac{M^2 a^{2\ell}\|X\| - \varepsilon}{M^2 a^{2\ell}}\|v'\|^2 \leq \frac{-\varepsilon}{2M^2 a^{2\ell}}\|v'\|^2.$$

Since $v' \neq 0$, we obtain $(Xu', u') < 0$, and therefore, (5.19) implies $(X(n)u, u)_{\mathbb{C}^{(n+1)r}} < 0$. Thus, $X(n) = (X_{ij})_{ij=0}^n$ is negative definite on M. By (5.16) this leads to (5.14).

Finally, we have

$$\nu_0\big((X_{ij})_{ij=0}^n\big) = (n+1)r - \nu_+\big((X_{ij})_{ij=0}^n\big) - \nu_-\big((X_{ij})_{ij=0}^n\big),$$

$$\nu_+\big((X_{ij})_{ij=0}^n\big) \leq (n+1)r - \nu_-\big((X_{ij})_{ij=0}^n\big),$$

and

$$\nu_-\big((X_{ij})_{ij=0}^n\big) \leq (n+1)r - \nu_+\big((X_{ij})_{ij=0}^n\big).$$

By (5.9) and (5.14) this implies

$$\nu\big((X_{ij})_{ij=0}^n\big) \leq (n+1)r - p(n+1) - (r-p)(n-\ell) = (r-p)(\ell+1)$$

$$\nu_+\big((X_{ij})_{ij=0}^n\big) \leq (n+1)r - (r-p)(n-\ell) \leq pn + r + (r-p)\ell,$$

and

$$\nu_-\big((X_{ij})_{ij=0}^n\big) \leq (n+1)r - p(n+1) = (r-p)(n+1).$$

These inequalities and (5.9) and (5.14) imply that $\lim_{n\to\infty} n^{-1}\nu_0\big((X_{ij})_{ij=0}^n\big) = 0$, $\lim_{n\to\infty} n^{-1}\nu_+\big((X_{ij})_{ij=0}^n\big) = p$, and $\lim_{n\to\infty} n^{-1}\nu_-\big((X_{ij})_{ij=0}^n\big) = r - p$. Thus, X has an asymptotic inertia (ν_+,ν_0,ν_-) given by

$$(\nu_+,\nu_0,\nu_-) = (p,0,r-p).$$

By (5.5) this equality imply (5.2). This proves the theorem in the case when the matrices $X_{n,n}$ have constant inertia.

We now consider the general case. It follows from (5.1) that

$$X_{n,n} - A_n^* X_{n+1,n+1} A_n \geq \varepsilon I_r \qquad (n = 0,1,\ldots).$$

Thus, by Corollary 2.6, there is a positive integer N such that the sequence $\{X_{n+N,n+N}\}_{n=0}^\infty$ has constant inertia. Denote $X_N = (X_{i+N,j+N})_{ij=0}^\infty$ and $G_N = (\delta_{i,j+1}A_{j+N})_{ij=0}^\infty$. Inequality (5.1) leads to

$$X_N - G_N^* X_N G_N \geq \varepsilon I.$$

By the previous part of the proof, it follows that $I - G_N$ is Fredholm, and that X_N has an asymptotic inertia (ν'_+,ν'_0,ν'_-) given by

$$(5.21) \qquad (\nu'_+,\nu'_0,\nu'_-) = \big(r + \mathrm{index}(I - G_N), 0, -\mathrm{index}(I - G_N)\big).$$

This implies that $I - G$ is Fredholm with

$$(5.22) \qquad \mathrm{index}(I - G) = \mathrm{index}(I - G_N).$$

Furthermore, note that we have

$$\Big|\nu_\pm\big((X_{ij})_{ij=0}^n\big) - \nu_\pm\big((X_{ij})_{ij=N}^n\big)\Big| \leq Nr \qquad (n = N, N+1,\ldots).$$

Hence,

$$
\begin{aligned}
\nu_\pm &= \lim_{n\to\infty} n^{-1}\nu_\pm\big((X_{ij})_{ij=0}^n\big) = \lim_{n\to\infty} n^{-1}\nu_\pm\big((X_{ij})_{ij=N}^n\big) \\
&= \lim_{n\to\infty} (n-N)^{-1}\nu_\pm\big((X_{i+N,j+N})_{ij=N}^n\big) = \nu'_\pm.
\end{aligned}
$$

(5.23)

These two limits imply that X has an asymptotic inertia $(\nu_+,\nu_0,\nu_-) = (\nu'_+,\nu'_0,\nu'_-)$. Combining this with (5.21)–(5.22) we obtain (5.2). $\square$

6. REFERENCES

[BG1] A. Ben-Artzi and I. Gohberg, Inertia theorems for nonstationary discrete systems and dichotomy, *Linear Algebra and Its Application* **120**: 95–138 (1989).

[BG2] A. Ben-Artzi and I. Gohberg, Band matrices and dichotomies, *Operator Theory: Advances and Applications* **50**: 137–170 (1991).

[BGK] A. Ben-Artzi, I. Gohberg, and M. A. Kaashoek, Invertibility and dichotomy of singular difference equations, *Operator Theory: Advances and Applications* **48**: 157–184 (1990).

[BO] C. de Boor, Dichotomies for band matrices, *SIAM J. Numer. Anal.* **17**(6) (1980).

[CS] Ch. V. Coffman and J. J. Schäffer, Dichotomies for linear difference equations, *Math. Annalen* **172**: 139–166 (1967).

[GKvS] I. Gohberg, M. A. Kaashoek and F. van Schagen, Non-compact integral operators with semi-separable kernels and their discrete analogous: inversion and Fredholm properties, *Integral Equations and Operator Theory* **7**: 642–703 (1984).

A. Ben-Artzi
Department of Mathematics
Indiana University
Swain East 222
Bloomington, IN 47405, U.S.A.

I. Gohberg
School of Mathematical Sciences
The Raymond and Beverly Sackler Faculty of Exact Sciences
Tel-Aviv University
Tel-Aviv, Ramat-Aviv 69978, Israel

Operator Theory:
Advances and Applications, Vol. 56
© 1992 Birkhäuser Verlag Basel

INTERPOLATION FOR UPPER TRIANGULAR OPERATORS

Patrick Dewilde and Harry Dym

In this paper the classical interpolation problems of Nevanlinna-Pick and Carathéodory-Fejér, as well as mixtures of the two, are solved in the general setting of upper triangular operators. Herein, the "points" at which the interpolation is carried out are themselves (diagonal) operators, and the components of all the intervening operators in their natural matrix representation may be finite or infinite dimensional. Moreover, we consider both contractive and strictly contractive solutions. A number of classical and new interpolation problems emerge as special cases.

CONTENTS

Patrick Dewilde wishes to acknowledge with thanks the support of the Commission of the EEC under the ESPRIT BRA Project NANA (3280) and the Meyerhoff Foundation for a visiting fellowship at the Weizmann Institute.

Harry Dym wishes to thank Renee and Jay Weiss for endowing the chair which supports his research at the Weizmann Institute, and the STW for a visitors' fellowship at Delft University of Technology under the JEMNA program.

1. INTRODUCTION

In this paper we shall use the machinery developed in [ADD] to formulate and study analogues of the interpolation problems of Nevanlinna-Pick and Carathéodory-Fejér, as well as mixtures of the two, in the setting of upper triangular operators.

We begin by recalling and somewhat extending that setting. To this end, let $\mathcal{X} = \mathcal{X}(\ell_{\mathcal{N}}^2; \ell_{\mathcal{M}}^2)$ denote the set of bounded linear operators from the space

$$\ell_{\mathcal{N}}^2 = \bigoplus_{i=-\infty}^{\infty} \mathcal{N}_i$$

of "square summable" sequences $f = (\ldots, f_{-1}, f_0, f_1, \ldots)$ with components f_i in a complex separable Hilbert space $\mathcal{N}_i = \mathcal{N}$ into the space

$$\ell_{\mathcal{M}}^2 = \bigoplus_{j=-\infty}^{\infty} \mathcal{M}_j$$

of "square summable" sequences $g = (\cdots g_{-1}, g_0, g_1, \ldots)$ with components g_j in a complex separable Hilbert space $\mathcal{M}_j = \mathcal{M}$. The spaces $\ell_{\mathcal{N}}^2$ and $\ell_{\mathcal{M}}^2$ are taken with the standard norms. Thus, for example, $f \in \ell_{\mathcal{N}}^2$ if and only if

$$\|f\|_{\ell_{\mathcal{N}}^2}^2 = \sum_{i=-\infty}^{\infty} \|f_i\|_{\mathcal{N}}^2 < \infty .$$

Let Z be the shift operator in $\ell_{\mathcal{N}}^2$:

$$(Zf)_i = f_{i+1} , \qquad i = \ldots, -1, 0, 1, \ldots ,$$

and let

$$\pi : u \in \mathcal{N} \longrightarrow f \in \ell_{\mathcal{N}}^2 , \quad \text{where} \quad \begin{cases} f_0 = u \\ f_i = 0 , \quad i \neq 0 \end{cases}$$

with adjoint

$$\pi^* : f \in \ell_{\mathcal{N}}^2 \longrightarrow f_0 \in \mathcal{N} ,$$

so that

$$\pi^* Z^i f = f_i .$$

It is readily seen that Z is unitary on $\ell_{\mathcal{N}}^2$: $ZZ^* = Z^*Z = I$, and that

$$\pi^* Z^j \pi = \begin{cases} I \ (\text{on } \mathcal{N}) \ \text{if} \ j = 0 \\ 0 \ (\text{on } \mathcal{N}) \ \text{if} \ j \neq 0 . \end{cases}$$

The latter implies in particular that

$$(Z^s\pi)^*(Z^t\pi) = 0 \quad \text{if} \quad s \neq t$$

and hence that

$$P_{ab} = \sum_{s=a}^{b} Z^s\pi\pi^*Z^{s*}$$

is a selfadjoint projection:

$$P_{ab} = P_{ab}^* = (P_{ab})^2 \ .$$

Moreover, P_{ab} converges strongly to the identity

$$I = \sum_{s=-\infty}^{\infty} Z^s\pi\pi^*Z^{*s} = \sum_{s=-\infty}^{\infty} Z^{*s}\pi\pi^*Z^s$$

in $\ell^2_{\mathcal{N}}$ as $a \downarrow -\infty$ and $b \uparrow \infty$.

In order to keep the typography simple, we shall continue to use the symbol π for the mapping which sends $g \in \ell^2_{\mathcal{M}}$ into $g_0 \in \mathcal{M}$, π^* for its adjoint and the symbol Z for the shift in $\ell^2_{\mathcal{M}}$. We shall also usually write $\mathcal{X}$ instead of the more cumbersome $\mathcal{X}(\ell^2_{\mathcal{N}}; \ell^2_{\mathcal{M}})$.

Next, keeping the preceding conventions in mind, we define the space of upper triangular operators

$$\mathcal{U} = \{A \in \mathcal{X} : \ \pi^*Z^iAZ^{*j}\pi = 0 \ \text{ for } \ i > j\} \ ,$$

the space of lower triangular operators

$$\mathcal{L} = \{A \in \mathcal{X} : \ \pi^*Z^iAZ^{*j}\pi = 0 \ \text{ for } \ i < j\} \ ,$$

and the space of diagonal operators

$$\mathcal{D} = \mathcal{U} \cap \mathcal{L} \ .$$

Here too, $\mathcal{U}$ is short for $\mathcal{U}(\ell^2_{\mathcal{N}}; \ell^2_{\mathcal{M}})$. Similar remarks apply to $\mathcal{L}$ and $\mathcal{D}$. Moreover, in the term $\pi^*Z^iAZ^{*j}\pi$, the π and the Z^* which sit on the right of A map $\mathcal{N}$ into $\ell^2_{\mathcal{N}}$ and $\ell^2_{\mathcal{N}}$ into itself, respectively, whereas the Z and the π^* which sit on the left of A map $\ell^2_{\mathcal{M}}$ into itself and $\ell^2_{\mathcal{M}}$ into $\mathcal{M}$, respectively. Thus

$$\pi^*Z^iAZ^{*j}\pi \ \text{ maps } \ \mathcal{N} \ \text{ into } \ \mathcal{M} \ .$$

The terminology upper triangular is consistent with picturing $A \in \mathcal{X}$ as a doubly infinite matrix of operators

$$\begin{bmatrix} \vdots & \vdots & \vdots & \vdots & \\ \cdots & A_{-1,-1} & A_{-1,0} & A_{-1,1} & A_{-1,2} & \cdots \\ \cdots & A_{0,-1} & \boxed{A_{00}} & A_{01} & A_{02} & \cdots \\ & A_{1,-1} & A_{10} & A_{11} & A_{12} & \cdots \\ \vdots & \vdots & \vdots & \vdots & \end{bmatrix}$$

with

$$A_{ij} = \pi^* Z^i A Z^{*j} \pi \, ,$$

where the 00 element is distinguished by a surrounding square.

Correspondingly, $f \in \ell^2_{\mathcal{N}}$ should be pictured as a doubly infinite column vector

$$f = \begin{bmatrix} \vdots \\ f_{-1} \\ \boxed{f_0} \\ f_1 \\ \vdots \end{bmatrix} \, .$$

The formula

$$(Af)_i = \sum_{j=-\infty}^{\infty} A_{ij} f_j \, , \qquad i = \ldots, -1, 0, 1, \ldots \, ,$$

then fits the usual rules for matrix-vector multiplication.

In the future we shall, as we have already indicated, often use the symbols $\mathcal{X}$, $\mathcal{L}$, $\mathcal{U}$, $\mathcal{D}$, π and Z in a generic way without designating the domains and ranges of the indicated sets of operators. Thus $\mathcal{X}$, $\mathcal{L}$, $\mathcal{U}$ and $\mathcal{D}$ may involve operators from either $\ell^2_{\mathcal{N}}$ into $\ell^2_{\mathcal{M}}$, or from $\ell^2_{\mathcal{M}}$ into $\ell^2_{\mathcal{N}}$, or from $\ell^2_{\mathcal{N}}$ or $\ell^2_{\mathcal{M}}$ into itself. With this in mind, is readily checked that $Z \in \mathcal{U}$. In fact if $A = Z$, then

$$A_{ij} = \begin{cases} I & \text{if } j = i+1 \\ 0 & \text{if } j \neq i+1 \end{cases} ,$$

for an appropriately chosen identity operator I, corresponding to the picture

$$Z = \begin{bmatrix} \ddots & \ddots & & & 0 \\ & 0 & I & & \\ & & \boxed{0} & I & \\ 0 & & & 0 & \ddots \\ & & & & \ddots \end{bmatrix} \, .$$

The space $\mathcal{U}$ [resp. $\mathcal{L}$] plays the role of bounded analytic functions inside the open [resp. outside the closed] unit disc, whereas $\mathcal{D}$ plays the role of the constants. A number of elementary properties of these spaces are established in Section 2 of [ADD] for the special case $\mathcal{M} = \mathcal{N}$. But most of the arguments go through even if $\mathcal{M} \neq \mathcal{N}$.

Next we recall the transformation

$$A^{(j)} = Z^{*j} A Z^j$$

for $A \in \mathcal{X}$ and every integer $j = \ldots, -1, 0, 1, \ldots$; it takes each of the spaces $\mathcal{L}$, $\mathcal{D}$ and $\mathcal{U}$ into themselves. It is readily checked that if also $B \in \mathcal{X}$ and ran $B \subset$ dom A so that AB is well defined, then

$$(AB)^{(j)} = A^{(j)} B^{(j)} \quad \text{and} \quad A^{(j+k)} = \{A^{(j)}\}^{(k)} .$$

In particular,

$$\left(A^{(j)}\right)_{st} = A_{s-j,t-j} ,$$

which is the same as to say that the entries in $A^{(j)}$ are translates of j units in the southeast direction along the diagonals of the entries in A.

If $F \in \mathcal{U}$, then, by Lemma 2.9 of [ADD] (which is easily adapted to the present more general setting), there exists a unique set of operators $F_{[j]} \in \mathcal{D}$, $j = 0, 1, \ldots$, such that

$$F - \sum_{k=0}^{n-1} Z^j F_{[j]} \in Z^n \mathcal{U} .$$

Correspondingly (still for $F \in \mathcal{U}$), we define the operator transform

$$\hat{F}(W) = \sum_{j=0}^{\infty} W^{[j]} F_{[j]} , \tag{1.1}$$

where $W^{[0]} = I$ and

$$W^{[j]} = W W^{(1)} \cdots W^{(j-1)} , \quad \text{for} \quad j \geq 1 ,$$

for every choice of $W \in \mathcal{X}$ for which the multiplications are meaningful and the indicated sum converges. More precisely, if F maps $\ell_{\mathcal{N}}^2$ into $\ell_{\mathcal{M}}^2$, then W will always be an operator from $\ell_{\mathcal{M}}^2$ into itself. In Section 3 of [ADD] it is shown that the notation

$$\ell_W = \lim_{n \uparrow \infty} \|W^{[n]}\|^{1/n}$$

is meaningful for every $W \in \mathcal{X}$ by identifying the limit with

$$r_{sp}(ZW^*) = r_{sp}(WZ^*) \, ,$$

the spectral radius of the bounded operator WZ^*. It follows by an easy estimate (which is also spelled out in Section 3 of [ADD]) that the indicated sum converges for every $F \in \mathcal{U}$ when $\ell_W < 1$.

In the sequel we shall be particularly interested in the diagonal transform which maps

$$F \in \mathcal{U} \longrightarrow \hat{F}(W) \in \mathcal{D}$$

for $W \in \mathcal{D}$ with $\ell_W < 1$. This set of diagonal evaluations suffices to specify F:

THEOREM 1.1.   *If F and G belong to $\mathcal{U}$ and if $\hat{F}(W) = \hat{G}(W)$ for every $W \in \mathcal{D}$ with $\ell_W < 1$, then $F = G$.*

PROOF. This is immediate from Theorem 3.2 of [ADD].   ∎

The following rule will prove to be very useful:

THEOREM 1.2.   *If $F \in \mathcal{U}$, $G \in \mathcal{U}$ and ran $G \subset$ dom F so that the multiplication FG is meaningful, then $FG \in \mathcal{U}$ and*

$$(FG)^{\wedge}(W) = \{\hat{F}(W)G\}^{\wedge}(W) \tag{1.2}$$

for every $W \in \mathcal{D}$ with $\ell_W < 1$. Moreover, if also $G \in \mathcal{D}$, then

$$(FG)^{\wedge}(W) = \hat{F}(W)G \, . \tag{1.3}$$

PROOF. See Lemmas 3.3 and 3.7 of [ADD].   ∎

COROLLARY.   *If $F \in \mathcal{U}$ is invertible in $\mathcal{U}$, then*

$$\hat{F}(W) \neq 0$$

for every $W \in \mathcal{D}$ with $\ell_W < 1$.

More generally, for every $F \in \mathcal{U}$ and every $V \in \mathcal{D}$ with $\ell_V < 1$ there exists a unique set of diagonal operators

$$F_{\{j\}} = F_{\{j\}}(V)$$

such that

$$F - \sum_{j=0}^{n} (Z - V)^j F_{\{j\}} \in (Z - V)^{n+1}\mathcal{U} \tag{1.4}$$

For $n = 0, 1, \ldots$. This may be verified by repeated application of Theorem 3.3 of [ADD]. These coefficients will be discussed in more detail in Section 8. For the moment, suffice it to note that

$$\hat{F}(V) = F_{\{0\}}(V) \ . \tag{1.5}$$

Now, upon letting

$$\mathcal{S} = \{F \in \mathcal{U} : \|F\| \leq 1\} \ ,$$

we are ready to state the analogues of the NP (Nevanlinna-Pick) problem and the CF (Carathéodory-Fejér) problem in the setting of this paper, in terms of operators on (and between) the spaces $\ell^2_{\mathcal{B}}$, $\ell^2_{\mathcal{M}}$, $\ell^2_{\mathcal{N}}$ based on separable Hilbert spaces $\mathcal{B}$, $\mathcal{M}$ and $\mathcal{N}$, respectively, and the diagonal transform.

NP PROBLEM: *Let*

$$V_j \in \mathcal{D}(\ell^2_{\mathcal{B}}; \ell^2_{\mathcal{B}}) \ , \quad j = 1, \ldots, n \ ,$$

$$\xi_j \in \mathcal{D}(\ell^2_{\mathcal{B}}; \ell^2_{\mathcal{M}}) \ , \quad j = 1, \ldots, n \ ,$$

$$\eta_j \in \mathcal{D}(\ell^2_{\mathcal{B}}; \ell^2_{\mathcal{N}}) \ , \quad j = 1, \ldots, n \ ,$$

be a given set of operators such that

$$\ell_{V_j} < 1 \ , \quad j = 1, \ldots, n \ ,$$

and the $n \times n$ *operator matrix* Δ *with components*

$$\Delta_{ij} = \left(\xi_i^* \xi_j \rho_{V_j}^{-1}\right)^{\wedge}(V_i) \ , \quad i, j = 1, \ldots, n \ ,$$

is uniformly positive on $(\ell^2_{\mathcal{B}})^n$. *Find necessary and sufficient conditions on the data to ensure the existence of an* $S \in \mathcal{S}(\ell^2_{\mathcal{N}}; \ell^2_{\mathcal{M}})$ *such that*

$$(\xi_j^* S)^{\wedge}(V_j) = \eta_j^*$$

for $j = 1, \ldots, n$.

CF PROBLEM: *Let*

$$V \in \mathcal{D}(\ell^2_{\mathcal{B}}; \ell^2_{\mathcal{B}}) \ , \quad \xi \in \mathcal{D}(\ell^2_{\mathcal{B}}; \ell^2_{\mathcal{M}}) \ , \quad \eta_j \subset \mathcal{D}(\ell^2_{\mathcal{B}}; \ell^2_{\mathcal{N}}) \ , \quad j = 0, \ldots, n \ ,$$

be a given set of operators such that

$$\ell_V < 1$$

and the $(n+1) \times (n+1)$ *operator matrix* Δ *with components*

$$\Delta_{ij} = \left(\xi^* \xi (\rho_V^{-1} Z)^j \rho_V^{-1}\right)_{\{i\}}(V) \ , \quad i, j = 0, \ldots, n \ ,$$

is uniformly positive on $(\ell_{\mathcal{B}}^2)^{(n+1)}$. Find necessary and sufficient conditions on the data to ensure the existence of an $S \in \mathcal{S}(\ell_{\mathcal{N}}^2; \ell_{\mathcal{M}}^2)$ such that

$$(\xi^* S)_{\{j\}}(V) = \eta_j^* ,$$

for $j = 0, \ldots, n$.

We shall use reproducing kernel Hilbert space methods to solve both of these problems as well as to obtain linear fractional representations for the solutions much as in [D1] and [D2], except that now the calculations are more formidable. They are carried out in the space $\mathcal{X}_2 \subset \mathcal{X}$ of Hilbert Schmidt operators on $\ell_{\mathcal{N}}^2$. It is well known that $\mathcal{X}_2$ is a Hilbert space with respect to the inner product

$$\langle A, B \rangle_{HS} = \text{trace } B^* A ,$$

as are each of the subspaces

$$\mathcal{U}_2 = \mathcal{U} \cap \mathcal{X}_2 , \quad \mathcal{L}_2 = \mathcal{L} \cap \mathcal{X}_2 \quad \text{and} \quad \mathcal{D}_2 = \mathcal{D} \cap \mathcal{X}_2 .$$

A number of basic facts about these spaces, some associated reproducing kernel spaces and some useful formulas (which are adapted from [ADD]) are summarized in Section 2. For the moment it suffices to note that the symbol $\underline{\underline{p}}$ denotes the orthogonal projection of $\mathcal{X}_2$ onto $\mathcal{U}_2$.

An important role will be played by the class $\mathcal{A}$ of matrix operators

$$\Theta = \begin{bmatrix} \Theta_{11} & \Theta_{12} \\ \Theta_{21} & \Theta_{22} \end{bmatrix}$$

for which

(1) $\Theta_{11} \in \mathcal{U}(\ell_{\mathcal{M}}^2; \ell_{\mathcal{M}}^2)$, $\Theta_{12} \in \mathcal{U}(\ell_{\mathcal{N}}^2; \ell_{\mathcal{M}}^2)$, $\Theta_{21} \in \mathcal{U}(\ell_{\mathcal{M}}^2; \ell_{\mathcal{N}}^2)$, and $\Theta_{22} \in \mathcal{U}(\ell_{\mathcal{N}}^2; \ell_{\mathcal{N}}^2)$.

(2) Θ_{22} is invertible in $\mathcal{U}(\ell_{\mathcal{N}}^2; \ell_{\mathcal{N}}^2)$

(3) $\Theta^* J \Theta = J$ and $\Theta J \Theta^* = J$, where

$$J = \begin{bmatrix} I_{\ell_{\mathcal{M}}^2} & 0 \\ 0 & -I_{\ell_{\mathcal{N}}^2} \end{bmatrix} . \tag{1.6}$$

The operators in $\mathcal{A}$ will be referred to as upper lossless chain scattering (ULCS) operators. We shall show that under suitable conditions on the data, there is an associated $\Theta \in \mathcal{A}$ which will play a crucial role in establishing the existence and representation of solutions to the interpolation problems under study. The details concerning the existence of such ULCS operators Θ are carried out in a number of sections. More precisely:

Section 3 summarizes some general facts from the theory of operator colligations and characteristic functions in a form which is useful for the problems at hand. In particular, it is shown that the characteristic function Θ of a certain operator colligation M is J unitary if and only if M is $\tilde{J}$ unitary and, moreover, that this will happen if and only if a second associated operator N is unitary. In the applications at hand M and N are only partially specified and the question of interest reduces to finding out when the partially specified N admits a unitary extension. In Sections 4 and 6 these results are tailored to fit the interpolation problems of interest. Some explicit formulas for Θ are presented in Section 5, under various extra assumptions. Sections 7 and 8 then treat the NP and CF problems, respectively. The mixed interpolation problem is discussed briefly in Section 9. A number of examples are presented in Section 10. Two of these are block Toeplitz and serve as a good introduction to Section 11, wherein this case is discussed in more detail, and some seemingly new interpolation problems are solved in the context of matrix valued Schur functions. The preceding analysis is adapted to a more general setting wherein the "coordinate" spaces $\mathcal{M}_i$, $\mathcal{N}_i$ and $\mathcal{B}_i$ of $\ell^2_{\mathcal{M}}$, $\ell^2_{\mathcal{N}}$ and $\ell^2_{\mathcal{B}}$, respectively, are allowed to vary with i in Section 12.

It turns out that the basic condition for the solvability of both the NF and the CF problem is that the solution Λ of the operator equation

$$\Lambda - V Z^* \Lambda Z V^* = \alpha\alpha^* - \beta\beta^* \tag{1.7}$$

be positive semidefinite. In (1.7) Z is a shift operator and α, β and V are operators which depend upon the data of the problem.

For the NF problem,

$$V = \begin{bmatrix} V_1 & & 0 \\ & \ddots & \\ 0 & & V_n \end{bmatrix}, \quad \alpha = \begin{bmatrix} \xi_1^* \\ \vdots \\ \xi_n^* \end{bmatrix} \quad \text{and} \quad \beta = \begin{bmatrix} \eta_1^* \\ \vdots \\ \eta_n^* \end{bmatrix}, \tag{1.8}$$

whereas for the CF analogue,

$$V = \begin{bmatrix} V & 0 & \cdots & 0 & 0 \\ I & V & & 0 & 0 \\ 0 & I & & \vdots & \vdots \\ \vdots & \vdots & & V & 0 \\ 0 & 0 & \cdots & I & V \end{bmatrix}, \quad \alpha = \begin{bmatrix} \xi \\ 0 \\ \vdots \\ 0 \end{bmatrix} \quad \text{and} \quad \beta = \begin{bmatrix} \eta_0^* \\ \vdots \\ \eta_n^* \end{bmatrix}. \tag{1.9}$$

The operator Λ plays the role of the Pick matrix in the present setting.

Both the NP problem and the CF problem, as well as mixtures of the two, can be fit into a general framework which basically amounts to resolving the following

FUNDAMENTAL INTERPOLATION PROBLEM: *Let*

$$F_i = \begin{bmatrix} G_i \\ H_i \end{bmatrix} , \qquad i = 1, \ldots, n ,$$

be the i'th block component of the operator

$$F = [F_1 \ \ F_2 \cdots F_n] = \begin{bmatrix} \alpha^* \\ \beta^* \end{bmatrix} (I - \boldsymbol{Z} \boldsymbol{V}^*)^{-1}$$

which is based on a given set of "data" operators

$$\boldsymbol{V} \in \{\mathcal{D}(\ell_B^2; \ell_B^2)\}^{n \times n} , \qquad \alpha \in \{\mathcal{D}(\ell_{\mathcal{M}}^2; \ell_B^2)\}^{n \times 1} , \qquad \beta \in \{\mathcal{D}(\ell_{\mathcal{N}}^2; \ell_B^2)\}^{n \times 1} ,$$

wherein it is always assumed that

$$\ell_V = r_{sp}(\boldsymbol{Z} \boldsymbol{V}^*) < 1 .$$

When does there exist an $S \in \mathcal{S}(\ell_{\mathcal{N}}^2; \ell_{\mathcal{M}}^2)$ such that

$$\underline{\underline{p}} S^* G_i E = H_i E \tag{1.10}$$

for $i = 1, \ldots, n$ and every choice of $E \in \mathcal{D}_2(\ell_B^2; \ell_B^2)$?

For the classical counterpart of this general one-sided interpolation problem, see e.g., Section 7 of [D2].

It turns out that a necessary condition for the existence of such an S is that the "Pick" operator Λ, which is the unique solution of (1.7), be positive semidefinite on $(\ell_B^2)^n$.

Just as in classical interpolation problems, the issue of the existence of an S which satisfies (1.10) is more subtle. In the sequel we shall first show that if

I. Λ *is uniformly positive on $(\ell_B^2)^n$,*

and a second technical condition:

II. *The kernel condition (4.16) is met,*

then there exists an operator

$$\Theta = \begin{bmatrix} \Theta_{11} & \Theta_{12} \\ \Theta_{21} & \Theta_{22} \end{bmatrix}$$

in $\mathcal{A}$ such that

$$F \Lambda^{-1} \hat{F}(\boldsymbol{W})^* = \{J - \Theta J \hat{\Theta}(\boldsymbol{W})^*\} \begin{bmatrix} I - ZW_1^* & 0 \\ 0 & I - ZW_2^* \end{bmatrix}^{-1}$$

for every choice of

$$W = \mathrm{diag}(W_1, W_2)$$

with $W_1 \in \mathcal{D}(\ell^2_{\mathcal{M}}; \ell^2_{\mathcal{M}})$, $W_2 \in \mathcal{D}(\ell^2_{\mathcal{N}}; \ell^2_{\mathcal{N}})$ *and*

$$r_{sp}(\boldsymbol{Z}\,\boldsymbol{W}^*) < 1 \ .$$

(The notation is discussed in more detail in the next section; see especially (2.5).)

The main conclusion is completely analogous to the conclusion for the conventional setting (see e.g., Lemma 7.1 and Theorem 7.2 of [D2]):

THEOREM 1.3. (THE FUNDAMENTAL INTERPOLATION THEOREM) *If* I *and* II *hold, then the following are equivalent for* $S \in \mathcal{S}$:

(1)
$$\underline{p}S^*G_i = H_i \ , \qquad for \ \ i = 1,\ldots,n \ . \tag{1.11}$$

(2)
$$S = T_{\Theta}[S_L] := (\Theta_{11}S_L + \Theta_{12})(\Theta_{21}S_L + \Theta_{22})^{-1} \tag{1.12}$$

for some choice of $S_L \in \mathcal{S}$.

Moreover, in (2), S is strictly contractive if and only if S_L is.

Theorem 1.3 is established in Theorem 6.3 and Lemma 4.1. An important auxiliary fact, which is established in Theorem 4.3, is:

If $\mathcal{B}$ is finite dimensional, then the kernel condition (4.16) is automatically met.

Subsequently we shall show by a limiting argument that if the operator

$$\Delta = \sum_{t=0}^{\infty} V^{[t]}(\alpha\alpha^*)^{(t)} V^{[t]^*} \tag{1.13}$$

is uniformly positive on $(\ell^2_{\mathcal{B}})^n$ and if also $\mathcal{B}$ is finite dimensional, then the condition $\Lambda \geq 0$ is also sufficient for the existence of an $S \in \mathcal{S}$ which satisfies (1.11).

The condition

III. Δ *is uniformly positive on* $(\ell^2_{\mathcal{B}})^n$,

insures that the operators G_i, $i = 1,\ldots,n$ are linearly independent in a strong enough way. In fact it guarantees that

$$\left\{ \sum_{i=1}^{n} G_i E_i : E_i \in \mathcal{D}_2(\ell^2_{\mathcal{B}}; \ell^2_{\mathcal{B}}) \right\}$$

is a Hilbert space with respect to the Hilbert-Schmidt inner product, as is spelled out in Theorem 4.7.

It is perhaps well to emphasize that $\Delta \geq \Lambda$ and hence that III needs no special mention when I is in force.

This paper has had a long and arduous birth, with many preliminary drafts of ever increasing generality and scope (punctuated by periods of total abandonment because of other obligations). The NP interpolation problem in the present general setting was first presented at Oberwolfach in October 1989. At that time the basic formulation of both the problem (which is fairly natural once the general transform theory developed in [ADD] is available) and its solution were pretty clear, however, there were still some gaps in the proof. These were completely filled for a simpler case with finite dimensional coordinate spaces and strictly contractive interpolants in the Summer of 1990. The results were presented in a tutorial lecture at the SIAM Conference on Linear Algebra in San Francisco in November 1990. Preprints of an expository article which contained these results were distributed at the Conference. The article itself [De] appeared in the Spring of 1991. In the Summer of 1991 we received two preprints: [BGK1] and [BGK2]. The former solves an NP interpolation problem similar to the one alluded to above. It too focuses on finite dimensional coordinate spaces and strictly contractive interpolants, but uses other methods. The latter presents an application to control theory. An earlier paper by the affiliated firm of [GKW] solves a Nehari problem for semi-infinite matrices of operators, which can in turn be used to solve a CF problem in this context for strictly contractive interpolants in the more transparent case of $V = 0$. The present analysis is, as we have indicated above, quite comprehensive. It covers a broad class of interpolation problems, infinite dimensional coordinate spaces, and does not restrict the interpolants to be strictly contractive. It also includes an extensive set of formulas and examples, and devotes special attention to the block Toeplitz case and to flexible coordinate spaces.

The authors wish to thank D. Alpay for his careful reading of, and useful comments on, a preliminary draft on NP interpolation, which preceded this paper. We also wish to thank I. Gohberg for raising a question which motivated Section 11, and J. Rovnyak, M. A. Dritschel, and A. Dijksma for useful discussions on unitary extensions.

2. PRELIMINARIES

In this section we shall briefly review some important facts about the space $\mathcal{X}_2(\ell^2_{\mathcal{N}}; \ell^2_{\mathcal{M}})$ of Hilbert-Schmidt operators from $\ell^2_{\mathcal{N}}$ into $\ell^2_{\mathcal{M}}$, and some reproducing kernel subspaces of $\mathcal{U}_2(\ell^2_{\mathcal{N}}; \ell^2_{\mathcal{M}})$.

We begin with a list of general properties of Hilbert-Schmidt operators. All of these and much more can be found in the monograph [GK] of Gohberg and Krein for $\mathcal{M} = \mathcal{N}$, but are easily adapted to the case $\mathcal{M} \neq \mathcal{N}$.

(1) If $A \in \mathcal{X}_2(\ell_{\mathcal{N}}^2; \ell_{\mathcal{M}}^2)$ and $B \in \mathcal{X}(\ell_{\mathcal{B}}^2; \ell_{\mathcal{N}}^2)$ [resp. $\mathcal{X}(\ell_{\mathcal{M}}^2; \ell_{\mathcal{B}}^2)$], then AB belongs to $\mathcal{X}_2(\ell_{\mathcal{M}}^2; \ell_{\mathcal{B}}^2)$ [resp. BA belongs to $\mathcal{X}(\ell_{\mathcal{N}}^2; \ell_{\mathcal{B}}^2)$].

(2) If $\{F_n\} \in \mathcal{X}_2$ and $F_n \to F$ in $\mathcal{X}$ as $n \uparrow \infty$, then $F \in \mathcal{X}_2$.

(3) If $F \in \mathcal{X}_2(\ell_{\mathcal{N}}^2; \ell_{\mathcal{M}}^2)$, then $F^* \in \mathcal{X}_2(\ell_{\mathcal{M}}^2; \ell_{\mathcal{N}}^2)$, where F^* denotes the adjoint of F as an operator from $\ell_{\mathcal{N}}^2$ to $\ell_{\mathcal{M}}^2$.

(4) If $F \in \mathcal{X}_2$, then
$$\|F\|_{HS}^2 = \langle F, F \rangle_{HS} = \|F^*\|_{HS}^2.$$

(5) If E and $F \in \mathcal{X}_2(\ell_{\mathcal{N}}^2; \ell_{\mathcal{M}}^2)$ and $G \in \mathcal{X}(\ell_{\mathcal{N}}^2; \ell_{\mathcal{N}}^2)$, then

$$\langle E, FG \rangle_{HS} = \langle EG^*, F \rangle_{HS},$$

whereas, if $G \in \mathcal{X}(\ell_{\mathcal{M}}^2; \ell_{\mathcal{M}}^2)$, then

$$\langle E, GF \rangle_{HS} = \langle G^* E, F \rangle_{HS}.$$

Let $\underline{p}$ denote the orthogonal projection of the Hilbert space $\mathcal{X}_2$ onto the Hilbert space $\mathcal{U}_2$ and let
$$\rho_V = I - Z V^*. \tag{2.1}$$
If $V \in \mathcal{D}$ with $\ell_V < 1$, then $\rho_V^{-1} \in \mathcal{U}$.

THEOREM 2.1. *If $F \in \mathcal{U}_2(\ell_{\mathcal{N}}^2; \ell_{\mathcal{B}}^2)$, $G \in \mathcal{D}_2(\ell_{\mathcal{N}}^2; \ell_{\mathcal{B}}^2)$, and $V \in \mathcal{D}(\ell_{\mathcal{B}}^2; \ell_{\mathcal{B}}^2)$ with $\ell_V < 1$, then:*

(1) $\hat{F}(V) \in \mathcal{D}_2(\ell_{\mathcal{N}}^2; \ell_{\mathcal{B}}^2)$.

(2) $\rho_V^{-1} G \in \mathcal{U}_2(\ell_{\mathcal{N}}^2; \ell_{\mathcal{B}}^2)$.

(3)
$$\langle F, \rho_V^{-1} G \rangle_{HS} = \mathrm{trace}\{G^* \hat{F}(V)\}. \tag{2.2}$$

PROOF. See Lemma 7.4 and Theorem 7.2 of [ADD]. ∎

The second and third conclusions of Theorem 2.1 serve to identify ρ_V^{-1} as a reproducing kernel for the Hilbert space $\mathcal{U}_2$.

In the sequel we shall often deal with matrices

$$F = [F_{ij}], \qquad i = 1, \ldots, m; \ j = 1, \ldots, n,$$

of operators $F_{ij} \in \mathcal{U}$. Such an F maps an n-fold copy $\ell_{\mathcal{N}}^2 \oplus \cdots \oplus \ell_{\mathcal{N}}^2$ of $\ell_{\mathcal{N}}^2$ into an m-fold copy of $\ell_{\mathcal{B}}^2$ by the usual rules of matrix multiplication. Let

$$\mathbf{Z} = \mathrm{diag}(Z, \ldots, Z) \qquad (m \text{ times})$$

and

$$\boldsymbol{W} = \mathrm{diag}(W, \ldots, W) \qquad (m \text{ times}).$$

Then, in a selfevident notation,

$$F = \left[\sum_{t=0}^{\infty} Z^t (F_{ij})_{[t]} \right]$$

$$= \sum_{t=0}^{\infty} \boldsymbol{Z}^t \left[(F_{ij})_{[t]} \right]$$

$$= \sum_{t=0}^{\infty} \boldsymbol{Z}^t F_{[t]} \ .$$

Accordingly, we define

$$\hat{F}(\boldsymbol{W}) = \sum_{t=0}^{\infty} \boldsymbol{W}^{[t]} F_{[t]} = [\hat{F}_{ij}(W)] \ ,$$

when $\ell_{\boldsymbol{W}} < 1$, with a selfevident adaptation when the entries in $\boldsymbol{W}$ are not all the same size (as in Theorem 2.3 below).

THEOREM 2.2. *If* $F \in \mathcal{U}(\ell_{\mathcal{N}}^2; \ell_{\mathcal{B}}^2)$, $G \in \mathcal{D}_2(\ell_{\mathcal{N}}^2; \ell_{\mathcal{B}}^2)$ *and* $V \in \mathcal{D}(\ell_{\mathcal{B}}^2; \ell_{\mathcal{B}}^2)$ *with* $\ell_V < 1$, *then*

$$\underline{p}F^* \rho_V^{-1} G = \hat{F}(V)^* \rho_V^{-1} G \ . \tag{2.3}$$

PROOF. See Theorem 7.3 of [ADD].   ∎

Let $\mathcal{U}_2^{2\times 1}$ [resp., $\mathcal{L}_2^{2\times 1}$, $\mathcal{D}_2^{2\times 1}$] denote the set of block columns $\underline{\underline{F}} = \begin{bmatrix} F_1 \\ F_2 \end{bmatrix}$ with $F_1 \in \mathcal{U}_2(\ell_{\mathcal{B}}^2; \ell_{\mathcal{M}}^2)$ and $F_2 \in \mathcal{U}_2(\ell_{\mathcal{B}}^2; \ell_{\mathcal{N}}^2)$, [resp., $\mathcal{L}_2$, $\mathcal{D}_2$]. $\mathcal{U}_2^{2\times 1}$ is a Hilbert space with inner product

$$\langle \underline{\underline{F}}, \underline{\underline{G}} \rangle_{HS} = \langle F_1, G_1 \rangle_{HS} + \langle F_2, G_2 \rangle_{HS}$$

$$= \mathrm{trace}\{\underline{\underline{G}}^* \underline{\underline{F}}\}.$$

The corresponding (indefinite) J inner product based on J as in (1.6) is then given by the rule:

$$\langle J\underline{\underline{F}}, \underline{\underline{G}} \rangle_{HS} = \langle F_1, G_1 \rangle_{HS} - \langle F_2, G_2 \rangle_{HS}.$$

Next, for every operator Θ in $\mathcal{A}$, we define

$$\mathcal{H}(\Theta) = \mathcal{U}_2^{2\times 1} \boxminus \Theta \mathcal{U}_2^{2\times 1},$$

where $\boxminus$ indicates that the orthogonal complement is taken with respect to the J inner product.

The spaces $\mathcal{H}(\Theta)$ are analogues of a class of reproducing kernel Hilbert spaces which were first introduced by de Branges [dB1] in a more classical setting.

THEOREM 2.3. *If $\Theta \in \mathcal{A}$, then $\mathcal{H}(\Theta)$ is a reproducing kernel Hilbert space with respect to the J inner product in $\mathcal{X}_2^{2 \times 1}$. Its reproducing kernel*

$$K_W = \{J - \Theta J \hat{\Theta}(W)^*\} \begin{bmatrix} \rho_{W_1} & 0 \\ 0 & \rho_{W_2} \end{bmatrix}^{-1} \tag{2.4}$$

for $W = \mathrm{diag}(W_1, W_2)$ with $W_1 \in \mathcal{D}(\ell_{\mathcal{M}}^2; \ell_{\mathcal{M}}^2)$, $W_2 \in \mathcal{D}(\ell_{\mathcal{N}}^2; \ell_{\mathcal{N}}^2)$ and $\ell_W < 1$.

PROOF. This is easily adapted Theorem 7.6 in [ADD].   ∎

Here
$$\hat{\Theta}(W) \;=\; \begin{bmatrix} \hat{\Theta}_{11}(W_1) & \hat{\Theta}_{12}(W_1) \\ \hat{\Theta}_{21}(W_2) & \hat{\Theta}_{22}(W_2) \end{bmatrix}. \tag{2.5}$$

Finally, seems well to recall some plications of condition I [resp. III] on the operator Λ [resp. Δ].

THEOREM 2.4. *Let $\Lambda \in \{\mathcal{D}(\ell_{\mathcal{B}}^2; \ell_{\mathcal{B}}^2)\}^{n \times n}$. Then the following are equivalent:*

(1)   Condition I is in force.

(2)   There exists an $\epsilon > 0$ such that

$$\langle \Lambda f, f \rangle \geq \epsilon \langle f, f \rangle$$

for every $f \in (\ell_{\mathcal{B}}^2)^n$.

(3)   There exists an $\epsilon > 0$ such that

$$\langle \Lambda E, E \rangle_{HS} \geq \epsilon \langle E, E \rangle_{HS}$$

for every $E \in \{\mathcal{D}_2(\ell_{\mathcal{B}}^2; \ell_{\mathcal{B}}^2)\}^{n \times 1}$. Moreover, if the inequality in (2) holds for some $\epsilon > 0$, then so does the inequality in (3), and vice versa.

(4) $\Lambda^{-1} \in \{\mathcal{D}(\ell_{\mathcal{B}}^2; \ell_{\mathcal{B}}^2)\}^{n \times n}$.

PROOF. (1) $\Longleftrightarrow$ (2) is the definition of uniform positivity.

(2) $\Longleftrightarrow$ (3) (and more) is established in the corollary to Lemma 8.4 of [ADD].

(2) $\Longleftrightarrow$ (4). It is well known that (2) holds if and only if Λ admits a bounded inverse, i.e., if and only if $\Lambda^{-1} \in \{\mathcal{X}(\ell_{\mathcal{B}}^2; \ell_{\mathcal{B}}^2)\}^{n \times n}$. If $n = 1$, it is easily checked that

$\Lambda^{-1} \in \mathcal{D}$. The case for general n then follows readily by induction with the help of the well known formula (see e.g., (0.8) of [D1]) for the inverse of a matrix in terms of Schur complements. ∎

3. COLLIGATIONS AND CHARACTERISTIC FUNCTIONS

In this section we shall show that a certain partially specified operator admits a $\tilde{J}$ unitary extension if and only if a certain associated operator admits a unitary extension. Conditions for such a unitary extension to exist are also formulated. When these conditions are met, the characteristic function associated with the extension will be J unitary. In the next section we shall further specialize these results to a form which will be useful for solving the interpolation problems under study. For the moment we shall work in the general setting of a matrix M of bounded linear operators M_{ij} from a Hilbert space $\mathcal{H}_j$ into a Hilbert space $\mathcal{H}_i$, $i,j = 1,2,3$, and we set

$$J = \begin{bmatrix} I_{\mathcal{H}_2} & 0 \\ 0 & -I_{\mathcal{H}_3} \end{bmatrix}, \quad \text{and} \quad \tilde{J} = \begin{bmatrix} I_{\mathcal{H}_1} & 0 \\ 0 & J \end{bmatrix}.$$

THEOREM 3.1. *Let* $\mathcal{H}_i$, $i = 1,2,3$, *be Hilbert spaces and let*

$$M_1 = \begin{bmatrix} M_{11} \\ M_{21} \\ M_{31} \end{bmatrix}$$

be a given operator from $\mathcal{H}_1$ *into* $\mathcal{H}_1 \oplus \mathcal{H}_2 \oplus \mathcal{H}_3$ *such that*

$$M_1^* \tilde{J} M_1 = I_{\mathcal{H}_1}.$$

Then M_1 *admits a* $\tilde{J}$ *unitary extension*

$$M = \begin{bmatrix} M_{11} & M_{12} & M_{13} \\ M_{21} & M_{22} & M_{23} \\ M_{31} & M_{32} & M_{33} \end{bmatrix}$$

with entries M_{ij} *which are bounded linear operators from* $\mathcal{H}_j$ *into* $\mathcal{H}_i$ *if and only if it admits such a* $\tilde{J}$ *unitary extension with* $M_{32} = 0$.

PROOF. Suppose first that M_1 admits a general $\tilde{J}$ unitary extension M of the indicated form without any additional constraints on M_{32}. Then it is readily checked that

$$M_{13}^* M_{13} + M_{23}^* M_{23} - M_{33}^* M_{33} = -I,$$
$$M_{31} M_{31}^* + M_{32} M_{32}^* - M_{33} M_{33}^* = -I,$$

and hence that

$$M_{33}^* M_{33} \geq I + M_{23}^* M_{23} \geq I$$

and

$$M_{33} M_{33}^* \geq I + M_{32} M_{32}^* \geq I \ .$$

Therefore both M_{33} and M_{33}^* have closed range and zero kernel. Thus they are both one to one maps of $\mathcal{H}_3$ onto itself and so are invertible with bounded inverses, as are the positive operators $M_{33}^* M_{33}$ and $(M_{33}^* M_{33})^{-1}$. By the last line of inequalities, the operator

$$K^* = -M_{33}^{-1} M_{32}$$

is subject to the bound

$$I - K^* K \geq (M_{33}^* M_{33})^{-1} \geq \epsilon I$$

for some $\epsilon > 0$. This proves that K is strictly contractive: $\|K\| < 1$. The same holds for K^* and therefore the Halmos extension

$$H(K) = \begin{bmatrix} I_{\mathcal{H}_2} & K \\ K^* & I_{\mathcal{H}_3} \end{bmatrix} \begin{bmatrix} (I - KK^*)^{-\frac{1}{2}} & 0 \\ 0 & (I - K^* K)^{-\frac{1}{2}} \end{bmatrix}$$

is a well defined J unitary operator from $\mathcal{H}_2 \oplus \mathcal{H}_3$ onto itself. Therefore

$$M \begin{bmatrix} I_{\mathcal{H}_1} & 0 \\ 0 & H(K) \end{bmatrix}$$

is $\tilde{J}$ unitary (since both factors are) with 32 entry equal to zero, by the choice of K.

This completes the proof, since the converse is selfevident.   ∎

THEOREM 3.2. *Let $\mathcal{H}_i$, $i = 1, 2, 3$, and M_1 be as in Theorem 3.1. Then M_1 admits a $\tilde{J}$ unitary extension from $\mathcal{H}_1 \oplus \mathcal{H}_2 \oplus \mathcal{H}_3$ onto itself if and only if there exists a pair of bounded linear operators M_{12} from $\mathcal{H}_2$ into $\mathcal{H}_1$ and M_{22} from $\mathcal{H}_2$ into itself such that the operator matrix*

$$M = \begin{bmatrix} M_{11} & M_{12} & M_{13} \\ M_{21} & M_{22} & M_{23} \\ M_{31} & 0 & M_{33} \end{bmatrix}$$

with the third block column specified in terms of the first by the formulas

$$M_{13} = M_{11} M_{31}^* (I + M_{31} M_{31}^*)^{-\frac{1}{2}} \tag{3.1}$$

$$M_{23} = M_{21} M_{31}^* (I + M_{31} M_{31}^*)^{-\frac{1}{2}} \tag{3.2}$$

$$M_{33} = (I + M_{31} M_{31}^*)^{\frac{1}{2}} \tag{3.3}$$

is $\tilde{J}$ unitary.

PROOF. If M_1 admits a $\tilde{J}$ unitary extension M to $\mathcal{H}_1 \oplus \mathcal{H}_2 \oplus \mathcal{H}_3$, then, by the preceding theorem it admits an extension with $M_{32} = 0$. But, for such an extension, the identity

$$M\tilde{J} \begin{bmatrix} M_{31}^* \\ 0 \\ M_{33}^* \end{bmatrix} = \begin{bmatrix} 0 \\ 0 \\ -I \end{bmatrix}$$

determines the entries in the third block column in terms of the entries of the first block column M_1, upto a right unitary factor U from $\mathcal{H}_3$ onto itself:

$$M_{13} = M_{13} M_{31}^* M_{33}^{-*}$$

$$M_{23} = M_{21} M_{31}^* M_{33}^{-*}$$

$$M_{33} = (I + M_{31} M_{31}^*)^{-\frac{1}{2}} U \ .$$

Formulas (3.1)–(3.3) are obtained by multiplying this $\tilde{J}$ unitary matrix on the right by the $\tilde{J}$ unitary matrix $I_{\mathcal{H}_1} \oplus I_{\mathcal{H}_2} \oplus U^*$. This proves that if M_1 admits a $\tilde{J}$ unitary extension to $\mathcal{H}_1 \oplus \mathcal{H}_2 \oplus \mathcal{H}_3$, then there exist a choice of M_{12} and M_{22} such that the operator matrix with first and third column as specified in the statement of the theorem and $M_{32} = 0$ is $\tilde{J}$ unitary.

This completes the proof, since the converse is selfevident. ∎

THEOREM 3.3. *Let $\mathcal{H}_i$, $i = 1, 2, 3$ be Hilbert spaces and let*

$$M = \begin{bmatrix} M_{11} & M_{12} & M_{13} \\ M_{21} & M_{22} & M_{23} \\ M_{31} & 0 & M_{33} \end{bmatrix}$$

be an operator matrix with entries M_{ij} which are bounded linear operators from $\mathcal{H}_j$ into $\mathcal{H}_i$ such that

$$M_{11}^* M_{11} + M_{21}^* M_{21} - M_{31}^* M_{31} = I_{\mathcal{H}_1}$$

and the third column is specified in terms of the first by formulas (3.1) to (3.3). Further, let N be a bounded linear invertible operator from $\mathcal{H}_1$ onto itself such that

$$N^*(I + M_{31}^* M_{31})N = I \ ,$$

and let

$$N = \begin{bmatrix} N_{11} & M_{12} \\ N_{21} & M_{22} \end{bmatrix}$$

be the operator matrix with

$$N_{11} = M_{11}N$$

$$N_{21} = M_{21}N \ .$$

Then

(1) $\quad M^* \tilde{J} M = \tilde{J} \iff N^* N = I$ *on* $\mathcal{H}_1 \oplus \mathcal{H}_2.$

(2) $\quad M \tilde{J} M^* = \tilde{J} \iff N N^* = I$ *on* $\mathcal{H}_1 \oplus \mathcal{H}_2.$

PROOF. The proof is largely computational. It is broken into steps.

STEP 1. $M^* \tilde{J} M = \tilde{J}$ *if and only if the following three equations are satisfied.*

$$M_{11}^* M_{11} + M_{21}^* M_{21} - M_{31}^* M_{31} = I_{\mathcal{H}_1} \tag{3.4}$$

$$M_{12}^* M_{11} + M_{22}^* M_{21} = 0 \tag{3.5}$$

$$M_{12}^* M_{12} + M_{22}^* M_{22} = I_{\mathcal{H}_2} \ . \tag{3.6}$$

PROOF OF STEP 1. By direct calculation it is readily seen that $M^* \tilde{J} M = \tilde{J}$ if and only if (3.4), (3.5) and (3.6) and the following supplementary three equations are satisfied:

$$M_{13}^* M_{11} + M_{23}^* M_{21} - M_{33}^* M_{31} = 0 \tag{3.7}$$

$$M_{13}^* M_{12} + M_{23}^* M_{22} = 0 \tag{3.8}$$

$$M_{13}^* M_{13} + M_{23}^* M_{23} - M_{33}^* M_{33} = -I_{\mathcal{H}_3} \ . \tag{3.9}$$

To complete the proof it therefore suffices to show that (because of the special choice of M_{13}, M_{23} and M_{33}) the last three equations are automatically satisfied whenever the first three are. To this end observe that the left hand side of (3.7) is equal to

$$(I + M_{31} M_{31}^*)^{-\frac{1}{2}} M_{31}(M_{11}^* M_{11} + M_{21}^* M_{21} - I - M_{31}^* M_{31}) \ ,$$

whereas the left hand side of (3.8) is equal to

$$(I + M_{31} M_{31}^*)^{-\frac{1}{2}} M_{31}(M_{11}^* M_{12} + M_{21}^* M_{22}) \ .$$

Thus (3.4) $\implies$ (3.7) and (3.5) $\implies$ (3.8). Finally, the left hand side of (3.9) can be reexpressed as

$$\{(M_{13}^* M_{11} + M_{23}^* M_{21})M_{31}^* - M_{33}^*(I + M_{31}M_{31}^*)\}(I + M_{31}M_{31}^*)^{-\frac{1}{2}}$$

$$= \{(M_{13}^* M_{11} + M_{23}^* M_{21} - M_{33}^* M_{31})M_{31}^* - M_{33}^*\}(I + M_{31}M_{31}^*)^{-\frac{1}{2}}$$

$$= -M_{33}^*(I + M_{31}M_{31}^*)^{-\frac{1}{2}}$$

$$= -I$$

by (3.7) and the definition of M_{33}. Since (3.7) follows from (3.4) by the first argument in this step, the proof of the step is complete.

STEP 2. $M^* \check{J} M = \check{J}$ *if and only if* N *is isometric.*

PROOF OF STEP 2. Clearly N is isometric if and only if the following three equations are satisfied:

$$N_{11}^* N_{11} + N_{21}^* N_{21} = I_{\mathcal{H}_1} \tag{3.10}$$

$$M_{12}^* N_{11} + M_{22}^* N_{21} = 0 \tag{3.11}$$

$$M_{12}^* M_{12} + M_{22}^* M_{22} = I_{\mathcal{H}_2} . \tag{3.12}$$

In view of Step 1, it suffices to show that these three equations hold if and only if (3.4), (3.5) and (3.6) hold.

Suppose first that (3.4) holds. Then

$$N_{11}^* N_{11} + N_{21}^* N_{21} = N^* \{M_{11}^* M_{11} + M_{21}^* M_{21}\} N$$

$$= N^* \{I + M_{31}^* M_{31}\} N$$

$$= I .$$

Thus (3.4) $\implies$ (3.10). On the other hand, if (3.10) holds, then

$$N^* \{M_{11}^* M_{11} + M_{21}^* M_{21} - (I + M_{31}^* M_{31})\} N$$

$$= N_{11}^* N_{11} + N_{21}^* N_{21} - I$$

$$= 0 .$$

Therefore, since N is invertible, the term inside the curly bracket must also vanish. This proves that (3.10) $\implies$ (3.4).

It is easily checked that (3.5) $\implies$ (3.11) and (3.6) $\implies$ (3.12) to complete the proof.

STEP 3. $M \tilde{J} M^* = \tilde{J}$ *if and only if the following three equations are satisfied:*

$$M_{11} M_{11}^* + M_{12} M_{12}^* - M_{13} M_{13}^* = I_{\mathcal{H}_1} \tag{3.13}$$

$$M_{21} M_{11}^* + M_{22} M_{12}^* - M_{23} M_{13}^* = 0 \tag{3.14}$$

$$M_{21} M_{21}^* + M_{22} M_{22}^* - M_{23} M_{23}^* = I_{\mathcal{H}_2} \; . \tag{3.15}$$

PROOF OF STEP 3. By direct computation it is readily seen that $M \tilde{J} M^* = \tilde{J}$ if and only if (3.13), (3.14), (3.15) and the following supplementary three equations are satisfied:

$$M_{31} M_{11}^* - M_{33} M_{13}^* = 0 \tag{3.16}$$

$$M_{31} M_{21}^* - M_{33} M_{23}^* = 0 \tag{3.17}$$

$$M_{31} M_{31}^* - M_{33} M_{33}^* = I_{\mathcal{H}_3} \; . \tag{3.18}$$

However, because of the special choice of M_{13}, M_{23} and M_{33}, the last three equations are automatically satisfied, and therefore the assertion is selfevident.

STEP 4. $M \tilde{J} M^* = \tilde{J}$ *if and only if* N *is coisometric.*

PROOF. N is coisometric if and only if the following three sets of equations are satisfied:

$$N_{11} N_{11}^* + M_{12} M_{12}^* = I_{\mathcal{H}_1} \tag{3.19}$$

$$N_{21} N_{11}^* + M_{22} M_{12}^* = 0 \tag{3.20}$$

$$N_{21} N_{21}^* + M_{22} M_{22}^* = I_{\mathcal{H}_2} \; . \tag{3.21}$$

Therefore, in view of Step 3, it suffices to show that this last set of three equations hold if and only if (3.13), (3.14) and (3.15) hold. More precisely, we shall show that (3.13) $\Longleftrightarrow$ (3.19), (3.14) $\Longleftrightarrow$ (3.20) and (3.15) $\Longleftrightarrow$ (3.21) by identifying the left hand sides of the indicated pairs of equations. This rests on the special choice of M_{13} and M_{23}. Thus, the left hand side of (3.13) is equal to

$$M_{11} \{ I - M_{31}^* (I + M_{31} M_{31}^*)^{-1} M_{31} \} M_{11}^* + M_{12} M_{12}^*$$

$$= M_{11} (I - M_{31} M_{31}^*)^{-1} M_{11}^* + M_{12} M_{12}^*$$

$$= M_{11} N \{ N^* (I - M_{31} M_{31}^*) N \}^{-1} N^* M_{11}^* + M_{12} M_{12}^*$$

$$= N_{11} N_{11}^* + M_{12} M_{12}^* \; ,$$

which is equal to the left hand side of (3.19), and hence (3.13) $\implies$ (3.19).

Similarly,

$$M_{21}M_{11}^* - M_{23}M_{13}^* = M_{21}\{I - M_{31}^*(I + M_{31}M_{31}^*)^{-1}M_{31}\}M_{11}^*$$

$$= M_{21}(I - M_{31}^*M_{31})^{-1}M_{11}^*$$

$$= N_{21}N_{11}^*$$

and

$$M_{21}M_{21}^* - M_{23}M_{23}^* = N_{21}N_{21}^*$$

and therefore (3.14) $\iff$ (3.20) and (3.15) $\iff$ (3.21), as asserted. $\blacksquare$

In the sequel we shall be interested in establishing the existence of a $\tilde{J}$ unitary operator M of the form given in the preceding theorem starting from a given set of operators M_{11}, M_{21} and M_{31} which satisfy (3.4). Since M_{13}, M_{23} and M_{33} are also specified in the statement of the theorem, the existence of such a $\tilde{J}$ unitary M reduces to the existence of an appropriate choice of M_{12} and M_{22}. The point of the theorem is to show that this is equivalent to choosing M_{12} and M_{22} so as to make the operator N with a specified isometric first column $\begin{bmatrix} N_{11} \\ N_{21} \end{bmatrix}$ unitary. This easier problem is discussed in the next theorem. The claimed isometry is due to the fact that (3.4) is equivalent to (3.10), as was shown in the proof of Step 2.

THEOREM 3.4. *Let M_1 be a $\tilde{J}$ isometric map of $\mathcal{H}_1$ into $\mathcal{H}_1 \oplus \mathcal{H}_2 \oplus \mathcal{H}_3$. Then M_1 admits a $\tilde{J}$ unitary extension to all of $\mathcal{H}_1 \oplus \mathcal{H}_2 \oplus \mathcal{H}_3$ if and only if*

$$\dim \ker[M_{11}^* \ \ M_{21}^*] \ = \ \dim \mathcal{H}_2 \ .$$

PROOF. By Theorems 3.1–3.3, M_1 admits a J unitary extension to all of $\mathcal{H}_1 \oplus \mathcal{H}_2 \oplus \mathcal{H}_3$ if and only if the isometry

$$N_1 \ = \ \begin{bmatrix} N_{11} \\ N_{21} \end{bmatrix}$$

from $\mathcal{H}_1$ into $\mathcal{H}_1 \oplus \mathcal{H}_2$ which is specified in Theorem 3.3 admits a unitary extension to all of $\mathcal{H}_1 \oplus \mathcal{H}_2$. But, as is well known (see e.g., Akhiezer and Glazman [AG]), this is possible if and only if N_1 has equal deficiency indices:

$$\dim (\text{dom } N_1)^{\perp} = \dim (\text{ran } N_1)^{\perp} \ ,$$

i.e., if and only if

$$\dim \mathcal{H}_2 = \dim \ker N_1^* \ .$$

This completes the proof, since

$$\dim \ker N_1^* = \dim \ker[M_{11}^* \quad M_{31}^*] .$$

LEMMA 3.1. *If, in the setting of Theorem 3.1, M is $\tilde{J}$ unitary, then*

$$\|M_{11} - M_{13}M_{33}^{-1}M_{31}\| \le 1 .$$

PROOF. Let

$$A = \begin{bmatrix} M_{11} & M_{12} \\ M_{21} & M_{22} \end{bmatrix} , \qquad B = \begin{bmatrix} M_{13} \\ M_{23} \end{bmatrix} ,$$

$$C = [M_{31} \quad M_{32}] \quad \text{and} \quad D = M_{33} .$$

Then, by Lemma 5.2 of [ADD], D is invertible with a bounded inverse and

$$\|A - BD^{-1}C\| \le 1 .$$

Therefore in each of the four block entries the 2×2 operator matrix $A - BD^{-1}C$ is contractive. The rest is easy, since the 11 block entry of $A - BD^{-1}C$ is equal to $M_{11} - M_{13}M_{33}^{-1}M_{31}$. ∎

THEOREM 3.5. *Let M be defined as in Theorem 3.1, let U be any unitary operator on $\mathcal{H}_1$ such that $I - UM_{11}$ is a bounded invertible operator on $\mathcal{H}_1$ with bounded inverse and let*

$$\Theta = \begin{bmatrix} M_{22} & M_{23} \\ M_{32} & M_{33} \end{bmatrix} + \begin{bmatrix} M_{21} \\ M_{31} \end{bmatrix} (I - UM_{11})^{-1}U[M_{12} \quad M_{13}] . \tag{3.22}$$

Then:

(1) $M^* \tilde{J} M = \tilde{J} \implies \Theta^* J \Theta = J$.

(2) $M \tilde{J} M^* = \tilde{J} \implies \Theta J \Theta^* = J$.

PROOF. The proof is well known but is incorporated for the sake of completeness; see e.g., Brodsky [Br] for the underlying idea.

It is convenient to rewrite M and Θ in the block forms

$$M = \begin{bmatrix} A & B \\ C & D \end{bmatrix}$$

and

$$\Theta = D + C(I - UA)^{-1}UB ,$$

where

$$A = M_{11}, \qquad\qquad B = [M_{12} \ \ M_{13}]$$

$$C = \begin{bmatrix} M_{21} \\ M_{31} \end{bmatrix} \quad \text{and} \quad D = \begin{bmatrix} M_{22} & M_{23} \\ M_{32} & M_{33} \end{bmatrix}.$$

Suppose now that $M^*\tilde{J}M = \tilde{J}$. Then clearly

$$A^*A + C^*JC = I$$

$$B^*A + D^*JC = 0$$

$$B^*B + D^*JD = J.$$

Therefore

$$\begin{aligned}
\Theta^*J\Theta &= D^*JD + B^*U^*(I - A^*U^*)^{-1}C^*JD \\
&\quad + D^*JC(I - UA)^{-1}UB \\
&\quad + B^*U^*(I - A^*U^*)^{-1}C^*JC(I - UA)^{-1}UB \\
&= J - B^*B - B^*U^*(I - A^*U^*)^{-1}A^*B \\
&\quad - B^*A(I - UA)^{-1}UB \\
&\quad + B^*U^*(I - A^*U^*)^{-1}(I - A^*A)(I - UA)^{-1}UB \\
&= J - B^*U^*(I - A^*U^*)^{-1}K(I - UA)^{-1}UB,
\end{aligned}$$

where

$$\begin{aligned}
K =&(I - A^*U^*)U^*U(I - UA) + A^*U^*(I - UA) \\
&+ (I - A^*U^*)UA + A^*A - I
\end{aligned}$$

is readily seen to vanish, since U is unitary. This completes the proof of (1).

Suppose next that $M\tilde{J}M^* = \tilde{J}$. Then

$$AA^* + BJB^* = I$$

$$CA^* + DJB^* = 0$$

$$CC^* + DJD^* = J.$$

Therefore

$$\Theta J \Theta^* = DJD^* + C(I - UA)^{-1}UBJD^*$$
$$+ DJB^*U^*(I - A^*U^*)^{-1}C^*$$
$$+ C(I - UA)^{-1}UBJB^*U^*(I - A^*U^*)^{-1}C^*$$
$$= J - CC^* - C(I - UA)^{-1}UAC^* - CA^*U^*(I - A^*U^*)^{-1}C^*$$
$$+ C(I - UA)^{-1}U(I - AA^*)U^*(I - A^*U^*)^{-1}C^*$$
$$= J - C(I - UA)^{-1}ULU^*(I - A^*U^*)^{-1}C^*$$

with

$$L = U^*(I - UA)(I - A^*U^*)U + A(I - A^*U^*)U$$
$$+ U^*(I - UA)A^* + AA^* - I$$
$$= 0 \ ,$$

much as before. This completes the proof of (2). ∎

4. TOWARDS INTERPOLATION

In this section we shall begin to specialize the general formulas of the preceding section to a form which will be useful in applications to interpolation. We begin with a general result.

THEOREM 4.1. *Let*
$$\begin{bmatrix} A & B \\ C & D \end{bmatrix}$$
be a $\tilde{J}$ coisometric matrix of operators with components

$$A = [A_{ij}], \quad i,j = 1,\ldots,n, \quad with \quad A_{ij} \in \mathcal{D}(\ell_{\mathcal{B}}^2; \ell_{\mathcal{B}}^2) \ ,$$

$$B = [B_{ij}], \quad i = 1,\ldots,n, \quad j = 1,2, \quad with \quad B_{i1} \in \mathcal{D}(\ell_{\mathcal{M}}^2; \ell_{\mathcal{B}}^2) \ ,$$

$$and \quad B_{i2} \in \mathcal{D}(\ell_{\mathcal{N}}^2; \ell_{\mathcal{B}}^2) \quad for \quad i = 1,\ldots,n \ ,$$

$$C = [C_{ij}], \quad i = 1,2, \quad j = 1,\ldots,n, \quad with \quad C_{1j} \in \mathcal{D}(\ell_{\mathcal{B}}^2; \ell_{\mathcal{M}}^2) \ ,$$

$$and \quad C_{2j} \in \mathcal{D}(\ell_{\mathcal{B}}^2; \ell_{\mathcal{N}}^2) \quad for \quad j = 1,\ldots,n \ ,$$

$$D = [D_{ij}], \quad i,j = 1,2, \quad with \quad D_{ij} \in \mathcal{D}(\mathcal{H}_j; \mathcal{H}_i)$$

$$\mathcal{H}_1 = \ell_{\mathcal{M}}^2 \quad and \quad \mathcal{H}_2 = \ell_{\mathcal{N}}^2 \ ,$$

and let

$$\Theta = D + GZB , \tag{4.1}$$

where

$$G = C(I - \boldsymbol{Z}A)^{-1} , \tag{4.2}$$

$$\boldsymbol{Z} = \mathrm{diag}\,(Z,\ldots,Z) ,$$

(with as many components as needed to make the formula meaningful) and it is further assumed that

$$r_{sp}(\boldsymbol{Z}A) < 1 .$$

Then

$$G\hat{G}(\boldsymbol{W})^* = \{J - \Theta J\hat{\Theta}(\boldsymbol{W})^*\}(I - \boldsymbol{Z}\boldsymbol{W}^*)^{-1} \tag{4.3}$$

for every choice of

$$\boldsymbol{W} = \mathrm{diag}\,(W_1, W_2)$$

with

$$W_1 \in \mathcal{D}(\ell^2_{\mathcal{M}}; \ell^2_{\mathcal{M}}) , \quad W_2 \in \mathcal{D}(\ell^2_{\mathcal{N}}; \ell^2_{\mathcal{N}}) \ \ and \ \ r_{sp}(\boldsymbol{Z}\boldsymbol{W}) < 1 .$$

(In the last line $\boldsymbol{Z} = \mathrm{diag}(Z, Z)$, where the first Z is based on $I_{\mathcal{M}}$ and the second on $I_{\mathcal{N}}$.)

PROOF. By (1.2),

$$\begin{aligned}
\hat{\Theta}(\boldsymbol{W}) &= D + (G\boldsymbol{Z})^{\wedge}(\boldsymbol{W})B \\
&= D + \{\hat{G}(\boldsymbol{W})\boldsymbol{Z}\}^{\wedge}(\boldsymbol{W})B \\
&= D + \{\boldsymbol{Z}\,\hat{G}(\boldsymbol{W})^{(1)}\}^{\wedge}(\boldsymbol{W})B \\
&= D + \boldsymbol{W}\hat{G}(\boldsymbol{W})^{(1)}B .
\end{aligned}$$

Thus

$$\begin{aligned}
\Theta J\hat{\Theta}(\boldsymbol{W})^* &= (D + G\boldsymbol{Z}B)J(D^* + B^*\{\hat{G}(\boldsymbol{W})^{(1)}\}^*\boldsymbol{W}^*) \\
&= DJD^* + G\boldsymbol{Z}BJD^* + DJB^*\{\hat{G}(\boldsymbol{W})^{(1)}\}^*\boldsymbol{W}^* \\
&\quad + G\boldsymbol{Z}BJB^*\{\hat{G}(\boldsymbol{W})^{(1)}\}^*\boldsymbol{W}^* \\
&= J - CC^* - G\boldsymbol{Z}AC^* - CA^*\{\hat{G}(\boldsymbol{W})^{(1)}\}^*\boldsymbol{W}^* \\
&\quad + G\boldsymbol{Z}\{\hat{G}(\boldsymbol{W})^{(1)}\}^*\boldsymbol{W}^* \\
&\quad - G\boldsymbol{Z}AA^*\{\hat{G}(\boldsymbol{W})^{(1)}\}^*\boldsymbol{W}^* \\
&= J - \textcircled{1} - \textcircled{2} - \textcircled{3} + \textcircled{4} - \textcircled{5} ,
\end{aligned}$$

in a selfevident notation.

Now, since

$$C + GZA = C\{I + (I - ZA)^{-1}ZA\}$$
$$= C(I - ZA)^{-1}$$
$$= G \, ,$$

it follows readily that

$$① \; + \; ② \; = GC^* \, ,$$
$$③ \; + \; ⑤ \; = GA^*\{\hat{G}(W)^{(1)}\}^*W^*$$

and hence that

$$\Theta J\hat{\Theta}(W)^* = J - GC^* - GA^*\{\hat{G}(W)^{(1)}\}^*W^* + G\hat{G}(W)^*ZW^* \, .$$

At the same time it follows from the formula

$$G = C(I - ZA)^{-1}$$
$$= C\sum_{j=0}^{\infty} Z^j\{A^{*[j]}\}^*$$
$$= \sum_{j=0}^{\infty} Z^j C^{(j)}\{A^{*[j]}\}^*$$

that

$$\hat{G}(W) = \sum_{j=0}^{\infty} W^{[j]}C^{(j)}\{A^{*[j]}\}^*$$
$$= C + W\hat{G}(W)^{(1)}A \, .$$

Therefore, for every W of the stated form, we have

$$\Theta J\hat{\Theta}(W)^* = J - G\hat{G}(W)^*(I - ZW^*) \, ,$$

which is clearly equivalent to the asserted formula.   ∎

THEOREM 4.2. *If, in the setting of the last theorem, the given operator matrix is $\tilde{J}$ unitary (and not just $\tilde{J}$ coisometric) and $r_{sp}(ZA) < 1$, then the operator Θ defined by (4.1) belongs to the class $\mathcal{A}$ of ULCS operators.*

PROOF. It is readily seen that the block entries of Θ belong to the requisite $\mathcal{U}$ spaces. By Theorem 3.5, Θ is J unitary and therefore, by Lemma 5.2 of [ADD], Θ_{22} is invertible in $\mathcal{X}(\ell^2_{\mathcal{N}}; \ell^2_{\mathcal{N}})$. It thus remains to show that $\Theta_{22}^{-1} \in \mathcal{U}(\ell^2_{\mathcal{N}}; \ell^2_{\mathcal{N}})$.

Since $r_{sp}(\boldsymbol{Z}\,A) < 1$, the "state variable" $x \in (\ell^2_{\mathcal{B}})^n$ in the system of equations

$$\begin{bmatrix} A & B \\ C & D \end{bmatrix} \begin{bmatrix} x \\ u \end{bmatrix} = \begin{bmatrix} \boldsymbol{Z}^* x \\ y \end{bmatrix}$$

can be eliminated to find an explicit expression for the "output"

$$y = \begin{bmatrix} a_2 \\ b_2 \end{bmatrix} \quad \text{in terms of the "input"} \quad u = \begin{bmatrix} a_1 \\ b_1 \end{bmatrix} ,$$

wherein $a_i \in \ell^2_{\mathcal{M}}$ and $b_i \in \ell^2_{\mathcal{N}}$ for $i = 1, 2$. More precisely, from the first row equation,

$$x = (I - \boldsymbol{Z}\,A)^{-1} \boldsymbol{Z} B u$$

and then, from the second,

$$y = \{D + C(I - \boldsymbol{Z}\,A)^{-1} \boldsymbol{Z}\,B\} u$$

$$= \Theta u \ .$$

In particular,

$$b_2 = \Theta_{21} a_1 + \Theta_{22} b_1 \ .$$

Since the operators A, B, C and D are matrices of diagonal operators, the original system of equations holds in each coordinate:

$$\begin{bmatrix} (A)_{jj} & (B)_{jj} \\ (C)_{jj} & (D)_{jj} \end{bmatrix} \begin{bmatrix} x_j \\ u_j \end{bmatrix} = \begin{bmatrix} x_{j-1} \\ y_j \end{bmatrix}$$

for $j = \ldots, -1, 0, 1, \ldots$. Therefore, since the operator matrix on the right is $(\tilde{J})_{jj}$ unitary, it is readily checked that

$$\|x_{j-1}\|^2 + \|(a_2)_j\|^2 - \|(b_2)_j\|^2 = \|x_j\|^2 + \|(a_1)_j\|^2 - \|(b_1)_j\|^2 \ ,$$

for $j = \ldots, -1, 0, 1, \ldots$. Thus

$$\|x_t\|^2 + \sum_{j=t+1}^{\infty} \|(a_2)_j\|^2 + \sum_{j=t+1}^{\infty} \|(b_1)_j\|^2 = \sum_{j=t+1}^{\infty} \|(a_1)_j\|^2 + \sum_{j=t+1}^{\infty} \|(b_2)_j\|^2$$

for every integer t. Now, if $a_1 = 0$, then

$$b_1 = \Theta_{22}^{-1} b_2 \ .$$

Moreover, if also $(b_2)_j = 0$ for $j > t$, then, by the preceding conservation of energy law, $(b_1)_j = 0$ for $j > t$. In other words, in terms of the orthogonal projector P_t on $\ell^2_{\mathcal{N}}$ which is defined by the diagonal operator with components

$$(P_t)_{jj} = \begin{cases} I_{\mathcal{N}} & \text{for} \quad j \le t \\ 0 & \text{for} \quad j > t \, , \end{cases}$$

and the complementary projection

$$P_t = I - Q_t \, ,$$

we have shown that

$$b_2 = P_t b_2 \implies Q_t b_1 = 0 \, .$$

But this in turns implies that for any $b_2 \in \ell^2_{\mathcal{N}}$,

$$Q_t \Theta_{22}^{-1} b_2 = Q_t \Theta_{22}^{-1} P_t b_2 + Q_t \Theta_{22}^{-1} Q_t b_2$$

$$= Q_t \Theta_{22}^{-1} Q_t b_2$$

and hence that

$$Q_t \Theta_{22}^{-1} = Q_t \Theta_{22}^{-1} Q_t$$

for every integer t. But this is the same as to say that $\Theta_{22}^{-1} \in \mathcal{U}$. $\blacksquare$

We remark that from a physical point of view it is more natural to work with row sequences on the left of the operator, rather than columns on the right, because then the traditional picture of causality fits with upper triangular operators.

In terms of our previous notation, the $A, \ldots, D$ in the last two theorems, can be identified as

$$A = M_{11} \, , \qquad B = [M_{12} \quad M_{13}] \, ,$$

$$C = \begin{bmatrix} M_{21} \\ M_{31} \end{bmatrix} \quad \text{and} \quad D = \begin{bmatrix} M_{22} & M_{23} \\ M_{23} & M_{33} \end{bmatrix} \, .$$

In the subsequent applications, we shall always specify the first block column of M in terms of a given set of blocks

$$V \in \{\mathcal{D}(\ell^2_B; \ell^2_B)\}^{n \times n} \, , \qquad \alpha \in \{\mathcal{D}(\ell^2_{\mathcal{M}}; \ell^2_B)\}^{n \times 1} \quad \text{and} \quad \beta \in \{\mathcal{D}(\ell^2_{\mathcal{N}}; \ell^2_B)\}^{n \times 1} \, ,$$

which come from the data and an operator

$$\Lambda = \sum_{j=0}^{\infty} V^{[j]} (\alpha\alpha^* - \beta\beta^*)^{(j)} V^{[j]*} \, , \tag{4.4}$$

which will play the role of the Pick matrix. It will always be assumed that (in a selfevident adaptation of the usual notation)

$$\ell_V := \lim_{n \uparrow \infty} \| V^{[n]} \|^{1/n} < 1 \, ,$$

so that the indicated sum converges. Since

$$\Lambda = \alpha\alpha^* - \beta\beta^* + \sum_{j=0}^{\infty} V^{[j+1]}(\alpha\alpha^* - \beta\beta^*)^{(j+1)} V^{[j+1]*}$$

and

$$V^{[j+1]} = VV^{(1)} \cdots V^{(j)} = V(V^{[j]})^{(1)}$$

it is readily checked that Λ is the unique solution of the equation

$$\Lambda - V\Lambda^{(1)}V^* = \alpha\alpha^* - \beta\beta^* \, . \tag{4.5}$$

The latter can also be expressed in the more familiar Stein equation form

$$\Lambda - VZ^*\Lambda ZV^* = [\alpha \ \beta] J [\alpha \ \beta]^* \, .$$

If Λ is uniformly positive on $(\ell_B^2)^n$, then Λ^{-1} also belongs to $\mathcal{D}^{n \times n}$ for $\mathcal{D} = \mathcal{D}(\ell_B^2; \ell_B^2)$ and (4.5) can be rewritten as

$$\Lambda^{-\frac{1}{2}} V\Lambda^{(1)}V^*\Lambda^{-\frac{1}{2}} + \Lambda^{-\frac{1}{2}}\alpha\alpha^*\Lambda^{-\frac{1}{2}} - \Lambda^{-\frac{1}{2}}\beta\beta^*\Lambda^{-\frac{1}{2}} = I \, , \tag{4.6}$$

which is of the form (3.4) with

$$M_{11} = (\Lambda^{\frac{1}{2}})^{(1)} V^*\Lambda^{-\frac{1}{2}} \tag{4.7}$$

$$M_{21} = \alpha^*\Lambda^{-\frac{1}{2}} \tag{4.8}$$

$$M_{31} = \beta^*\Lambda^{-\frac{1}{2}} \, . \tag{4.9}$$

By Theorem 3.2, the specified first block column M_1 will admit a $\tilde{J}$ unitary extension to $\mathcal{H}_1 \oplus \mathcal{H}_2 \oplus \mathcal{H}_3$ if and only if it admits one with $M_{32} = 0$ and

$$M_{13} = M_{11} M_{31}^* (I + M_{31} M_{31}^*)^{-\frac{1}{2}}$$
$$= (\Lambda^{\frac{1}{2}})^{(1)} V^*\Lambda^{-1}\beta(I + \beta^*\Lambda^{-1}\beta)^{-\frac{1}{2}} \, , \tag{4.10}$$

$$M_{23} = M_{21} M_{31}^* (I + M_{31} M_{31}^*)^{-\frac{1}{2}}$$
$$= \alpha^*\Lambda^{-1}\beta(I + \beta^*\Lambda^{-1}\beta)^{-\frac{1}{2}} \, , \tag{4.11}$$

$$M_{33} = (I + M_{31} M_{31}^*)^{\frac{1}{2}}$$
$$= (I + \beta^*\Lambda^{-1}\beta)^{\frac{1}{2}} \, . \tag{4.12}$$

Now choose

$$N = \Lambda^{\frac{1}{2}}(\Lambda + \beta\beta^*)^{-\frac{1}{2}} \tag{4.13}$$

so that

$$N_{11} = (\Lambda^{\frac{1}{2}})^{(1)}V^*(\Lambda + \beta\beta^*)^{-\frac{1}{2}} \tag{4.14}$$

and

$$N_{21} = \alpha^*(\Lambda + \beta\beta^*)^{-\frac{1}{2}} . \tag{4.15}$$

Since

$$\ker[M_{11}^* \quad M_{21}^*] = \ker[V(\Lambda^{\frac{1}{2}})^{(1)} \quad \alpha] ,$$

it follows from Theorem 3.4, that the given column M_1 with entries specified by (4.7) to (4.9) admits a $\tilde{J}$ unitary extension M with blocks of diagonal operators as components if and only if there exist a pair of operators

$$M_{12} \in \{\mathcal{D}(\ell^2_{\mathcal{M}}; \ell^2_{\mathcal{B}})\}^{n \times 1} \quad \text{and} \quad M_{22} \in \mathcal{D}(\ell^2_{\mathcal{M}}; \ell^2_{\mathcal{N}})$$

such that

$$\begin{bmatrix} M_{12} \\ M_{22} \end{bmatrix} \text{ is an isometric mapping of } \ell^2_{\mathcal{M}} \text{ onto } \ker[V(\Lambda^{\frac{1}{2}})^{(1)} \quad \alpha] . \tag{4.16}$$

We shall refer to this condition as the kernel condition. Therein, the indicated kernel is a subspace of $(\ell^2_{\mathcal{B}})^n \oplus \ell^2_{\mathcal{M}}$. It is also important to bear in mind that the operators M_{12} and M_{22} are required to have block diagonal form and hence that this condition really corresponds to infinitely many conditions, one for the action of the operator on each coordinate entry of the corresponding ℓ^2 space:

The kernel condition is always met if $\mathcal{B}$ is finite dimensional:

THEOREM 4.3. *If $\Lambda + \beta\beta^*$ is uniformly positive on $(\ell^2_{\mathcal{B}})^n$ and if $\mathcal{B}$ is finite dimensional, then the kernel condition (4.16) is satisfied, i.e., there exist a pair of operators $M_{12} \in \{\mathcal{D}(\ell^2_{\mathcal{M}}; \ell^2_{\mathcal{B}})\}^{n \times 1}$ and $M_{22} \in \mathcal{D}(\ell^2_{\mathcal{M}}; \ell^2_{\mathcal{M}})$ such that the operator matrix*

$$N = \begin{bmatrix} N_{11} & M_{12} \\ N_{21} & M_{22} \end{bmatrix} ,$$

with first block column N_1 specified by (4.14) and (4.15), is unitary on $(\ell^2_{\mathcal{B}})^n \oplus \ell^2_{\mathcal{M}}$.

PROOF. The condition on $\Lambda + \beta\beta^*$ ensures that the specified operator N_1 is well defined. N_1 is isometric and, since all the operators in question are diagonal operators or matrices of diagonal operators, it will admit a unitary extension of the requisite form if and only if each of its components $(N_1)_{jj}$, $j = 0, \pm 1, \ldots$, which are isometric mappings of

$$\mathcal{B}_j^n \longrightarrow \mathcal{B}_j^n \oplus \mathcal{M}_j \ ,$$

admits a unitary extension to $\mathcal{B}_j^n \oplus \mathcal{M}_j$. Here $\mathcal{B}_j = \mathcal{B}$ [resp. $\mathcal{M}_j = \mathcal{M}$] designates the j-th coordinate entry in $\ell_\mathcal{B}^2$ [resp. $\ell_\mathcal{M}^2$]. But $(N_1)_{jj}$ will admit such a unitary extension if and only if

$$\dim \ \{\mathrm{dom}(N_1)_{jj}\}^\perp = \dim \ \{\mathrm{ran}(N_1)_{jj}\}^\perp \ ,$$

i.e., if and only if

$$\dim \ \mathcal{M} = \dim \ \{\mathrm{ran}(N_1)_{jj}\}^\perp \ .$$

But this will always be so if $\mathcal{B}$ is finite dimensional.   ∎

The last assertion in the proof may fail if $\mathcal{B}$ is infinite dimensional. For example, if $\mathcal{B}$ is infinite dimensional with orthonormal basis $\varphi_0, \varphi_1, \ldots$, then the operator T which is defined by the rule

$$T\varphi_0 \ = \ \frac{1}{\sqrt{2}} \begin{bmatrix} \varphi_1 \\ 1 \end{bmatrix} \ ,$$

$$T\varphi_j \ = \ \begin{bmatrix} \varphi_{j+1} \\ 0 \end{bmatrix} \ , \qquad j = 1, 2, \ldots \ ,$$

is an isometric mapping of $\mathcal{B}$ into $\mathcal{B} \oplus \mathbb{C}$ for which

$$1 = \dim \ (\mathrm{dom} \ T)^\perp \neq \dim \ (\mathrm{ran} \ T)^\perp = 2 \ .$$

For ease of future applications, we now summarize the main implications of the preceding analysis in the next theorem. To avoid excessive duplication, however, we shall first review the basic setting which is common to all the interpolation problems considered in the first eleven sections of this paper.

BASIC SETTING. *Let*

$$V \in \{\mathcal{D}(\ell_\mathcal{B}^2; \ell_\mathcal{B}^2)\}^{n \times n} \ , \qquad \alpha \in \{\mathcal{D}(\ell_\mathcal{M}^2; \ell_\mathcal{B}^2)\}^{n \times 1} \quad and \quad \beta \in \{\mathcal{D}(\ell_\mathcal{N}^2; \ell_\mathcal{B}^2)\}^{n \times 1}$$

and suppose that $r_{sp}(Z V^*) < 1$ so that (4.5) has a unique solution $\Lambda \in \mathcal{D}^{n \times n}$. Let

$$F \ = \ \begin{bmatrix} \alpha^* \\ \beta^* \end{bmatrix} (I - Z V^*)^{-1}$$

and

$$\mathcal{F} = \{F\underline{E} : \underline{E} \in \mathcal{D}_2^{n \times 1} \quad with \quad \mathcal{D}_2 = \mathcal{D}_2(\ell_\mathcal{B}^2; \ell_\mathcal{B}^2)\}$$

endowed with the inner product

$$\langle F\underline{D}, \ F\underline{E}\rangle_{\mathcal{F}} = \langle JF\underline{D}, \ F\underline{E}\rangle_{HS} \ ,$$

with J as in (1.6). Further, let e_j be the block column operator in $\mathcal{D}^{n\times 1}$ with an I in the jth place and 0's elsewhere:

$$e_1 \ = \ \begin{bmatrix} I \\ 0 \\ \vdots \\ 0 \end{bmatrix}, \dots, e_n \ = \ \begin{bmatrix} 0 \\ \vdots \\ 0 \\ I \end{bmatrix} \ ,$$

and let

$$\Lambda_{ij} = e_i^* \Lambda e_j \quad and \quad F_i = Fe_i = \begin{bmatrix} G_i \\ H_i \end{bmatrix} \ ,$$

with components

$$G_i \in \mathcal{U}(\ell_{\mathcal{B}}^2; \ell_{\mathcal{M}}^2) \quad and \quad H_i \in \mathcal{U}(\ell_{\mathcal{B}}^2; \ell_{\mathcal{N}}^2) \ .$$

THEOREM 4.4. *If, in the basic setting described above,*

I. Λ *is uniformly positive on* $(\ell_{\mathcal{B}}^2)^n$,

then

(1) $\langle JF_jD, F_iE\rangle_{HS} = \text{trace } E^*\Lambda_{ij}D$
for $i,j = 1,\dots,n$ and every choice of D and E in $\mathcal{D}_2$.

(2) $\mathcal{F}$ *is a reproducing kernel Hilbert space with reproducing kernel*

$$K_{\boldsymbol{W}} = F\Lambda^{-1}\hat{F}(\boldsymbol{W})^*$$

for every choice of $\boldsymbol{W} = \text{diag}(W_1, W_2)$ with $W_1 \in \mathcal{D}(\ell_{\mathcal{M}}^2; \ell_{\mathcal{M}}^2)$, $W_2 \in \mathcal{D}(\ell_{\mathcal{N}}^2; \ell_{\mathcal{N}}^2)$ and $\ell_{\boldsymbol{W}} < 1$.

If also

II. *The kernel condition (4.16) is satisfied,*

then the operator matrix Θ with entries

$$\Theta_{11} = M_{22} + \alpha^*(I - \boldsymbol{Z}\boldsymbol{V}^*)^{-1}\Lambda^{-\frac{1}{2}}\boldsymbol{Z}M_{12} \ , \tag{4.17}$$

$$\Theta_{21} = \beta^*(I - \boldsymbol{Z}\boldsymbol{V}^*)^{-1}\Lambda^{-\frac{1}{2}}\boldsymbol{Z}M_{12} \ , \tag{4.18}$$

$$\Theta_{12} = \alpha^*(I - \boldsymbol{Z}\boldsymbol{V}^*)^{-1}\Lambda^{-1}\beta(I + \beta^*\Lambda^{-1}\beta)^{-\frac{1}{2}} \ , \tag{4.19}$$

$$\Theta_{22} = \{I + \beta^*(I - \boldsymbol{Z}\boldsymbol{V}^*)^{-1}\Lambda^{-1}\beta\}(I + \beta^*\Lambda^{-1}\beta)^{-\frac{1}{2}} \tag{4.20}$$

belongs to $\mathcal{A}$ and the reproducing kernel

$$F\Lambda^{-1}\hat{F}(W)^* = \{J - \Theta J\hat{\Theta}(W)^*\}(I - ZW^*)^{-1} \tag{4.21}$$

for every choice of $W = \mathrm{diag}(W_1, W_2)$ as in (2).

Moreover, if Θ is any ULCS operator in $\mathcal{A}$ which satisfies (4.21) (and not just the specific Θ exhibited in (4.17)–(4.20)), then

$$\langle J\Theta\underline{E}\, ,\, F_j D\rangle_{HS} = 0 \tag{4.22}$$

for every choice of $D \in \mathcal{D}_2(\ell_{\mathcal{B}}^2; \ell_{\mathcal{B}}^2)$ and

$$\underline{E} = \begin{bmatrix} E_1 \\ E_2 \end{bmatrix}$$

with $E_1 \in \mathcal{U}_2(\ell_{\mathcal{B}}^2; \ell_{\mathcal{M}}^2)$ and $E_2 \in \mathcal{U}_2(\ell_{\mathcal{B}}^2; \ell_{\mathcal{N}}^2)$.

PROOF. To begin with, it is readily checked with the help of (2.2) that

$$\langle JF_j D,\ F_i E\rangle_{HS} = \langle F_i^* JF_j D,\ E\rangle_{HS}$$

$$= \mathrm{trace}\ E^* \{\underline{p}F_i^* JF_j D\}^{\wedge}(0)$$

for every choice of D and E in $\mathcal{D}_2$.

On the other hand

$$F^* JF = (I - ZV^*)^{-*}(\alpha\alpha^* - \beta\beta^*)(I - ZV^*)^{-1}$$

$$= \sum_{t=0}^{\infty}(I - ZV^*)^{-*}Z^t(\alpha\alpha^* - \beta\beta^*)^{(t)}V^{[t]^*}$$

and therefore

$$\underline{p}F_i^* JF_j D = \sum_{t=0}^{\infty}\underline{p}e_i^*(I - ZV^*)^{-*}Z^t(\alpha\alpha^* - \beta\beta^*)^{(t)}V^{[t]^*}e_j D$$

$$= \sum_{t=0}^{\infty} e_i^* \sum_{s=0}^{t} Z^{t-s}(V^{[s]})^{(t-s)}(\alpha\alpha^* - \beta\beta^*)^{(t)}V^{[t]^*}e_j D\ .$$

Thus

$$\{\underline{p}F_i^* JF_j D\}^{\wedge}(0) = \sum_{t=0}^{\infty} e_i^* V^{[t]}(\alpha\alpha^* - \beta\beta^*)^{(t)}V^{[t]^*}e_j D$$

$$= \Lambda_{ij}D\ .$$

The asserted formula now drops out easily upon substituting the last evaluation into the first identity in the proof.

Next, to establish (2), observe first that for every choice of $\underline{D} \in \{\mathcal{D}_2(\ell^2_{\mathcal{B}}; \ell^2_{\mathcal{B}})\}^{n \times 1}$ and

$$\underline{E} = \begin{bmatrix} E_1 \\ E_2 \end{bmatrix}$$

with $E_1 \in \mathcal{D}_2(\ell^2_{\mathcal{B}}; \ell^2_{\mathcal{M}})$ and $E_2 \in \mathcal{D}_2(\ell^2_{\mathcal{B}}; \ell^2_{\mathcal{N}})$, $K_{\boldsymbol{W}}\underline{E}$ clearly belongs to $\mathcal{F}$ and, in view of (1),

$$\langle JF\underline{D}, \ K_{\boldsymbol{W}}\underline{E} \rangle_{HS} = \langle JF\underline{D}, \ F\Lambda^{-1}\hat{F}(\boldsymbol{W})^*\underline{E} \rangle_{HS}$$

$$= \text{trace } \underline{E}^*\hat{F}(\boldsymbol{W})\Lambda^{-1}\Lambda\underline{D}$$

$$= \text{trace } \underline{E}^*\hat{F}(\boldsymbol{W})\underline{D} \ .$$

To check that $\mathcal{F}$ is complete, let $F\underline{D}_j$ be a Cauchy sequence in $\mathcal{F}$. Then, since

$$\langle JF(\underline{D}_j - \underline{D}_k), \ F(\underline{D}_j - \underline{D}_k) \rangle_{HS} = \text{trace}(\underline{D}_j - \underline{D}_k)^*\Lambda(\underline{D}_j - \underline{D}_k)$$

$$= \langle \Lambda(\underline{D}_j - \underline{D}_k), \ \underline{D}_j - \underline{D}_k \rangle_{HS}$$

$$\geq \epsilon \langle \underline{D}_j - \underline{D}_k, \ \underline{D}_j - \underline{D}_k \rangle_{HS}$$

for some $\epsilon > 0$ by the presumed uniform positivity of Λ, it follows that $\{\underline{D}_j\}$ is a Cauchy sequence in the Hilbert space $\mathcal{D}_2^{n \times 1}$. Therefore $\underline{D}_j$ tends to a limit $\underline{D}$ in $\mathcal{D}_2^{n \times 1}$ as $j \to \infty$ and consequently $F\underline{D}_j \to F\underline{D}$. Thus $\mathcal{F}$ is complete, as needed. For another argument, see the proof of Theorem 4.7.

Suppose next that the kernel condition (4.16) is met. Then, by Theorem 3.4, there exists a choice of $M_{12} \in \mathcal{D}^{n \times 1}$ and $M_{22} \in \mathcal{D}$ such that the operator matrix

$$M = \begin{bmatrix} M_{11} & M_{12} & M_{13} \\ M_{21} & M_{22} & M_{23} \\ M_{31} & 0 & M_{33} \end{bmatrix}$$

with first and last block column specified by (4.7) – (4.12) is $\tilde{J}$ unitary. Thus, as

$$r_{sp}(\boldsymbol{Z}M_{11}) = r_{sp}(\Lambda^{\frac{1}{2}}\boldsymbol{Z}\boldsymbol{V}^*\Lambda^{-\frac{1}{2}})$$

$$= r_{sp}(\boldsymbol{Z}\boldsymbol{V}^*) < 1 \ ,$$

it follows from Theorem 4.2 that the corresponding characteristic operator function Θ of the operator colligation M (as defined by (4.1)) belongs to $\mathcal{A}$. By direct calculation, its block entries are given by (4.17) - (4.20) when U is taken equal to $\boldsymbol{Z}$ in (3.22). Moreover, upon identifying the entries in Theorem 4.1 in terms of the current choice of the M_{ij}, it turns out that

$$G = C(I - \mathbf{Z} A)^{-1}$$

$$= \begin{bmatrix} \alpha^* \\ \beta^* \end{bmatrix} (I - \mathbf{Z} V^*)^{-1} \Lambda^{-\frac{1}{2}}$$

$$= F\Lambda^{-\frac{1}{2}}$$

and hence, in view of (4.3), that

$$F\Lambda^{-1}\hat{F}(\mathbf{W})^* = \{J - \Theta J \hat{\Theta}(\mathbf{W})^*\}(I - \mathbf{Z} \mathbf{W}^*)^{-1}$$

for every choice of $\mathbf{W}$ of the stated form. This completes the proof of (4.21).

Finally, to prove (4.22), observe first that

$$\text{trace } \underline{\underline{H}}^* J \hat{F}_j(\mathbf{W})E = \langle JF_j E, \{J - \Theta J \hat{\Theta}(\mathbf{W})^*\}(I - \mathbf{Z} \mathbf{W}^*)^{-1}\underline{\underline{H}}\rangle_{HS}$$

$$= \langle F_j E, (I - \mathbf{Z} \mathbf{W}^*)^{-1}\underline{\underline{H}}\rangle_{HS} - \langle JF_j E, \Theta J \hat{\Theta}(\mathbf{W})^*(I - \mathbf{Z} \mathbf{W}^*)^{-1}\underline{\underline{H}}\rangle_{HS}$$

for every choice of $E \in \mathcal{D}_2(\ell_B^2; \ell_B^2)$,

$$\underline{\underline{H}} = \begin{bmatrix} H_1 \\ H_2 \end{bmatrix} \quad \text{and} \quad \mathbf{W} = \text{diag}(W_1, W_2)$$

with $H_1 \in \mathcal{D}_2(\ell_B^2; \ell_M^2)$, $H_2 \in \mathcal{D}_2(\ell_B^2; \ell_N^2)$, $W_1 \in \mathcal{D}(\ell_M^2; \ell_M^2)$, $W_2 \in \mathcal{D}(\ell_N^2; \ell_N^2)$ and $\ell_{\mathbf{W}} < 1$. Therefore, since the first inner product on the right in the last line of the equality is equal to the left hand side by Theorem 2.1,

$$0 = \langle JF_j E, \Theta J \hat{\Theta}(\mathbf{W})^*(I - \mathbf{Z} \mathbf{W}^*)^{-1}\underline{\underline{H}}\rangle_{HS}$$

$$= \langle \hat{\Theta}(\mathbf{W})J\underline{p}\Theta^* JF_j E, (I - \mathbf{Z} \mathbf{W}^*)^{-1}\underline{\underline{H}}\rangle_{HS}$$

$$= \text{trace } \underline{\underline{H}}^*(\hat{\Theta}(\mathbf{W})J\underline{p}\Theta^* JF_j E\}^\wedge(\mathbf{W}) .$$

But this in turn implies that

$$0 = \{\hat{\Theta}(\mathbf{W})J\underline{p}\Theta^* JF_j E\}^\wedge(\mathbf{W})$$

and hence by Theorem 1.2,

$$0 = \{\Theta J\underline{p}\Theta^* JF_j E\}^\wedge(\mathbf{W}) .$$

Thus, by Theorem 1.1 and the arbitrariness of $\mathbf{W}$,

$$\Theta J\underline{p}\Theta^* JF_j E = 0$$

which in turn implies that

$$\underline{p}\Theta^* J F_j E = 0 \ , \tag{4.23}$$

since Θ is J unitary. But this is equivalent to (4.22). This completes the proof. ∎

USAGE: *From now on we shall refer to any $\Theta \in \mathcal{A}$ which satisfies (4.21) with F prescribed as in the basic setting, as an associated ULCS operator.*

LEMMA 4.1. *If $\Theta \in \mathcal{A}$, then*

$$S = T_\Theta[S_L]$$

is a strictly contractive operator in $S(\ell^2_{\mathcal{N}}; \ell^2_{\mathcal{M}})$ if and only if S_L is.

PROOF. Let

$$S = T_\Theta[S_L]$$

for some choice of $S_L \in \mathcal{S}$. Then,

$$I - SS^* = 2(\Theta_{21}S_L + \Theta_{22})^{-*}(I - S_L^* S_L)(\Theta_{21}S_L + \Theta_{22})^{-1} \ .$$

Then, since Θ_{22} is invertible in $\mathcal{U}$ and $\|\Theta_{22}^{-1}\Theta_{21}\| < 1$ by Lemma 5.2 of [ADD], it is readily seen that S is strictly contractive if and only if S_L is. ∎

THEOREM 4.5. *Suppose that in the basic setting of Theorem 4.4, conditions I and II are fulfilled, and that $\Theta \in \mathcal{A}$ is any associated ULCS operator. Then*

$$\underline{p}S^* G_i E = H_i E \ , \tag{4.24}$$

for $i = 1, \ldots, n$ and every choice of $E \in \mathcal{D}_2$ and S of the form

$$S = T_\Theta[S_L] := (\Theta_{11}S_L + \Theta_{12})(\Theta_{21}S_L + \Theta_{22})^{-1} \tag{4.25}$$

with $S_L \in \mathcal{S}$.

PROOF. To simplify the typography, we shall write

$$\Theta = \begin{bmatrix} A & B \\ C & D \end{bmatrix} \ .$$

Then, formula (4.23) implies that

$$\underline{p}(A^* G_i - C^* H_i)E = 0 \tag{4.26}$$

and

$$\underline{p}(B^* G_i - D^* H_i)E = 0 \tag{4.27}$$

for $i = 1, \ldots, n$ and every choice of $E \in \mathcal{D}_2$. Since D is invertible in $\mathcal{U}$, it follows readily from (4.27) that

$$\underline{p}D^*\underline{p}(D^{-*}B^*G_i - H_i)E = 0$$

and hence that

$$\underline{p}(D^{-*}B^*G_i - H_i)E = \underline{p}D^{-*}\underline{p}D^*\underline{p}(D^{-*}B^*G_i - H_i)E$$

$$= 0 \ .$$

This proves (4.25) for

$$S = T_\Theta[0] = BD^{-1} \ .$$

Next, for general $S = T_\Theta[S_L]$, we have,

$$S - BD^{-1} = \{AS_L + B - BD^{-1}(CS_L + D)\}(CS_L + D)^{-1}$$

$$= (A - BD^{-1}C)Q$$

with

$$Q = S_L(CS_L + D)^{-1}$$

in $\mathcal{U}$. Thus, upon setting $S_0 = BD^{-1}$ for short,

$$\underline{p}(S^*G_i - H_i)E = \underline{p}\{S_0^*G_i - H_i\}E + \underline{p}Q^*(A - S_0C)^*G_iE$$

$$= \underline{p}Q^*\underline{p}\{A^*G_iE - C^*S_0^*G_iE\}$$

$$= \underline{p}Q^*\underline{p}\{C^*H_iE - C^*S_0^*G_iE\} \quad \text{(by (4.26))}$$

$$= \underline{p}Q^*C^*\underline{p}(H_i - S_0^*G_i)E \ ,$$

since $C \in \mathcal{U}$. But this vanishes by (4.24).　　∎

COROLLARY 1.　*If, in the basic setting of Theorem 4.4, conditions I and II are in force and $\Theta \in \mathcal{A}$ is any associated ULCS operator, then $S = T_\Theta[S_L]$ is a strictly contractive solution of (4.24) for every choice of S_L in $\mathcal{S}(\ell_\mathcal{N}^2; \ell_\mathcal{M}^2)$ which is strictly contractive.*

PROOF.　This is immediate from the preceding theorem and Lemma 4.1.　　∎

The next theorem is a partial converse to the last corollary. Before proceeding to the statement and the proof, it is perhaps well to recall condition III from Section 1:

III.　Δ *is uniformly positive on* $(\ell_\mathcal{B}^2)^n$,

and that I $\Rightarrow$ III, always.

THEOREM 4.6. *If, in the basic setting of Theorem 4.4, there exists an $S \in \mathcal{S}(\ell^2_{\mathcal{N}}; \ell^2_{\mathcal{M}})$ which solves (4.24), then Λ is positive semidefinite on $(\ell^2_B)^n$.*

If S is strictly contractive, then III $\Rightarrow$ I.

PROOF. Let $S \in \mathcal{S}$ be a solution (4.24). Then, for every choice of $D_1, \ldots, D_n$ in $\mathcal{D}_2(\ell^2_B; \ell^2_B)$,

$$
\text{trace} \sum_{i,j=1}^{n} D_i^* \Lambda_{ij} D_j = \sum_{i,j=1}^{n} \langle J F_j D_j, F_i D_i \rangle_{HS}
$$

$$
= \sum_{i,j=1}^{n} \langle G_j D_j, G_i D_i \rangle_{HS}
$$

$$
- \sum_{i,j=1}^{n} \langle p S^* G_j D_j, \underline{p} S^* G_i D_i \rangle_{HS}
$$

$$
= \left\| \sum_{j=1}^{n} G_j D_j \right\|^2_{HS} - \left\| \underline{p} S^* \sum_{j=1}^{n} G_j D_j \right\|^2_{HS}
$$

$$
\geq (1 - \|S\|^2) \left\| \sum_{j=1}^{n} G_j D_j \right\|^2_{HS},
$$

which suffices to prove the first assertion. If S is strictly contractive and III is in force, then the bound can be refined because the last term on the right is equal to

$$
(1 - \|S\|^2)\text{trace} \sum_{j=1}^{n} D_i^* \Delta_{ij} D_j \geq (1 - \|S\|^2)\epsilon \, \text{trace} \sum_{j=1}^{n} D_i^* D_j
$$

for some $\epsilon > 0$ (because of III). Therefore I is also in force.   ∎

We conclude this section with a brief discussion on the significance of condition III.

LEMMA 4.2. *In the basic setting of Theorem 4.4, the following are equivalent:*

(1)   The elements G_i, $i = 1, \ldots, n$, are right linearly independent over $\mathcal{D}_2$: if $\sum_{j=1}^{n} G_j D_j = 0$ for some choice of $D_1, \ldots, D_n$ in $\mathcal{D}_2$, then $D_j = 0$ for $j = 1, \ldots, n$.

(2) $\ker \Delta = 0$, when Δ is restricted to $\mathcal{D}_2^{n \times 1}$.

PROOF. Suppose first that (1) holds and let $\underline{D} \in \mathcal{D}_2^{n \times 1}$ be a block column vector in the kernel of Δ with components $D_1, \ldots, D_n$. Then

$$0 = \text{trace} \left\{ \sum_{i=1}^{n} D_i^* \sum_{j=1}^{n} \Delta_{ij} D_j \right\}$$

$$= \left\| \sum_{j=1}^{n} G_j D_j \right\|_{HS}^2 ,$$

which in turn implies that

$$\sum_{j=1}^{n} G_j D_j = 0 \tag{4.28}$$

and hence by (1) that $\underline{\underline{D}} = 0$. Thus (1) implies (2).

Suppose next that (2) is in force and that (4.28) holds for some choice of $D_1, \ldots, D_n$ in $\mathcal{D}_2$. Then

$$\text{trace } E^* \sum_{j=1}^{n} \Delta_{ij} D_j = \langle \sum_{j=1}^{n} G_j D_j, G_i E \rangle_{HS} = 0$$

for $i = 1, \ldots, n$ and every choice of $E \in \mathcal{D}_2(\ell_B^2; \ell_B^2)$. Thus $D_1 = \cdots = D_n = 0$ and so (2) implies (1). ∎

THEOREM 4.7. *In the basic setting of Theorem 4.4, the following are equivalent:*

(1)   *Condition III is in force.*

(2)   *The operators G_i are linearly independent over $\mathcal{D}_2$ and the space*

$$\mathcal{M} = \left\{ \sum_{i=1}^{n} G_i D_i : \ D_i \in \mathcal{D}_2 \ \text{ for } \ i = 1, \ldots, n \right\}$$

is a closed subspace of $\mathcal{U}_2$.

PROOF. Suppose first that (1) holds. Then, since

$$\left\| \sum_{j=1}^{n} G_j D_j \right\|_{HS}^2 = \langle \Delta \underline{\underline{D}}, \underline{\underline{D}} \rangle_{HS}$$

for every choice of

$$\underline{\underline{D}} = \begin{bmatrix} D_1 \\ \vdots \\ D_n \end{bmatrix}$$

in $\mathcal{D}_2^{n \times 1}$, it follows readily that there exist a pair of positive numbers α and β such that

$$\alpha \|\underline{D}\|_{HS} \leq \left\| \sum_{j=1}^{n} G_j D_j \right\|_{HS} \leq \beta \|\underline{D}\|_{HS} \, .$$

Therefore (2) holds.

Conversely, if (2) holds, then the transformation which sends

$$\underline{D} \in \mathcal{D}_2^{n \times 1} \longrightarrow \sum_{j=1}^{n} G_j D_j \in \mathcal{M}$$

is a one to one bounded linear mapping of the Hilbert space $\mathcal{D}_2^{n \times 1}$ onto the Hilbert space $\mathcal{M}$. Therefore, by the open mapping theorem, it has a bounded inverse (see e.g., Theorem 5.10 of [Ru]). Thus

$$\|\underline{D}\|_{HS}^2 \leq \gamma \left\| \sum_{j=1}^{n} G_j D_j \right\|_{HS}^2 = \gamma \langle \Delta \underline{D}, \underline{D} \rangle_{HS}$$

form some $\gamma > 0$. This proves (1). ∎

5. EXPLICIT FORMULAS FOR Θ

In general, even if conditions I and II are met, it does not seem possible to find nice explicit formulas for Θ_{11} and Θ_{21} in the setting of Theorem 4.4. A number of exceptions are considered in this section.

CASE 1. *$n = 1$ and α is invertible in $\mathcal{D}$.*

In this instance it is readily seen that

$$\ker[V(\Lambda^{\frac{1}{2}})^{(1)} \quad \alpha] = \operatorname{ran} \begin{bmatrix} I \\ -\alpha^{-1} V(\Lambda^{\frac{1}{2}})^{(1)} \end{bmatrix} \, .$$

Thus, to meet (3.5) (or equivalently (3.11)), we shall choose

$$M_{12} = Q$$

$$M_{22} = -\alpha^{-1} V(\Lambda^{\frac{1}{2}})^{(1)} Q$$

for some $Q \in \mathcal{D}$ which, because of (3.12), is subject to the constraint

$$Q^* \{ I + (\Lambda^{\frac{1}{2}})^{(1)} V^* (\alpha \alpha^*)^{-1} V (\Lambda^{\frac{1}{2}})^{(1)} \} Q = I \, . \tag{5.1}$$

LEMMA 5.1. *If $n = 1$ and $\alpha\alpha^*$ is invertible in $\mathcal{D}$, then*

$$\{\Lambda^{(1)} + \Lambda^{(1)}V^*(\alpha\alpha^*)^{-1}V\Lambda^{(1)}\}\{(\Lambda^{-1})^{(1)} - V^*(\Lambda + \beta\beta^*)^{-1}V\} = I,$$

$$\{(\Lambda^{-1})^{(1)} - V^*(\Lambda + \beta\beta^*)^{-1}V\}\{\Lambda^{(1)} + \Lambda^{(1)}V^*(\alpha\alpha^*)^{-1}V\Lambda^{(1)}\} = I,$$

and thus $(\Lambda^{-1})^{(1)} - V^*(\Lambda + \beta\beta^*)^{-1}V$ is positive definite and invertible in $\mathcal{D}$.

PROOF. It suffices to prove the first identity (because the second is just the adjoint of the first) and this is a straightforward calculation with the help of (4.5). ∎

It turns out that for the particular case

$$Q = (\Lambda^{\frac{1}{2}})^{(1)}\{\Lambda^{(1)} + \Lambda^{(1)}V^*(\alpha\alpha^*)^{-1}V\Lambda^{(1)}\}^{-\frac{1}{2}},$$

$M_{12} = Q$ and $M_{22} = -\alpha^{-1}V(\Lambda^{\frac{1}{2}})^{(1)}Q$ satisfy (3.11)–(3.12) and, as follows with the help of Lemma 5.1, (3.19)–(3.21). Thus N is unitary and hence, by Theorems 3.1 and 3.4, the operator Θ with components

$$\Theta_{11} = \{-\alpha^{-1}V\Lambda^{(1)} + \alpha^*\rho_V^{-1}Z\}\{\Lambda^{(1)} + \Lambda^{(1)}V^*(\alpha\alpha^*)^{-1}V\Lambda^{(1)}\}^{-\frac{1}{2}} \tag{5.2}$$

$$\Theta_{21} = \beta^*\rho_V^{-1}Z\{\Lambda^{(1)} + \Lambda^{(1)}V^*(\alpha\alpha^*)^{-1}V\Lambda^{(1)}\}^{-\frac{1}{2}} \tag{5.3}$$

and Θ_{12}, Θ_{22} as given by (4.19) and (4.20), respectively, belong to $\mathcal{U}^{2\times2}$ and is J unitary.

In the special case that $\alpha = I$, this leads to the simpler formula

$$\Theta = \begin{bmatrix} \rho_V^{-1}Z - V\Lambda^{(1)} & \rho_V^{-1}\Lambda^{-1}\beta \\ \beta^*\rho_V^{-1}Z & I + \beta^*\rho_V^{-1}\Lambda^{-1}\beta \end{bmatrix} \times \begin{bmatrix} \Lambda^{(1)} + \Lambda^{(1)}V^*V\Lambda^{(1)} & 0 \\ 0 & I + \beta^*\Lambda^{-1}\beta \end{bmatrix}^{-\frac{1}{2}}. \tag{5.4}$$

We remark that much of the preceding analysis goes through under the less stringent assumption that only $\alpha\alpha^*$ is invertible: For the choices

$$M_{12} = Q,$$

$$M_{22} = -\alpha^*(\alpha\alpha^*)^{-1}V(\Lambda^{\frac{1}{2}})^{(1)}Q,$$

with Q as above, the operator N is isometric. However, it is not coisometric: (3.19) and (3.20) hold, but

$$N_{21}N_{21}^* + M_{22}M_{22}^* = \alpha^*(\alpha\alpha^*)^{-1}\alpha,$$

which is equal to I if and only if $\alpha^*(\alpha\alpha^*)^{-1}$ is a left inverse of α as well as a right inverse, i.e., if and only if α is invertible. (The problem is that the mapping defined by this choice of M_{12} and M_{22} does not map onto the requisite kernel unless α is invertible.)

CASE 2. *V is invertible in $\mathcal{D}^{n \times n}$.*

In this instance

$$\ker[V(\Lambda^{\frac{1}{2}})^{(1)} \quad \alpha] = \operatorname{ran} \begin{bmatrix} -(\Lambda^{-\frac{1}{2}})^{(1)} V^{-1} \alpha \\ I \end{bmatrix}$$

and thus we shall choose

$$M_{12} = -(\Lambda^{-\frac{1}{2}})^{(1)} V^{-1} \alpha Q$$

and

$$M_{22} = Q$$

for some $Q \in \mathcal{D}$ which, because of (3.12), is subject to the constraint

$$Q^* \{ I + \alpha^* (V \Lambda^{(1)} V^*)^{-1} \alpha \} Q = I \ . \tag{5.5}$$

LEMMA 5.2. *If V is invertible (as well as Λ), then*

$$I + \alpha^* (V \Lambda^{(1)} V^*)^{-1} \alpha = \{ I - \alpha^* (\Lambda + \beta \beta^*)^{-1} \alpha \}^{-1} \ . \tag{5.6}$$

PROOF. Let $K = (\Lambda + \beta \beta^*)^{-\frac{1}{2}} \alpha$. Then, by (4.5),

$$I + \alpha^* (V \Lambda^{(1)} V^*)^{-1} \alpha = I + \alpha^* (\Lambda + \beta \beta^* - \alpha \alpha^*)^{-1} \alpha$$

$$= I + K^* (I - K K^*)^{-1} K$$

$$= I + (I - K^* K)^{-1} K^* K$$

$$= (I - K^* K)^{-1} \ .$$

But this is the same as the asserted identity. The indicated inverses exist because of the presumed invertibility of Λ and V.   ∎

In view of Lemma 5.2, it is now clear that if V is invertible, then

$$Q = \{ I - \alpha^* (\Lambda + \beta \beta^*)^{-1} \alpha \}^{\frac{1}{2}} U$$

is well defined and satisfies (5.5) for every isometric $U \in \mathcal{D}^{n \times n}$. We shall choose $U = I$.

To complete the proof that N is unitary, it remains to check that (3.19)–(3.21) hold for $M_{12} = -(\Lambda^{-\frac{1}{2}})^{(1)} V^{-1} \alpha Q$ and $M_{22} = Q$ for this choice of Q. To carry out the requisite calculations it is useful to let

$$K = V \Lambda^{(1)} V^*$$

and to bear in mind that under the prevailing assumptions K is both invertible in $\mathcal{D}^{n \times n}$ and positive definite. Then

$$N_{11}N_{11}^* + M_{12}M_{12}^*$$

$$= (\Lambda^{\frac{1}{2}})^{(1)}V^*\{(K + \alpha\alpha^*)^{-1} + K^{-1}\alpha(I + \alpha^*K^{-1}\alpha)^{-1}\alpha^*K^{-1}\}V(\Lambda^{\frac{1}{2}})^{(1)}$$

$$= I \, ,$$

as needed, since the term in curly brackets is readily seen to be equal to K^{-1}.

Similarly,

$$N_{21}N_{11}^* + M_{22}M_{12}^* = \{\alpha^*(K + \alpha\alpha^*)^{-1}K - (I + \alpha^*K^{-1}\alpha)^{-1}\alpha^*\}V^{-*}(\Lambda^{-\frac{1}{2}})^{(1)}$$

$$= \alpha^*\{(K + \alpha\alpha^*)^{-1}K - (I + K^{-1}\alpha\alpha^*)^{-1}\}V^{-*}(\Lambda^{-\frac{1}{2}})^{(1)}$$

$$= 0 \, .$$

Finally, it remains to verify (3.21), but this turns out to be easy.

For this choice of M_{12}, M_{22} and Q it follows from (4.17) that

$$\Theta_{11}Q = I - \alpha^*(\Lambda + \beta\beta^*)^{-1}\alpha$$

$$- \alpha^*(I - ZV^*)^{-1}\Lambda^{-\frac{1}{2}}Z(\Lambda^{-\frac{1}{2}})^{(1)}V^{-1}\alpha\{I - \alpha^*(\Lambda + \beta\beta^*)^{-1}\alpha\}$$

$$= I - \alpha^*(\Lambda + \beta\beta^*)^{-1}\alpha$$

$$- \alpha^*(I - ZV^*)^{-1}\Lambda^{-1}ZV^{-1}\{I - \alpha\alpha^*(\Lambda + \beta\beta^*)^{-1}\}\alpha$$

$$= I - \alpha^*(\Lambda + \beta\beta^*)^{-1}\alpha - \alpha^*(I - ZV^*)^{-1}\Lambda^{-1}ZV^{-1}V\Lambda^{(1)}V^*(\Lambda + \beta\beta^*)^{-1}\alpha$$

$$= I - \alpha^*(\Lambda + \beta\beta^*)^{-1}\alpha - \alpha^*(I - ZV^*)^{-1}ZV^*(\Lambda + \beta\beta^*)^{-1}\alpha$$

$$= I - \alpha^*(I - ZV^*)^{-1}(\Lambda + \beta\beta^*)^{-1}\alpha \, ,$$

and hence that

$$\Theta_{11} = \{I - \alpha^*(I - ZV^*)^{-1}(\Lambda + \beta\beta^*)^{-1}\alpha\}\{I - \alpha^*(\Lambda + \beta\beta^*)^{-1}\alpha\}^{-\frac{1}{2}} \, .$$

Similarly, by (4.18),

$$\Theta_{21}Q = \beta^*(I - ZV^*)^{-1}\Lambda^{-\frac{1}{2}}Z(\Lambda^{-\frac{1}{2}})^{(1)}V^{-1}\alpha G^2$$

$$= \beta^*(I - ZV^*)^{-1}\Lambda^{-1}ZV^{-1}\alpha\{I - \alpha^*(\Lambda + \beta\beta^*)^{-1}\alpha\}$$

$$= \beta^*(I - ZV^*)^{-1}\Lambda^{-1}ZV^{-1}\{I - \alpha\alpha^*(\Lambda + \beta\beta^*)^{-1}\}\alpha$$

$$= \beta^*(I - ZV^*)^{-1}\Lambda^{-1}ZV^{-1}V\Lambda^{(1)}V^*(\Lambda + \beta\beta^*)^{-1}\alpha$$

$$= \beta^*(I - ZV^*)^{-1}ZV^*(\Lambda + \beta\beta^*)^{-1}\alpha \, ,$$

and thus

$$\Theta_{21} = \beta^*(I - \mathbf{Z}\mathbf{V}^*)^{-1}\mathbf{Z}\mathbf{V}^*(\Lambda + \beta\beta^*)^{-1}\alpha\{I - \alpha^*(\Lambda + \beta\beta^*)^{-1}\alpha\}^{-\frac{1}{2}} .$$

Formulas (4.19) and (4.20) for the entries in the second block column continue to remain in force.

CASE 3. $N_{11}^* = (I - N_{21}^* N_{21})^{\frac{1}{2}} U$ *for some unitary operator* $U \in \mathcal{D}^{n \times n}$.

In this instance it is readily checked that the choice

$$M_{12} = -U^* N_{21}^*$$

and

$$M_{22} = (I - N_{21} N_{21}^*)^{\frac{1}{2}}$$

serve to make M a $\tilde{J}$ unitary operator.

CASE 4. *In the setting (1.9) of the CF problem,* $\mathcal{M} = \mathcal{B}$ *and* ξ *is invertible.*

Let

$$Y = \begin{bmatrix} (-V)^n \\ \vdots \\ -V \\ I \end{bmatrix} , \quad \text{then} \quad VY = \begin{bmatrix} (-V)^{n+1} \\ 0 \\ \vdots \\ 0 \end{bmatrix} .$$

It is then readily checked that

$$\begin{bmatrix} M_{12} \\ M_{22} \end{bmatrix} = \begin{bmatrix} (\Lambda^{-\frac{1}{2}})^{(1)} Y Q \\ \xi^{-1}(-V)^{n+1} Q \end{bmatrix} ,$$

with

$$Q = \{Y^*(\Lambda^{-1})^{(1)} Y + (V^*)^{n+1}(\xi\xi^*)^{-1} V^{n+1}\}^{-\frac{1}{2}}$$

satisfies the kernel condition (4.16). The indicated inverse exists because $Y^*(\Lambda^{-1})^{(1)} Y$ is uniformly positive on $\ell_{\mathcal{B}}^2$. To verify this, it is convenient to write the latter in the form

$$Y^*(\Lambda^{-1})^{(1)} Y = [y^* \; I] \begin{bmatrix} K_{11} & K_{12} \\ K_{12}^* & K_{22} \end{bmatrix} \begin{bmatrix} y \\ I \end{bmatrix}$$

$$= [y^* \; I] \begin{bmatrix} I & 0 \\ K_{12}^* K_{11}^{-1} & I \end{bmatrix} \begin{bmatrix} K_{11} & 0 \\ 0 & K_{22} - K_{12}^* K_{11}^{-1} K_{12} \end{bmatrix} \begin{bmatrix} I & K_{11}^{-1} K_{12} \\ 0 & I \end{bmatrix} \begin{bmatrix} y \\ I \end{bmatrix}$$

$$\geq [y^* \; I] \begin{bmatrix} I & 0 \\ K_{12}^* K_{11}^{-1} & I \end{bmatrix} \begin{bmatrix} 0 & 0 \\ 0 & K_{22} - K_{12}^* K_{11}^{-1} K_{12} \end{bmatrix} \begin{bmatrix} I & K_{11}^{-1} K_{12} \\ 0 & I \end{bmatrix} \begin{bmatrix} y \\ I \end{bmatrix}$$

$$= K_{22} - K_{12}^* K_{11}^{-1} K_{12} .$$

This does the trick since the indicated Schur complement is uniformly postive on $\ell^2_{\mathcal{B}}$, as follows easily from the identity

$$K_{22} - K_{12}^* K_{11}^{-1} K_{12} = [0\ \ I] \begin{bmatrix} I & 0 \\ -K_{12}^* K_{11}^{-1} & I \end{bmatrix} \begin{bmatrix} K_{11} & K_{12} \\ K_{12}^* & K_{22} \end{bmatrix} \begin{bmatrix} I & -K_{11}^{-1} K_{12} \\ 0 & I \end{bmatrix} \begin{bmatrix} 0 \\ I \end{bmatrix} \quad (5.7)$$

and condition I.   ∎

In this setting, formulas (4.17)–(4.20) with M_{12} and M_{22} as just above yield an associated ULCS operator $\Theta \in \mathcal{A}$.

CASE 5. $\alpha\alpha^* - \beta\beta^*$ *is uniformly positive on* $(\ell^2_{\mathcal{B}})^n$.

It is well to notice that, since

$$\Lambda \geq \alpha\alpha^* - \beta\beta^* \ ,$$

the condition imposed in this case forces I to be satisfied.

We begin with an abstract theorem on $\tilde{J}$ unitary extensions M of a given first column in the general setting of Section 3, wherein M is partitioned as in the proof of Theorem 3.5.

THEOREM 5.1. *Let A and C be a given pair of bounded linear operators from $\mathcal{H}_1$ into $\mathcal{H}_1$ and $\mathcal{H}_1$ into $\mathcal{H}_2 \oplus \mathcal{H}_3$, respectively, such that*

$$A^* A + C^* J C = I$$

and $C^* J C$ is uniformly positive on $\mathcal{H}_1$, and let U be any unitary operator from $\mathcal{H}_1$ into itself. Then the matrix operator

$$M = \begin{bmatrix} A & B \\ C & D \end{bmatrix}$$

with components

$$B = (I - AA^*)^{\frac{1}{2}} U (I - A^* A)^{-\frac{1}{2}} C^* J$$

and

$$D = I - C(C^* J C)^{-\frac{1}{2}} (I + A^* U)(C^* J C)^{-\frac{1}{2}} C^* J$$

is $\tilde{J}$ unitary.

PROOF. The proof is by computation.   ∎

The indicated formulas were obtained by first imposing the requirement that M be $\tilde{J}$ isometric. This forces B and D to be solutions of the pair of equations

$$A^* B + C^* J D = 0 \tag{5.8}$$

$$B^* B + D^* J D = J \ . \tag{5.9}$$

For B and D of the special form

$$B = HC^*J \tag{5.10}$$

and

$$D = I + CKC^*J , \tag{5.11}$$

equations (5.8) and (5.9) reduce to the following more tractable pair of equations for H and K:

$$\{I + A^*H + C^*JCK\}C^*J = 0 \tag{5.12}$$

and

$$JCH^*HC^*J + (I + JCK^*C^*)J(I + CKC^*J) = J . \tag{5.13}$$

The former is solved by choosing

$$K = -(C^*JC)^{-1}(I + A^*H) , \tag{5.14}$$

which helps to reduce the latter to

$$JC^*\{H^*[I + A(C^*JC)^{-1}A^*]H - (C^*JC)^{-1}\}CJ = 0 .$$

Since

$$I + A(C^*JC)^{-1}A^* = I + A(I - A^*A)^{-1}A^*$$
$$= I + (I - AA^*)^{-1}AA^*$$
$$= (I - AA^*)^{-1} ,$$

this will be achieved by choosing H so that

$$H^*(I - AA^*)^{-1}H = (I - A^*A)^{-1} . \tag{5.15}$$

The choice

$$H = (I - AA^*)^{\frac{1}{2}}U(I - A^*A)^{-\frac{1}{2}} \tag{5.16}$$

is appropriate if C^*JC is uniformly positive, and leads easily to the formulas given in Theorem 5.1.

In the basic setting of Theorem 4.4,

$$A = M_{11} = (\Lambda^{\frac{1}{2}})^{(1)}V\Lambda^{-\frac{1}{2}}$$

and

$$C = \begin{bmatrix} \alpha^* \\ \beta^* \end{bmatrix} \Delta^{-\frac{1}{2}} .$$

Upon letting,

$$u = \begin{bmatrix} \alpha^* \\ \beta^* \end{bmatrix} ,$$

for short, and setting the unitary operator

$$U = (u^* J u)^{-\frac{1}{2}} \Delta^{\frac{1}{2}} (C^* J C)^{\frac{1}{2}} ,$$

the formulas for B and D in Theorem 5.1 can be reexpressed as

$$B = (I - AA^*)^{\frac{1}{2}} (u^* J u)^{-\frac{1}{2}} u^* J$$

and

$$D = I - u(u^* J u)^{-1} \{ I + V(\Lambda^{\frac{1}{2}})^{(1)} (I - AA^*)^{\frac{1}{2}} (u^* J u)^{-\frac{1}{2}} \} u^* J ,$$

while

$$C(I - \mathbf{Z} A)^{-1} = u(I - \mathbf{Z} V^*)^{-1} \Lambda^{-\frac{1}{2}} .$$

Thus, the corresponding ULCS operator

$$\Theta = D + C(I - \mathbf{Z} A)^{-1} \mathbf{Z} B \tag{5.17}$$

is now fully specified in terms of Λ and the interpolation data.

It is instructive to reexpress Θ in terms of the "elementary" Blaschke operator

$$U_{\mathbf{W}} = (I - \mathbf{Z} \mathbf{W}^*)^{-1} (\mathbf{Z} - \mathbf{W} L_{\mathbf{W}}) L_{\mathbf{W}}^{-\frac{1}{2}} Q , \tag{5.18}$$

which is a variant of the form considered in Section 4 of [ADD]. Here

$$\mathbf{W} = (u^* J u)^{-\frac{1}{2}} V \{ (u^* J u)^{\frac{1}{2}} \}^{(1)} \tag{5.19}$$

$$B_{\mathbf{W}} = \sum_{j=0}^{\infty} \mathbf{W}^{[j]} \mathbf{W}^{[j]*} = (u^* J u)^{-\frac{1}{2}} \Lambda (u^* J u)^{-\frac{1}{2}} \tag{5.20}$$

$$\begin{aligned} L_{\mathbf{W}} &= B_{\mathbf{W}}^{(1)} - B_{\mathbf{W}}^{(1)} \mathbf{W}^* B_{\mathbf{W}}^{-1} \mathbf{W} B_{\mathbf{W}}^{(1)} \\ &= \{ (u^* J u)^{-\frac{1}{2}} \}^{(1)} \{ \Lambda^{(1)} - \Lambda^{(1)} V^* \Lambda^{-1} V \Lambda^{(1)} \} \{ (u^* J u)^{-\frac{1}{2}} \}^{(1)} \end{aligned} \tag{5.21}$$

and

$$Q = L_{\mathbf{W}}^{-\frac{1}{2}} \{ (u^* J u)^{-\frac{1}{2}} \}^{(1)} (\Lambda^{\frac{1}{2}})^{(1)} (I - AA^*)^{\frac{1}{2}} \tag{5.22}$$

is a unitary "constant" factor.

THEOREM 5.2. *If, in the basic setting of Theorem 4.4, $\alpha\alpha^* - \beta\beta^*$ is uniformly positive on $(\ell_B^2)^n$, then the associated ULCS operator $\Theta \in \mathcal{A}$ given by (5.15) can be reexpressed as*

$$\Theta = I + u(u^*Ju)^{-\frac{1}{2}}(U_W - I)(u^*Ju)^{-\frac{1}{2}}u^*J \,, \tag{5.23}$$

where U_W is the elementary Blaschke operator factor defined by (5.18)-(5.22).

PROOF. Let

$$\gamma = u(u^*Ju)^{-\frac{1}{2}}$$

and

$$\Psi = I + \gamma(U_W - I)\gamma^*J \,.$$

Then

$$\Psi = \Psi_1 + \Psi_2 \,,$$

where

$$\Psi_1 = I + \gamma\{\hat{U}_W(0) - I\}\gamma^*J$$
$$= I - \gamma\{I + WL_W^{\frac{1}{2}}Q\}\gamma^*J$$

and

$$\Psi_2 = \gamma\{U_W + WL_W^{\frac{1}{2}}Q\}\gamma^*J$$
$$= \gamma(I - ZW^*)^{-1}Z(I - W^*WL_W)L_W^{-\frac{1}{2}}Q\gamma^*J$$
$$= u(I - ZV^*)^{-1}Z\{(u^*Ju)^{-\frac{1}{2}}\}^{(1)}(I - W^*WL_W)L_W^{-\frac{1}{2}}Q\gamma^*J \,.$$

But now as

$$(I - WW^*L_W)L_W^{-\frac{1}{2}} = (B_W^{-1})^{(1)}L_W^{\frac{1}{2}}$$

(see the verification of (4.9) in [ADD] for help with this, if need be),

$$\Psi_2 = u(I - ZV^*)^{-1}Z(\Lambda^{-1})^{(1)}\{(u^*Ju)^{\frac{1}{2}}\}^{(1)}L_W^{\frac{1}{2}}Q\gamma^*J$$
$$= u(I - ZV^*)^{-1}Z(\Lambda^{-\frac{1}{2}})^{(1)}(I - AA^*)^{\frac{1}{2}}(u^*Ju)^{-\frac{1}{2}}u^*J \,,$$

by the choice of Q. But this is exactly the same as

$$C(I - ZA)^{-1}ZB \,.$$

It remains to check that

$$D = \Psi_1 \,,$$

but that is easy. ∎

CASE 6. $\alpha\alpha^* - \beta\beta^*$ is uniformly negative on $(\ell_B^2)^n$.

THEOREM 5.3. If A, C and U are as in Theorem 5.1, except that now C^*JC is uniformly negative on $\mathcal{H}_1$, then the operator matrix

$$M = \begin{bmatrix} A & B \\ C & D \end{bmatrix}$$

with components

$$B = (AA^* - I)^{\frac{1}{2}} U (A^*A - I)^{-\frac{1}{2}} C^* J$$

and

$$D = I + C(-C^*JC)^{-\frac{1}{2}}(I + A^*U)(-C^*JC)^{-\frac{1}{2}} C^* J$$

is $\tilde{J}$ unitary.

PROOF. The proof is by computation. ∎

The formulas for B and D are obtained just as in the discussion following Theorem 5.1, except that now, in place of (5.14), we choose

$$H = (AA^* - I)^{\frac{1}{2}} U (A^*A - I)^{-\frac{1}{2}}$$

and K as in (5.14), but with this new choice of H.

CASE 7. $\alpha\alpha^* - \beta\beta^* = 0$.

We begin much as in the discussion of Case 5.

THEOREM 5.5. Let A and C be a pair of bounded linear operators from $\mathcal{H}_1$ into $\mathcal{H}_1$ and $\mathcal{H}_1$ into $\mathcal{H}_2 \oplus \mathcal{H}_3$, respectively, such that

$$A^*A + C^*JC = I$$

and

$$C^*JC = 0 \ .$$

Then the matrix operator

$$M = \begin{bmatrix} A & B \\ C & D \end{bmatrix}$$

with components

$$B = -AC^*J$$

and

$$D = I - \frac{1}{2}CC^*J$$

is $\tilde{J}$ unitary.

Again, this theorem is easily verified by direct computation. The formulas themselves may be obtained much as in Case 5, only easier.

If $C^* J C = 0$, then $A^* A = I$ and it is readily seen that

$$H = -A$$

is a solution of (5.12). Under this condition, (5.13) reduces to

$$JC\{I + K^* + K\}C^* J = 0 \ ,$$

which is solved by choosing

$$K = -I/2 \ .$$

Substituting into (5.10) and (5.11) yields the indicated formulas for B and D.

THEOREM 5.6. *If, in the basic setting of Theorem 4.4, I is in force and* $\alpha\alpha^* - \beta\beta^* = 0$, *then the associated ULCS operator* $\Theta \in \mathcal{A}$ *given by (5.15) can be expressed in the form*

$$\Theta = I - u(I - Z V^*)^{-1}\Lambda^{-1}u^* J \ . \tag{5.24}$$

PROOF. This is an easy calculation. ∎

6. ADMISSIBILITY AND MORE ON GENERAL INTERPOLATION

We shall say, just as in [ADD], that an operator

$$X = [A \ \ B]$$

with components $A \in \mathcal{U}(\ell^2_{\mathcal{M}}; \ell^2_{\mathcal{M}})$ and $B \in \mathcal{U}(\ell^2_{\mathcal{N}}; \ell^2_{\mathcal{M}})$ is admissible if

$$\sum_{i,j=1}^{k} \{\langle \hat{A}(W_j)^* \rho_{W_j}^{-1} D_j, \hat{A}(W_i)^* \rho_{W_i}^{-1} D_i\rangle_{HS} - \langle \hat{B}(W_j)^* \rho_{W_j}^{-1} D_j, \hat{B}(W_i)^* \rho_{W_i}^{-1} D_i\rangle_{HS}\}$$

$$\geq 0$$

for every choice of $W_1, \ldots, W_k$ in $\mathcal{D}(\ell^2_{\mathcal{M}}; \ell^2_{\mathcal{M}})$ with $\ell_{W_j} < 1$ and $D_1, \ldots, D_k$ in $\mathcal{D}_2(\ell^2_{\mathcal{M}}; \ell^2_{\mathcal{M}})$.

Let M_A, M_B and $M_X = [M_A \ \ M_B]$ denote the restriction of the operators A, B and X, respectively, to $\mathcal{U}_2$ spaces of the appropriate size. Then

$$M_A^* = \underline{p}M_{A^*} \ , \qquad M_B^* = \underline{p}M_{B^*} \quad \text{and} \quad M_X^* = \underline{p}M_{X^*} \ ,$$

where $\underline{\underline{p}}$ designates the orthogonal projection of $\mathcal{X}_2$ onto $\mathcal{U}_2$. By Theorem 8.5 of [ADD], X is admissible if and only if the operator

$$\Gamma_X = M_X J M_X^*$$

is positive semidefinite on $\ell^2_\mathcal{M}$.

THEOREM 6.1. *The operator* $X = [A \ \ B]$ *with components* $A \in \mathcal{U}(\ell^2_\mathcal{M}; \ell^2_\mathcal{M})$ *and* $B \in \mathcal{U}(\ell^2_\mathcal{N}; \ell^2_\mathcal{M})$ *is admissible if and only if*

$$B = AS$$

for some $S \in \mathcal{S}(\ell^2_\mathcal{N}; \ell^2_\mathcal{M})$.

PROOF. The operator M_A maps the Hilbert space $\mathcal{U}_2(\ell^2_\mathcal{M}; \ell^2_\mathcal{M})$ into itself, whereas M_B maps the Hilbert space $\mathcal{U}_2(\ell^2_\mathcal{N}; \ell^2_\mathcal{N})$ into the Hilbert space $\mathcal{U}_2(\ell^2_\mathcal{N}; \ell^2_\mathcal{M})$. Therefore, by a well known construction (see e.g., the lemma in 1.14 of Rosenblum and Rovnyak [RR]), the inequality

$$M_A M_A^* \geq M_B M_B^*$$

holds if and only if there exists an operator C from $\mathcal{U}_2(\ell^2_\mathcal{N}; \ell^2_\mathcal{N})$ into $\mathcal{U}_2(\ell^2_\mathcal{N}; \ell^2_\mathcal{M})$ with $\|C\| \leq 1$ such that

$$M_B = M_A C .$$

Since C maps $\mathcal{U}_2$ into $\mathcal{U}_2$ it must belong to $\mathcal{U}$, i.e.,

$$C \in \mathcal{S}(\ell^2_\mathcal{N}; \ell^2_\mathcal{M}) .$$

This completes the proof in one direction, since X (as was noted just above) is admissible if and only if the stated inequality holds.

The converse is established in Theorem 5.1 of [ADD] for the case $\mathcal{M} = \mathcal{N}$. But the same proof goes through in the present setting also.   ∎

To each admissible X, we can associate a reproducing kernel Hilbert space

$$\mathcal{B}(X) = \mathrm{ran} \ \Gamma_X^{\frac{1}{2}} ,$$

acting on $\mathcal{U}_2(\ell^2_\mathcal{M}; \ell^2_\mathcal{M})$. The details are spelled out in [ADD] (for the case $\mathcal{M} = \mathcal{N}$, but they are easily adapted to the present setting). In particular,

$$\Gamma_X \rho_W^{-1} E = X J \hat{X}(W)^* \rho_W^{-1} E \tag{6.1}$$

and

$$\langle F, \Gamma_X \rho_W^{-1} E \rangle_{\mathcal{B}(X)} = \mathrm{trace} \ E^* \hat{F}(W) , \tag{6.2}$$

for every choice of $F \in \mathcal{B}(X)$, $E \in \mathcal{D}_2(\ell^2_{\mathcal{M}}; \ell^2_{\mathcal{M}})$ and $W \in \mathcal{D}(\ell^2_{\mathcal{M}}; \ell^2_{\mathcal{M}})$ with $\ell_W < 1$. This serves to identify

$$\Lambda_W = X J \hat{X}(W)^* \rho_W^{-1} \tag{6.3}$$

as the reproducing kernel for $\mathcal{B}(X)$. It is also well to recall that

$$\langle \Gamma_X G, \Gamma_X H \rangle_{\mathcal{B}(X)} = \langle \Gamma_X G, H \rangle_{HS} \tag{6.4}$$

for G and H in $\mathcal{U}_2(\ell^2_{\mathcal{M}}; \ell^2_{\mathcal{M}})$.

THEOREM 6.2. *Suppose that in the basic setting of Theorem 4.4 assumptions I and II are in force and that $\Theta \in \mathcal{A}$ is an associated ULCS operator. Suppose further that there exists an $S \in \mathcal{S}(\ell^2_{\mathcal{N}}; \ell^2_{\mathcal{M}})$ such that*

$$\underline{p} S^* G_i E = H_i E \tag{6.5}$$

for $i = 1, \ldots, n$ and every choice of $E \in \mathcal{D}_2(\ell^2_{\mathcal{B}}; \ell^2_{\mathcal{B}})$. Then

$$Y = [I \quad -S]\Theta$$

is admissible. (The converse is also true, and will be established in Theorem 6.3.)

PROOF. Let

$$X = [I \quad -S], \quad \text{so that} \quad Y = X\Theta,$$

and set

$$L = \sum_{j=1}^{k} \rho_{W_j}^{-1} D_j$$

for some choice of $W_1, \ldots, W_k$ in $\mathcal{D}$ with $\ell_{W_j} < 1$ and $D_1, \ldots, D_k$ in $\mathcal{D}_2$. It suffices to show that

$$\langle M_Y J M_Y^* L, L \rangle_{HS} \geq 0$$

for every L of the indicated form. The proof rests on the decomposition

$$M_Y J M_Y^* = \Gamma_X - M_X \Gamma_\Theta M_X^*$$

in terms of the operators $\Gamma_X = M_X J M_X^*$ and $\Gamma_\Theta = J - M_\Theta J M_\Theta^*$, both of which are discussed fairly extensively in Section 8 of [ADD], and the evaluations

$$\langle \Gamma_X L, L \rangle_{HS} = \|\Gamma_X L\|^2_{\mathcal{B}(X)}$$

and

$$\langle M_X \Gamma_\Theta M_X^* L, L \rangle_{HS} = \|P \Gamma_X L\|^2_{\mathcal{B}(X)},$$

where P denotes the orthogonal projection of $\mathcal{B}(X)$ onto a closed subspace thereof, which will be identified in the course of the proof.

The asserted positivity of $M_Y J M_Y^*$ is then immediate, since

$$\|\Gamma_X L\|^2_{\mathcal{B}(X)} \geq \|P\Gamma_X L\|^2_{\mathcal{B}(X)} \ .$$

It remains therefore to obtain the indicated evaluations. This is somewhat lengthy and is carried out in steps.

The notation P_0 for the orthogonal projection of $\mathcal{X}_2(\ell^2_\mathcal{M}; \ell^2_\mathcal{B})$ onto $\mathcal{D}_2(\ell^2_\mathcal{M}; \ell^2_\mathcal{B})$ will prove convenient.

STEP 1.

$$F_i E = J M_X^* G_i E$$

and

$$X F_i E = \Gamma_X G_i E$$

for $i = 1, \ldots, n$ and every choice of $E \in \mathcal{D}_2$.

PROOF OF STEP 1. This is immediate from the definitions.

STEP 2.

$$\Gamma_\Theta M_X^* \rho_W^{-1} D = \sum_{s,t=1}^n F_s (\Lambda^{-1})_{st} P_0 G_t^* \Gamma_X \rho_W^{-1} D$$

for every choice of $W \in \mathcal{D}(\ell^2_\mathcal{M}; \ell^2_\mathcal{M})$ with $\ell_W < 1$ and $D \in \mathcal{D}_2(\ell^2_\mathcal{M}; \ell^2_\mathcal{M})$.

PROOF OF STEP 2. Let

$$R = \Gamma_\Theta M_X^* \rho_W^{-1} D$$

for some permitted choice of W and D. Then, since $\mathcal{H}(\Theta)$ and ran Γ_Θ coincide as sets (by Lemma 8.2 of [ADD]), R must belong to $\mathcal{H}(\Theta)$. Therefore, by Theorem 4.4,

$$R = \sum_{s=1}^n F_s \delta_s$$

for some choice of $\delta_1, \ldots, \delta_n$ in $\mathcal{D}_2$. But this in turn implies that

$$\langle JR, F_i E \rangle_{HS} = \langle J\{J - M_\Theta J M_\Theta^*\} M_X^* \rho_W^{-1} D, F_i E \rangle_{HS}$$

$$= \langle M_X^* \rho_W^{-1} D, F_i E \rangle_{HS}$$

$$= \langle \Gamma_X \rho_W^{-1} D, G_i E \rangle_{HS}$$

$$= \operatorname{trace} E^* P_0 G_i^* \Gamma_X \rho_W^{-1} D \ .$$

On the other hand,

$$\langle JR, F_i E\rangle_{HS} = \sum_{j=1}^{n} \langle JF_j\delta_j, F_i E\rangle_{HS}$$

$$= \text{trace } E^* \sum_{j=1}^{n} \Lambda_{ij}\delta_j \ ,$$

which, upon matching expressions, yields the identities

$$P_0 G_i^* \Gamma_X \rho_W^{-1} D = \sum_{j=1}^{n} \Lambda_{ij}\delta_j$$

for $i = 1, \ldots, n$. Thus

$$\delta_i = \sum_{j=1}^{n} (\Lambda^{-1})_{ij} P_0 G_j^* \Gamma_X \rho_W^{-1} D$$

for $j = 1, \ldots, n$ which, in conjunction with the formula $R = \Sigma F_i \delta_i$, serves to complete the proof of this step.

STEP 3.

$$\Gamma_X L \in \mathcal{B}(X) \quad and \quad \langle \Gamma_X L, L\rangle_{HS} = \|\Gamma_X L\|^2_{\mathcal{B}(X)} \ .$$

PROOF OF STEP 3. The asserted inclusion is immediate from Theorem 8.6 of [ADD] since $\rho_W^{-1} D \in \mathcal{U}_2$ for every choice of $W \in \mathcal{D}$ with $\ell_W < 1$ and $D \in \mathcal{D}_2$. The asserted identity may then be verified either by evaluating each side separately, or by invoking (3) of Theorem 8.1 of [ADD].

STEP 4.

$$\langle \Gamma_\Theta M_X^* L, M_X^* L\rangle_{HS} = \text{trace} \sum_{s,t=1}^{n} (P_0 G_s^* \Gamma_X L)^* (\Lambda^{-1})_{st} (P_0 G_t^* \Gamma_X L) \ .$$

PROOF OF STEP 4. By Step 2,

$$\langle \Gamma_\Theta M_X^* L, M_X^* L\rangle_{HS} = \sum_{s,t=1}^{n} \langle F_s Q_{st}, M_X^* L\rangle_{HS} \ ,$$

where

$$Q_{st} = (\Lambda^{-1})_{st} P_0 G_t^* \Gamma_X L \ ,$$

for short. By Step 1, the last inner product is equal to

$$\langle F_s Q_{st}, M_X^* L \rangle_{HS} = \langle J M_X^* G_s Q_{st}, M_X^* L \rangle_{HS}$$

$$= \langle G_s Q_{st}, \Gamma_X L \rangle_{HS}$$

$$= \{ \langle \Gamma_X L, G_s Q_{st} \rangle_{HS} \}^*$$

$$= \{ \text{trace } Q_{st}^* P_0 G_s^* \Gamma_X L \}^*$$

$$= \text{trace}(P_0 G_s^* \Gamma_X L)^* Q_{st} \ .$$

The desired result now drops out easily by combining terms.

STEP 5. *Let*

$$\mathcal{C} = \left\{ \sum_{i=1}^{n} \Gamma_X G_t E_t : \ E_t \in \mathcal{D}_2 \right\} \ .$$

Then $\mathcal{C}$ is a closed subspace of $\mathcal{B}(X)$ and, upon letting P denote the orthogonal projection of $\mathcal{B}(X)$ onto $\mathcal{C}$,

$$\langle \Gamma_\Theta M_X^* L, M_X^* L \rangle_{HS} = \| P \Gamma_X L \|_{\mathcal{B}(X)}^2 \ .$$

PROOF OF STEP 5. First, since

$$\langle \Gamma_X G_t E_t, \Gamma_X G_s E_s \rangle_{\mathcal{B}(X)} = \langle \Gamma_X G_t E_t, G_s E_s \rangle_{HS}$$

$$= \langle J M_X^* G_t E_t, M_X^* G_s E_s \rangle_{HS}$$

$$= \langle J F_t E_t, F_s E_s \rangle_{HS} \quad \text{(by Step 1)}$$

$$= \text{trace } E_s^* \Lambda_{st} E_t$$

and Λ is uniformly positive on $(\ell_B^2)^n$, it follows, just as in the proof of Theorem 4.7, that $\mathcal{C}$ is a closed subspace of $\mathcal{B}(X)$. Therefore,

$$P \Gamma_X L = \sum_{s=1}^{n} \Gamma_X G_s \epsilon_s$$

for some choice of ϵ_s in $\mathcal{D}_2$. These coefficients can be evaluated much as in the proof of Step 2:

$$\langle P \Gamma_X L, \Gamma_X G_t E \rangle_{\mathcal{B}(X)} = \sum_{s=1}^{n} \langle \Gamma_X G_s \epsilon_s, \Gamma_X G_t E \rangle_{\mathcal{B}(X)}$$

$$= \text{trace } E^* \sum_{s=1}^{n} \Lambda_{ts} \epsilon_s$$

and also

$$\langle P\Gamma_X L, \Gamma_X G_t E\rangle_{\mathcal{B}(X)} = \langle \Gamma_X L, \Gamma_X G_t E\rangle_{\mathcal{B}(X)}$$

$$= \langle \Gamma_X L, G_t E\rangle_{HS}$$

$$= \text{trace } E^* P_0 G_t^* \Gamma_X L$$

for every choice of $E \in \mathcal{D}_2$. Therefore

$$P_0 G_t^* \Gamma_X L = \sum_{s=1}^{n} \Lambda_{ts}\epsilon_s$$

for $s = 1, \ldots, n$, and hence

$$\epsilon_s = \sum_{t=1}^{n} (\Lambda^{-1})_{st} P_0 G_t^* \Gamma_X L$$

for $s = 1, \ldots, n$. Consequently

$$\|P\Gamma_X L\|_{\mathcal{B}(X)}^2 = \sum_{s,t=1}^{n} \langle \Gamma_X G_t \epsilon_t, \Gamma_X G_s \epsilon_s\rangle_{\mathcal{B}(X)}$$

$$= \text{trace } \sum_{s,t=1}^{n} \epsilon_s^* \Lambda_{st}\epsilon_t ,$$

which, with the help of Step 4, is readily seen to yield the asserted identity.	∎

THEOREM 6.3. *Suppose that in the basic setting of Theorem 4.4 conditions I and II are in force, that $\Theta \in \mathcal{A}$ is an associated ULCS operator, and that $S \in \mathcal{S}(\ell_N^2, \ell_M^2)$. Then the following are equivalent:*

(1) $S = T_\Theta[S_L]$
   for some choice of $S_L \in \mathcal{S}(\ell_N^2; \ell_M^2)$.

(2) $[I \quad - S]\Theta$ is admissible.

(3) $\underline{\underline{p}}S^* G_i E = H_i E$
for $i = 1, \ldots, n$ and every choice of $E \in \mathcal{D}_2(\ell_B^2; \ell_B^2)$.

PROOF. The proof is broken into steps. Therein we shall once again use $A, \ldots, D$ for the blocks of Θ, as in the proof of Theorem 4.5.

STEP 1. (1) $\Longleftrightarrow$ (2).

PROOF OF STEP 1. If $S = T_\Theta[S_L]$, then it is readily checked that

$$B - SD = -(A - SC)S_L$$

and hence that

$$[I \quad -S]\Theta = -(A - SC)[I \quad -S_L] \; .$$

(2) is now immediate from Theorem 6.1.

On the other hand, if (2) holds, then

$$[I \quad -S]\Theta = Q[I \quad -S_L]$$

for some choice of $Q \in \mathcal{U}$ and $S_L \in \mathcal{S}$, by another application of Theorem 6.1. Thus, by a straightforward calculation,

$$B - SD = -(A - SC)S_L \; ,$$

from which in turn it follows that

$$S(CS_L + D) = AS_L + B \; .$$

Since D is invertible in $\mathcal{U}$ and (by Lemma 5.2 of [ADD]) $\|D^{-1}C\| < 1$, $CS_L + D$ is invertible in $\mathcal{U}$ and

$$S = (AS_L + B)(CS_L + D)^{-1}$$

as claimed.

STEP 2. $(2) \Longleftrightarrow (3)$.

PROOF OF STEP 2. This is now easy because the work has already been done: $(3) \Longrightarrow (2)$ by Theorem 6.2 and $(1) \Longrightarrow (3)$ by Theorem 4.5. The desired result is now immediate from Step 1. ∎

For the counterpart of Theorem 6.3 in a more conventional setting, see e.g., Lemma 7.1 and Theorem 7.2 of [D2] and Theorem 2.4 of [AD].

7. NEVANLINNA-PICK INTERPOLATION

7.1. The NP problem

We turn now to the NP interpolation problem which was formulated in the first section. The conditions on the data will be expressed in terms of the $n \times n$ matrices

$$\Delta = [\Delta_{ij}] \quad \text{and} \quad \Lambda = [\Lambda_{ij}] \; , \qquad i, j = 1, \ldots, n \; ,$$

of the operators

$$\Delta_{ij} = \{\xi_i^* \xi_j \rho_{V_j}^{-1}\}^{\wedge}(V_i)$$

$$= \sum_{t=0}^{\infty} V_i^{[t]} (\xi_i^* \xi_j)^{(t)} V_j^{[t]*} \tag{7.1}$$

and

$$\Lambda_{ij} = \Delta_{ij} - \{\eta_i^* \eta_j \rho_{V_j}^{-1}\}^\wedge(V_i)$$

$$= \sum_{t=0}^{\infty} V_i^{[t]}(\xi_i^* \xi_j - \eta_i^* \eta_j)^{(t)} V_j^{[t]*} \ . \tag{7.2}$$

In the setting of the NP problem,

$$Fe_i = \begin{bmatrix} \alpha^* \\ \beta^* \end{bmatrix} (I - \mathbf{Z}\mathbf{V}^*)^{-1} e_i$$

$$= \begin{bmatrix} \xi_i \\ \eta_i \end{bmatrix} (I - ZV_i^*)^{-1} \ , \qquad i = 1,\ldots,n \ .$$

Thus the components of

$$F_i = Fe_i = \begin{bmatrix} G_i \\ H_i \end{bmatrix} \ , \qquad i = 1,\ldots,n \ ,$$

are given by the formulas

$$G_i = \xi_i \rho_{V_i}^{-1} \tag{7.3}$$

and

$$H_i = \eta \rho_{V_i}^{-1} \ , \tag{7.4}$$

respectively.

WARNING: *Throughout this section, all references to* Δ, Λ, F_i *and* G_i *will refer to the explicit choices made in (7.1) to (7.4).*

The basic observation which makes the general analysis carried out in Sections 4 and 6 applicable is:

THEOREM 7.1. *An operator* $S \in \mathcal{S}(\ell_\mathcal{N}^2; \ell_\mathcal{M}^2)$ *is a solution of the NP problem if and only if*

$$\underline{p}S^* G_i E = H_i E \tag{7.5}$$

for $i = 1,\ldots,n$, *and every choice of* $E \in \mathcal{D}_2(\ell_\mathcal{B}^2; \ell_\mathcal{B}^2)$. *(Here, of course,* G_i *and* H_i *are specified by (7.3) and (7.4), respectively.)*

PROOF. In the present setting

$$\underline{p}S^* G_i E = \underline{p}(\xi_i^* S)^* \rho_{V_i}^{-1} E$$

$$= (\xi_i^* S)^\wedge(V_i)^* \rho_{V_i}^{-1} E$$

and

$$H_i E = \eta_i \rho_{V_i}^{-1} E \ .$$

Thus, if (7.5) holds, then

$$(\xi_i^* S)^\wedge (V_i)^* \rho_{V_i}^{-1} E = \eta_i \rho_{V_i}^{-1} E$$

and hence the diagonal components of both sides must also match:

$$(\xi_i^* S)^\wedge (V_i)^* E = \eta_i E$$

for $i = 1, \ldots, n$ and every $E \in \mathcal{D}_2$. But this in turn implies that

$$(\xi_i^* S)^\wedge (V_i) = \eta_i^*$$

for $i = 1, \ldots, n$ and so exhibits S as a solution of the NP problem.

Conversely, if S is a solution of the NP problem, then it is easily verified that (7.5) holds via the evaluations in the first few lines of the proof. ∎

THEOREM 7.2. *If the NP problem is solvable, then the operator matrix Λ is positive semidefinite on $(\ell_B^2)^n$.*

In the other direction, if conditions I and II are in force, then the NP problem is solvable and, if $\Theta \in \mathcal{A}$ denotes an associated ULCS operator (e.g., as defined in (4.17)–(4.20)), then

$$\{T_\Theta[S_L]: \; S_L \in \mathcal{S}(\ell_\mathcal{N}^2; \ell_\mathcal{M}^2)\}$$

is a complete list of the set of solutions.

PROOF. The first assertion is immediate from Theorem 4.6.

In the other direction, the supplementary condition $\ell_{V_j} < 1$, $j = 1, \ldots, n$, which is imposed in the statement of the NP problem, implies that $r_{sp}(Z V^*) < 1$ and hence, by Theorem 4.4, conditions I and II guarantee the existence of an associated ULCS operator $\Theta \in \mathcal{A}$. Thus, by Theorems 6.3 and 7.1, the NP problem is solvable and the set of all solutions is equal to the indicated set of linear fractional transformations. ∎

THEOREM 7.3. *If $\mathcal{B}$ is finite dimensional, then the NP problem is solvable if and only if Λ is positive semidefinite on $(\ell_B^2)^n$.*

If Λ is uniformly positive, then

$$\{T_\Theta[S_L]: \; S_L \in \mathcal{S}(\ell_\mathcal{N}^2; \ell_\mathcal{M}^2)\}$$

is a complete list of the set of solutions.

PROOF. In view of Theorem 7.2, it remains only to show the sufficiency of the stated condition for the existence of a solution. To this end, fix $\epsilon > 0$ and, just as in

[D1], consider the new interpolation problem with η_j replaced by $(1+\epsilon)^{-\frac{1}{2}}\eta_j$, $j = 1, \ldots, n$ and set

$$\Lambda_{ij}^{\epsilon} = \sum_{t=0}^{\infty} V_i^{[t]} (\xi_i^* \xi_j - \frac{\eta_j^* \eta_j}{1+\epsilon})^{(t)} V_j^{[t]*}$$

$$= \frac{1}{1+\epsilon} \Lambda_{ij} + \frac{\epsilon}{1+\epsilon} \sum_{t=0}^{\infty} V_i^{[t]} (\xi_i^* \xi_j)^{(t)} V_j^{[t]*} \ .$$

Thus,

$$\Lambda^{\epsilon} = \frac{1}{1+\epsilon} \Lambda + \frac{\epsilon}{1+\epsilon} \Delta$$

$$\geq \frac{\epsilon}{1+\epsilon} \Delta > 0$$

meets condition I for every $\epsilon > 0$. The assumption that $\mathcal{B}$ is finite dimensional further guarantees that condition II is met for the perturbed CF problem for every $\epsilon > 0$ and hence, by Therem 7.2, there exists an $F^{\epsilon} \in \mathcal{S}(\ell_{\mathcal{N}}^2; \ell_{\mathcal{M}}^2)$ such that

$$(\xi_j^* F^{\epsilon})^{\wedge}(V_j) = (1 + \epsilon)^{-\frac{1}{2}} \eta_j^*$$

for $j = 1, \ldots, n$ and every $\epsilon > 0$.

The proof is now completed by letting $\epsilon \downarrow 0$ and invoking the fact that the unit ball in $\mathcal{X}$ is compact in the weak operator topology (see e.g., Lemma 2.3 on p.102 of [Be]). This means that there exists an $F \in \mathcal{X}$ with $\|F\| \leq 1$ such that

$$\lim_{\epsilon \downarrow 0} \langle F^{\epsilon} f, g \rangle_{\ell_{\mathcal{M}}^2} = \langle F f, g \rangle_{\ell_{\mathcal{M}}^2}$$

for every choice of $f \subset \ell_{\mathcal{N}}^2$ and $g \in \ell_{\mathcal{M}}^2$, as $\epsilon \downarrow 0$ through an appropriately chosen subsequence (which we shall not indicate in the notation). Since

$$\pi^* Z^i F^{\epsilon} Z^{*j} \pi = 0 \quad \text{for} \quad i > j$$

it follows readily that the same holds true for F, i.e., $F \in \mathcal{S}$.

Finally, it remains to check that F is an interpolant. To this end, it suffices to justify the evaluation

$$\text{trace } D^*(\xi_j^* F)^{\wedge}(V_j)E = \langle FE, \xi_j \rho_{V_j}^{-1} D \rangle_{HS}$$

$$= \lim_{\epsilon \downarrow 0} \langle F^{\epsilon} E, \xi_j \rho_{V_j}^{-1} D \rangle_{HS}$$

$$= \lim_{\epsilon \downarrow 0} (1 + \epsilon)^{-\frac{1}{2}} \text{ trace } D^* \eta_j^* E$$

$$= \text{trace } D^* \eta_j^* E , \qquad j = 1, \ldots, n ,$$

for a sufficiently rich class of D and E in $\mathcal{D}_2$. The delicate point is the interchange of limits in the passage from line 1 to line 2, but since D and E may be chosen to have finite rank, this presents no real difficulty. The details are left to the reader.   ∎

We remark that the condition that $\mathcal{B}$ be finite dimensional can obviously be relaxed. A glance at the proof indicates that what is really needed is that there exist a $\Theta \in \mathcal{A}$ for every choice of the perturbed data.

THEOREM 7.4. *If condition II is in force, then the NP problem admits a strictly contractive solution if and only if I holds. Moreover, in this instance*

$$\{T_\Theta[S_L]: \ S_L \in \mathcal{S}(\ell^2_{\mathcal{M}}; \ell^2_{\mathcal{N}}) \ \text{ and } \ \|S\| < 1\}$$

is equal to the set of all strictly contractive solutions of the NP problem.

PROOF. Suppose first that the NP problem admits a strictly contractive solution. Then, by Theorem 4.6, I must hold.

Conversely, if I holds, then, by Theorem 7.2, the NP problem is solvable and $\{T_\Theta[S_L]: \ S_L \in \mathcal{S}\}$ is a complete list of the set of all solutions to the NP problem. By Lemma 4.1, the set of strictly contractive solutions is obtained by restricting S_L to be strictly contractive.   ∎

COROLLARY. *If $\mathcal{B}$ is finite dimensional, then the NP problem admits a strictly contractive solution if and only if the operator Λ defined by (7.2) is uniformly positive on $(\ell^2_{\mathcal{B}})^n$. Moreover, in this instance,*

$$\{T_\Theta[S_L]: \ S_L \in \mathcal{S}(\ell^2_{\mathcal{N}}; \ell^2_{\mathcal{M}}) \ \text{ and } \ \|S\| < 1\}$$

is equal to the set of all strictly contractive solutions to the NP problem. (Θ may be defined by (4.17)-(4.20).)

PROOF. If $\mathcal{B}$ is finite dimensional, then, by Theorem 4.2, condition II is in force. Therefore Theorem 7.4 is applicable.   ∎

7.2. Another look at Δ

It is instructive to analyze the role of condition III in the more explicit setting of the NP problem. In particular, condition III, which is imposed in the statement of the NP problem, ensures that the operators $\xi_j \rho_{V_j}^{-1}$, $j = 1, \ldots, n$, are right linearly independent and that the set of linear combinations over $\mathcal{D}_2$ is a Hilbert space; see Lemma 4.2 and Theorem 4.7. It is not enough to simply choose the points $V_1, \ldots, V_n$ all different. Thus, for example, if $n = 2$ and if $(V_1)_{jj} = 0$ for $j \neq 0$ and $(V_2)_{jj} = 0$ for $j \neq 1$, then

$$\rho_{V_1}^{-1} = I + ZV_1 \, ,$$

$$\rho_{V_2}^{-1} = I + ZV_2 \, ,$$

and it is readily seen that there are many nonzero choices of D_1 and D_2 in $\mathcal{D}$ for which

$$\rho_{V_1}^{-1} D_1 + \rho_{V_2}^{-1} D_2 = 0 \; :$$

$D_1 = -D_2$ and $(D_1)_{jj} = 0$ for $j = 0$ and 1, but otherwise arbitrary, will do.

Condition III is also enough to ensure that the simple interpolation problem without norm constraints is solvable:

THEOREM 7.5. *Let* $\xi_j \in \mathcal{D}(\ell_B^2; \ell_\mathcal{M}^2)$, $\eta_j \in \mathcal{D}(\ell_B^2; \ell_\mathcal{N}^2)$ *and* $V_j \in \mathcal{D}(\ell_B^2; \ell_B^2)$ *with* $\ell_{V_j} < 1$ *for* $j = 1, \ldots, n$ *and suppose that the operator matrix* Δ *is uniformly positive on* $(\ell_B^2)^n$. *Then* $\Delta^{-1} \in \{\mathcal{D}(\ell_B^2; \ell_B^2)\}^{n \times n}$,

$$F = \sum_{i,j=1}^{n} \xi_i \rho_{V_i}^{-1} (\Delta^{-1})_{ij} \eta_j^*$$

belongs to $\mathcal{U}$ *for every choice of* $\eta_1, \ldots, \eta_n$ *in* $\mathcal{D}(\ell_B^2; \ell_\mathcal{N}^2)$ *and*

$$(\xi_t^* F)^\wedge(V_t) = \eta_t^*$$

for $t = 1, \ldots, n$.

PROOF. The first assertion is immediate from Theorem 2.4 applied to Δ. The rest is a straightforward calculation:

$$(\xi_t^* F)^\wedge(V_t) = \sum_{i,j=1}^{n} (\xi_t^* \xi_i \rho_{V_i}^{-1})^\wedge(V_t)(\Delta^{-1})_{ij} \eta_j^*$$

$$= \sum_{i,j=1}^{n} \Delta_{ti}(\Delta^{-1})_{ij} \eta_j^*$$

$$= \sum_{j=1}^{n} (\Delta \Delta^{-1})_{tj} \eta_j^*$$

$$= \eta_t^* \; . \qquad\blacksquare$$

8. CARATHÉODORY-FEJÉR INTERPOLATION

We turn now to the CF interpolation problem for upper triangular operators which was formulated in Section 1. It is convenient, however, to first prepare some needed evaluations in a preliminary subsection. The CF problem itself is then treated in a second subsection.

8.1. Preliminary evaluations

Let

$$\varphi_{j,V} = (\rho_V^{-1} Z)^j \rho_V^{-1} \, , \tag{8.1}$$

for $V \in \mathcal{D}$ with $\ell_V < 1$ and $j = 0, 1, \dots$. The operator (8.1) serves to evaluate the j'th coefficient in the expansion of an operator $F \in \mathcal{U}$ in powers of $Z - V$. It plays the same role as the function φ_j defined in the first line of (6.2) of [D1].

LEMMA 8.1. *If $V \in \mathcal{D}$ with $\ell_V < 1$ and if D and E are arbitrary elements in $\mathcal{D}_2$, then for any choice of $j, k = 0, 1, \dots$,*

$$\langle (Z - V)^j D, \varphi_{k,V} E \rangle_{HS} = \begin{cases} \text{trace } E^* D & \text{if} \quad j = k \\ 0 & \text{if} \quad j \neq k \end{cases}$$

PROOF. Clearly

$$(\rho_V^{-1} Z)^* = \{ (ZZ^* - ZV^*)^{-1} Z \}^*$$

$$= (Z - V)^{-1} \, .$$

Thus, if $j \geq k$, then

$$\langle (Z - V)^j D, \ \varphi_{k,V} E \rangle_{HS} = \langle (Z - V)^{j-k} D, \ \rho_V^{-1} E \rangle_{HS}$$

$$= \text{trace } E^* \{ (Z - V)^{j-k} D \}^\wedge (V)$$

$$= \begin{cases} \text{trace } E^* D & \text{if} \quad j = k \\ 0 & \text{if} \quad j > k \, . \end{cases}$$

On the other hand, if $j < k$, then

$$\langle (Z - V)^j D, (\rho_V^{-1} Z)^k \rho_V^{-1} E \rangle_{HS} = \langle D, (\rho_V^{-1} Z)^{k-j} \rho_V^{-1} E \rangle_{HS}$$

$$= 0 \, ,$$

since $D \in \mathcal{D}_2$, $(\rho_V^{-1} Z)^{k-j} \rho_V^{-1} E \in Z^{k-j} \mathcal{U}_2$ and these two spaces are orthogonal in $\mathcal{X}_2$. ∎

Before stating the next lemma, it is well recall that for every $F \in \mathcal{U}$ and every $V \in \mathcal{D}$ with $\ell_V < 1$ there exists a unique set of diagonal operators $F_{\{j\}} = F_{\{j\}}(V)$, $j = 0, 1, \dots$, such that

$$F - \sum_{j=0}^{n} (Z - V)^j F_{\{j\}} \in (Z - V)^{n+1} \mathcal{U} \, , \tag{8.2}$$

for $n = 0, 1, \ldots$. This may, as we have already noted in Section 1, be verified by repeated application of Theorem 3.3 of [ADD].

LEMMA 8.2. *If $V \in \mathcal{D}$ with $\ell_V < 1$ and $F \in \mathcal{U}$, then*

$$\langle FD, \ \varphi_{k,V} E \rangle_{HS} = \text{trace } E^* F_{\{k\}}(V) D$$

for $k = 0, 1, \ldots$, and every choice of D and E in $\mathcal{D}_2$.

PROOF. Fix k and choose $n \geq k$. Then, since

$$F = \sum_{j=0}^{n} (Z - V)^j F_{\{j\}} + (Z - V)^{n+1} G$$

for some choice of $G \in \mathcal{U}$ and

$$\langle (Z - V)^{n+1} GD, \ \varphi_{k,V} E \rangle_{HS} = \langle (Z - V)^{n+1-k} GD, \ \rho_V^{-1} E \rangle_{HS}$$

$$= \text{trace } E^* \{(Z - V)^{n+1-k} G\}^{\wedge}(V) D$$

$$= \text{trace } E^* \{(Z - V)^{\wedge}(V)(Z - V)^{n-k} G\}^{\wedge}(V) D$$

$$= 0 \ ,$$

the evaluation of interest is equal to

$$\sum_{j=0}^{n} \langle (Z - V)^j F_{\{j\}} D, \ \varphi_{k,V} E \rangle_{HS} \ .$$

But this in turn is easily evaluated by the last lemma.   ∎

LEMMA 8.3. *If $V \in \mathcal{D}$ with $\ell_V < 1$ and if $F \in \mathcal{U}$, $F_{\{j\}} = F_{\{j\}}(V)$ and $\varphi_j = \varphi_{j,V}$, then*

$$\underline{p} F^* \varphi_k D = \{ F_{\{0\}}^* \varphi_k + F_{\{1\}}^* \varphi_{k-1} + \cdots + F_{\{k\}}^* \varphi_0 \} D \tag{8.3}$$

for every choice of $D \in \mathcal{D}_2$.

PROOF. Since

$$F = \sum_{j=0}^{k} (Z - V)^j F_{\{j\}} + (Z - V)^{k+1} G$$

with $G \in \mathcal{U}$ and

$$Z^* - V^* = Z^* \rho_V \ ,$$

it is readily seen that

$$F^* \varphi_k D = \sum_{j=0}^{k} F_{\{j\}}^* (Z^* \rho_V)^j \varphi_k D + G^* (Z^* \rho_V)^{k+1} \varphi_k D$$

$$= \sum_{j=0}^{k} F_{\{j\}}^* \varphi_{k-j} D + G^* Z^* D \ .$$

The rest is plain since the sum belongs to $\mathcal{U}_2$, whereas $G^* Z^* D \in Z^* \mathcal{L}_2$, and these two spaces are orthogonal in $\mathcal{X}_2$. ∎

We remark that formula (8.3) is the natural analogue in the present setting of the evaluation given in Lemma 6.1 of [D1].

LEMMA 8.4. *If $V \in \mathcal{D}$ with $\ell_V < 1$ and $F \in \mathcal{U}$, then*

$$F = F_{\{0\}} + (Z - V)H$$

for some choice of $H \in \mathcal{U}$,

$$[(Z - V)F]_{\{j\}}(V) = \begin{cases} 0 & for \quad j = 0 \\ F_{\{j-1\}}(V) & for \quad j \geq 1 \end{cases}$$

and

$$H_{\{j\}}(V) = F_{\{j+1\}}(V) \ , \quad for \quad j = 0, 1, \ldots \ .$$

PROOF. This is an elementary consequence of the uniqueness of the coefficients in expansions of the form (8.2). ∎

The next result is the analogue of the classical convolution formula in the present setting.

THEOREM 8.1. *If $V \in \mathcal{D}$ with $\ell_V < 1$ and $F \in \mathcal{U}$, $G \in \mathcal{U}$ are expressed in the form*

$$F = F_{\{0\}} + (Z - V)F_{\{1\}} + \cdots$$

$$G = G_{\{0\}} + (Z - V)G_{\{1\}} + \cdots \ ,$$

then, in the corresponding representation of the product FG, the j-th coefficient

$$(FG)_{\{j\}} = \sum_{i=0}^{j} (F_{\{i\}}G)_{\{j-i\}} \ . \tag{8.4}$$

PROOF. The proof is by induction. By Theorem 1.2, formula (8.4) is valid for $j = 0$. Suppose next that it is valid for $j = k - 1$ for some $k > 1$ and consider the case $j = k$. By definition,

$$(FG)_{\{k\}} = \left\{ (Z - V)^{-k}[FG - \sum_{s=0}^{k-1}(Z - V)^s(FG)_{\{s\}}] \right\}^{\wedge}(V) \ .$$

By Theorem 3.3 of [ADD],

$$F = F_{\{0\}} + (Z - V)H$$

for some choice of $H \in \mathcal{U}$. Thus, in terms of $F_{\{0\}}$ and H,

$$FG - \sum_{s=0}^{k-1}(Z - V)^s(FG)_{\{s\}} = F_{\{0\}}G - \sum_{s=0}^{k-1}(Z - V)^s(F_{\{0\}}G)_{\{s\}}$$

$$+ (Z - V)HG - \sum_{s=0}^{k-1}(Z - V)^s((Z - V)HG)_{\{s\}} \ .$$

With the help of Lemma 8.4, the sum of the last two terms can be reexpressed as

$$(Z - V)HG - \sum_{s=1}^{k-1}(Z - V)^s(HG)_{\{s-1\}}$$

$$= (Z - V)\left[HG - \sum_{s=0}^{k-2}(Z - V)^s(HG)_{\{s\}} \right]$$

and hence

$$(FG)_{\{k\}} = (F_{\{0\}}G)_{\{k\}} + (HG)_{\{k-1\}} \ .$$

By the induction hypothesis, and the last lemma,

$$(HG)_{\{k-1\}} = \sum_{t=0}^{k-1}(H_{\{t\}}G)_{\{k-1-t\}}$$

$$= \sum_{t=0}^{k-1}(F_{\{t+1\}}G)_{\{k-1-t\}}$$

$$= \sum_{t=1}^{k}(F_{\{t\}}G)_{\{k-t\}} \ .$$

The validity of (8.4) for the case $j = k$ now drops out easily upon combining the last two formulas. ∎

8.2. The CF problem

Let V, α and β be defined in terms of the data of the CF problem as in (1.9), let

$$\Lambda = \sum_{t=0}^{\infty} V^{[t]}(\alpha\alpha^* - \beta\beta^*)^{(t)}V^{[t]*} \tag{8.5}$$

be the unique solution of the equation

$$\Lambda - V\Lambda^{(1)}V^* = \alpha\alpha^* - \beta\beta^* . \tag{8.6}$$

and for the rest of this section let

$$F = \begin{bmatrix} \alpha^* \\ \beta^* \end{bmatrix} (I - ZV^*)^{-1}$$

$$= \begin{bmatrix} \xi & 0 & \cdots & 0 \\ \eta_0 & \eta_1 & \cdots & \eta_n \end{bmatrix} \begin{bmatrix} \rho_V & -Z & 0 & \cdots & 0 & 0 \\ 0 & \rho_V & -Z & \cdots & 0 & 0 \\ \vdots & \vdots & \vdots & & \vdots & \vdots \\ 0 & 0 & 0 & \cdots & \rho_V & -Z \\ 0 & 0 & 0 & \cdots & 0 & \rho_V \end{bmatrix}^{-1} \tag{8.7}$$

$$= \begin{bmatrix} \xi & 0 & \cdots & 0 \\ \eta_0 & \eta_1 & \cdots & \eta_n \end{bmatrix} \begin{bmatrix} \varphi_0 & \varphi_1 & \cdots & \varphi_n \\ 0 & \varphi_0 & \cdots & \varphi_{n-1} \\ \vdots & \vdots & & \vdots \\ 0 & 0 & \cdots & \varphi_0 \end{bmatrix} ,$$

where φ_j is short for $\varphi_{j,V}$. The sum in (8.5) converges because

$$r_{sp}(ZV^*) = r_{sp}(ZV^*) = \ell_V$$

and

$$\ell_V < 1 ,$$

by assumption. The spectral radii match because the two operators ZV^* and ZV^* have the same spectrum, i.e., $\lambda I - ZV^*$ is invertible if and only if $\lambda I - ZV^*$ is invertible, as follows by expressing the former in terms of the latter much as the development of (8.7).

Formula (8.7) serves to identify the components of the i-th block entry of F:

$$F_i = \begin{bmatrix} G_i \\ H_i \end{bmatrix}$$

as

$$G_i = \xi\varphi_i \quad \text{and} \quad H_i = \sum_{s=0}^{i} \eta_s\varphi_{i-s} , \tag{8.8}$$

for $i = 0, \ldots, n$. Consequently, by (1) of Theorem 4.4 and repeated applications of Lemma 8.2, we can now evaluate the components of the $(n + 1) \times (n + 1)$ operator matrices Δ and Λ explicitly in terms of the CF data:

$$\Delta_{ij} = (\xi^* \xi \varphi_j)_{\{i\}}(V) \tag{8.9}$$

and

$$\Lambda_{ij} = \Delta_{ij} - \sum_{s,t=0}^{n} (\eta_{i-s}^* \eta_{j-t} \varphi_t)_{\{s\}}(V) , \tag{8.10}$$

for $i, j = 0, \ldots, n$.

THEOREM 8.2. *An operator* $S \in \mathcal{S}(\ell_{\mathcal{N}}^2; \ell_{\mathcal{M}}^2)$ *is a solution of the CF problem if and only if*

$$\underline{p} S^* G_i E = H_i E \tag{8.11}$$

for $i = 0, \ldots, n$, *and every choice of* $E \in \mathcal{D}_2(\ell_{\mathcal{B}}^2; \ell_{\mathcal{B}}^2)$, *where now* G_i *and* H_i *are specified by (8.8).*

PROOF. In the present setting,

$$\underline{p} S^* G_i E = \underline{p} S^* \xi \varphi_i E$$

$$= \underline{p}(\xi^* S)^* \varphi_i E$$

$$= \sum_{t=0}^{i} (\xi^* S)_{\{i-t\}}^* \varphi_t E$$

by Lemma 8.3, whereas

$$H_i E = \sum_{t=0}^{i} \eta_{i-t} \varphi_t E .$$

Thus, if (8.11) holds for $i = 0, \ldots, n$ and every choice of $E \in \mathcal{D}_2$, then it follows by a straightforward inductive argument that

$$(\xi^* S)_{\{j\}}(V) = \eta_j^* \tag{8.12}$$

for $j = 0, \ldots, n$. More precisely, the verification of (8.12) for $j = 0$ rests on the very same argument that was used in the proof of Theorem 7.1; the connecting link is (1.5). Next, if (8.12) is valid for $j = 0, \ldots, i - 1$, then

$$0 = \underline{p} S^* G_i E - H_i E$$

$$= \sum_{t=0}^{i} (\xi^* S)_{\{i-t\}}^* \varphi_t E - \sum_{t=0}^{i} \eta_{i-t} \varphi_t E$$

$$= (\xi^* S)_{\{i\}}^* \varphi_0 E - \eta_i \varphi_0 E$$

which serves to verify (8.12) for $j = i$ also. The point here is that, just as in the verification of (8.12) for $j = 0$, we are again dealing with diagonal operator coefficients on the left of $\varphi_0 E$. This completes the proof that (8.11) implies (8.12). The converse is selfevident. ∎

THEOREM 8.3. *If the CF problem is solvable then the operator matrix Λ defined by (8.10) is positive semidefinite on $(\ell_{\mathcal{B}}^2)^{n+1}$.*

In the other direction, if conditions I and II are in force, then the CF problem is solvable and, if $\Theta \in \mathcal{A}$ denotes an associated ULCS operator (e.g., as defined in (4.17)– (4.20)), then

$$\{T_\Theta[S_L]:\ S_L \in \mathcal{S}(\ell_{\mathcal{N}}^2; \ell_{\mathcal{M}}^2)\}$$

is a complete list of the set of solutions.

PROOF. The argument is a complete paraphrase of the proof of Theorem 7.2 with NP replaced by CF and Theorem 7.1 replaced by Theorem 8.2. The point is that Theorem 8.2 enables us to invoke the abstract theory developed in Sections 4 and 6. ∎

THEOREM 8.4. *If $\mathcal{B}$ is finite dimensional, then the CF problem is solvable if and only if Λ is positive semidefinite on $(\ell_{\mathcal{B}}^2)^{n+1}$.*

If Λ is uniformly positive, then

$$\{T_\Theta[S_L]:\ S_L \in \mathcal{S}(\ell_{\mathcal{N}}^2; \ell_{\mathcal{M}}^2)\}$$

is a complete list of the set of solutions.

PROOF. In view of Theorem 8.3, it remains only to show the sufficiency of the stated condition for the existence of a solution. To this end, fix $\epsilon > 0$ and consider the new interpolation problem with η_j replaced by $(1+\epsilon)^{-\frac{1}{2}}\eta_j$. Then, the corresponding Pick operator,

$$\Lambda^\epsilon = \frac{1}{1+\epsilon}\Lambda + \frac{\epsilon}{1+\epsilon}\Delta$$

$$\geq \frac{\epsilon}{1+\epsilon}\Delta$$

meets condition I for every $\epsilon > 0$. The assumption that $\mathcal{B}$ is finite dimensional further guarantees that condition II is met for the perturbed CF problem for every $\epsilon > 0$, and hence, by Theorem 8.3, there exists an $F^\epsilon \in \mathcal{S}(\ell_{\mathcal{N}}^2; \ell_{\mathcal{M}}^2)$ such that

$$(\xi^* F^\epsilon)_{\{j\}}(V) = (1+\epsilon)^{-\frac{1}{2}}\eta_j^*$$

for $j = 0, \ldots, n$ and every $\epsilon > 0$. The rest goes through just as in the proof of the corresponding fact in the proof of Theorem 7.3, except that the last four lines of equality must now be changed to

$$\mathrm{trace}\, D^*(\xi^* F)_{\{j\}}(V)E = \langle FE, \xi\varphi_j D\rangle_{HS}$$

$$= \lim_{\epsilon\downarrow 0}\langle F^\epsilon E, \xi\varphi_j D\rangle_{HS}$$

$$= \lim_{\epsilon\downarrow 0}\mathrm{trace}\, D^*(\xi^* F^\epsilon)_{\{j\}}(V)E$$

$$= \lim_{\epsilon\downarrow 0}(1 + \epsilon)^{-\frac{1}{2}}\mathrm{trace}\, D^*\eta_j^* E$$

$$= \mathrm{trace}\, D^*\eta_j^* E \; ,$$

for $j = 0, \ldots, n$. $\blacksquare$

Here too, the condition that $\mathcal{B}$ be finite dimensional can be relaxed. The basic requirement is only that Θ should belong to $\mathcal{A}$ for every choice of $\epsilon > 0$.

THEOREM 8.5. *If condition II is in force, then the CF problem admits a strictly contractive solution if and only if I holds. Moreover, in this instance*

$$\{T_\Theta[S_L]: \; S_L \in \mathcal{S}(\ell^2_{\mathcal{M}}; \ell^2_{\mathcal{N}}) \; and \; \|S\| < 1\}$$

is equal to the set of all strictly contractive solutions of the CF problem.

PROOF. Suppose first that the CF problem admits a strictly contractive solution. Then, by Theorems 4.6 and 4.5, I holds.

Conversely, if I holds, then, since II is assumed, the rest is immediate from Theorem 8.3 and Lemma 4.1. $\blacksquare$

COROLLARY. *If $\mathcal{B}$ is finite dimensional, then the CF problem admits a strictly contractive solution if and only if the operator Λ defined by (8.5) is uniformly positive on $(\ell^2_{\mathcal{B}})^{n+1}$. Moreover, in this instance,*

$$\left\{T_\Theta[S_L]: \; S_L \in \mathcal{S}(\ell^2_{\mathcal{N}}; \ell^2_{\mathcal{M}}) \; and \; \|S\| < 1\right\}$$

is equal to the set of all strictly contractive solutions to the CF problem. (Here Θ is any associated ULCS operator; it may be defined by (4.17)–(4.20).)

PROOF. If $\mathcal{B}$ is finite dimensional, then, by Theorem 4.3, condition II is in force. Therefore Theorem 8.5 is applicable. $\blacksquare$

9. MIXED INTERPOLATION PROBLEMS

The mixed interpolation problem is basically an array of CF problems, wherein the data now consists of k sets of CF data: A set of operators

$$V_j \in \mathcal{D}(\ell_B^2; \ell_B^2) \, , \quad \text{with} \quad \ell_{V_j} < 1 \, ,$$

$$\xi_j \in \mathcal{D}(\ell_B^2; \ell_{\mathcal{M}}^2)$$

$$\eta_{jt} \in \mathcal{D}(\ell_B^2; \ell_{\mathcal{N}}^2) \, , \qquad t = 1, \ldots, n_j \, , \quad j = 1, \ldots, k \, .$$

The objective is to establish conditions for the existence of an $S \in \mathcal{S}(\ell_{\mathcal{N}}^2; \ell_{\mathcal{M}}^2)$ such that

$$(\xi_j^* S)_{\{t-1\}}(V_j) = \eta_t^*$$

for $t = 1, \ldots, n_j$ and $j = 1, \ldots, k$.

The NP problem corresponds to the special case in which $n_j = 1$ for $j = 1, \ldots, k$, whereas the CF problem corresponds to the special case in which $k = 1$.

The mixed problem also fits into the general framework considered in Sections 4 and 6. Now

$$\begin{bmatrix} \alpha^* \\ \beta^* \end{bmatrix} = \left[\begin{array}{cccc|cccc|c|cccc} \xi_1 & 0 & \cdots & 0 & \xi_2 & 0 & \cdots & 0 & & \xi_k & 0 & \cdots & 0 \\ & & & & & & & & \cdots & & & & \\ \eta_{11} & & \cdots & \eta_{1n_1} & \eta_{21} & & \cdots & \eta_{2n_2} & & \eta_{k1} & & \cdots & \eta_{kn_k} \end{array} \right]$$

and

$$V = \mathrm{diag}(V_1, \ldots, V_k) \, ,$$

wherein V_j is an $n_j \times n_j$ operator matrix of the form exhibited in (1.9) but based on V_j:

$$V_j = \begin{bmatrix} V_j & & & \\ I & V_j & & \\ & \ddots & \ddots & \\ & & I & V_j \end{bmatrix} \, .$$

We shall also suppose that

$$\Delta = \sum_{t=0}^{\infty} V^{[t]}(\alpha \alpha^*)^{(t)} V^{[t]*}$$

is uniformly positive on $(\ell_B^2)^n$.

By our earlier analysis, the already imposed condition

$$\ell_{V_j} < 1 \, , \qquad j = 1, \ldots, k \, ,$$

guarantees that

$$r_{sp}(\mathbf{Z}\mathbf{V}^*) < 1$$

and hence that

$$F = \begin{bmatrix} \alpha^* \\ \beta^* \end{bmatrix} (I - \mathbf{Z}\mathbf{V}^*)^{-1}$$

is well defined. It inherits a natural decomposition of the form

$$F = \begin{bmatrix} F_{11} \cdots F_{1n_1} \ \vdots \ \cdots \ \vdots \ F_{k1} \cdots F_{kn_k} \end{bmatrix} ,$$

where

$$[F_{j1} \cdots F_{jn_j}] = \begin{bmatrix} \xi_j & 0 & \cdots & 0 \\ \eta_{j1} & & \cdots & \eta_{jn_j} \end{bmatrix} \left(I - \mathbf{Z}\mathbf{V}_j^* \right)^{-1} .$$

It is convenient to reindex the

$$n_1 + \cdots + n_k = n$$

entries in F by $F_1, \ldots, F_n$ and to set

$$F_i = \begin{bmatrix} G_i \\ H_i \end{bmatrix} .$$

THEOREM 9.1.   *Let $S \in \mathcal{S}(\ell_{\mathcal{N}}^2; \ell_{\mathcal{M}}^2)$. Then, in the setting described just above, the following are equivalent:*

(1) $\underline{p}S^* G_i E = H_i E$
for $i = 1, \ldots, \nu$, and every choice of $E \in \mathcal{D}_2(\ell_{\mathcal{B}}^2; \ell_{\mathcal{B}}^2)$.

(2) $(\xi_j^* S)_{\{t-1\}}(V_j) = \eta_{jt}^*$
for $t = 1, \ldots, n_j$ and $j = 1, \ldots, k$.

PROOF.   Suppose $\underline{p}S^* G_i E = H_i E$ and let $i = n_{j-1} + t$, $1 \le t \le n_j$, with the understanding that $n_0 = 0$. Then

$$G_i = \xi_j \varphi_{t-1,V_j}$$

$$H_i = \eta_{j1} \varphi_{t-1,V_j} + \eta_{j2} \varphi_{t-2,V_j} + \cdots + \eta_{jt} \varphi_{0,V_j} .$$

Thus, by Lemma 8.3,

$$\underline{p}S^* G_i E = \underline{p}(\xi_j^* S)^* \varphi_{t-1} E$$

$$= \left[(\xi_j^* S)_{\{0\}}^* \varphi_{t-1} + (\xi_j^* S)_{\{1\}}^* \varphi_{t-2} + \cdots + (\xi_j^* S)_{\{t-1\}}^* \varphi_0 \right] E$$

and, as follows from Theorem 8.2, this will match $H_i E$ for $i = n_{j-1} + t$ and every $E \in \mathcal{D}_2(\ell_{\mathcal{B}}^2; \ell_{\mathcal{B}}^2)$ if and only if

$$(\xi_j^* S)_{\{s-1\}}(V_j) = \eta_{js}^* , \qquad s = 1, \ldots, n_j .$$

Thus, as was to be expected, if

$$pS^* G_i E = H_i E$$

for $i = 1, \ldots, n$, then S is a solution of the mixed interpolation problem.

The converse is selfevident.   ■

It is now a routine matter to formulate the requisite theorems as before. We state the results for the sake of completeness but skip the proofs. In particular:

Theorems 8.3–8.5 and the corollary to the latter all hold with the words "CF problem" replaced by "mixed interpolation problem" and $n + 1$ replaced by n (because of the different ways of enumerating) providing that Δ and Λ are computed with respect to the data operators for the mixed interpolation problem.

Thus, for example, Theorem 8.4 translates to:

THEOREM 9.2. *If $\mathcal{B}$ is finite dimensional, then the mixed interpolation problem is solvable if and only if Λ is positive semidefinite on $(\ell_{\mathcal{B}}^2)^n$.*

If Λ is uniformly positive on $(\ell_{\mathcal{B}}^2)^n$, then there is an ULCS operator $\Theta \in \mathcal{A}$ associated with the data and

$$\{T_\Theta[S_l] : \ S_L \in \mathcal{S}(\ell_{\mathcal{N}}^2; \ell_{\mathcal{M}}^2)\}$$

is a complete list of the set of solutions.

10.   EXAMPLES

In this section we shall present some examples of the theory established to this point. They will not always be carried through to the end because typically the relevant calculations are lengthy and, more often than not, explicit formulas are either not attainable, or are very complicated.

EXAMPLE 1 (BLOCK TOEPLITZ).

Let $\mathcal{B} = \mathbb{C}^k$, $\mathcal{M} = \mathbb{C}^p$, $\mathcal{N} = \mathbb{C}^q$,

$$\begin{aligned}
\xi = \alpha^* &= \mathrm{diag}(x_j) \quad \text{with} \quad x_j = x \in \mathbb{C}^{p \times k} \\
\eta = \beta^* &= \mathrm{diag}(y_j) \quad \text{with} \quad y_j = y \in \mathbb{C}^{q \times k} \\
V &= \mathrm{diag}(\omega_j) \quad \ \ \text{with} \quad \omega_j = \omega \in \mathbb{C}^{k \times k} \ ,
\end{aligned}$$

and suppose that

$$\ell_V = r_{sp}(\omega) < 1 \ .$$

Then it is readily checked that

$$\Lambda = \operatorname{diag}(\lambda_j)$$

with

$$\lambda_j = \lambda = \sum_{t=0}^{\infty} \omega^t (x^* x - y^* y) \omega^{*t}$$

for $j = 0, \pm 1, \ldots$, and

$$\rho_V^{-1} = \begin{bmatrix} \ddots & & \ddots & & \ddots & \\ & \ddots & I_k & \omega^* & \omega^{*2} & \omega^{*3} \\ & & 0 & \boxed{I_k} & \omega^* & \omega^{*2} & \ddots \\ & & 0 & 0 & I_k & \omega^* & \ddots \\ & & & & \ddots & \ddots & \ddots \end{bmatrix}$$

Thus both Λ and ρ_V^{-1} are themselves block Toeplitz, and the condition that Λ be uniformly positive on $\ell_{\mathcal{B}}^2$ is equivalent to requiring that the $k \times k$ matrix

$$\lambda = \sum_{t=0}^{\infty} \omega^t (x^* x - y^* y) \omega^{*t} \tag{10.1}$$

be positive definite. In the special case $k = 1$,

$$r_{sp}(\omega) = |\omega| < 1$$

and the condition on λ reduces to the familiar constraint

$$\frac{x^* x - y^* y}{1 - |\omega|^2} > 0 \, ,$$

from classical NP interpolation theory. If $k > 1$, then (10.1) is of course more complicated.

In the present setting, the NP interpolation problem is to find an $S \in \mathcal{S}(\ell_{\mathcal{N}}^2; \ell_{\mathcal{M}}^2)$ such that

$$(x^* S)^{\wedge}(V) = \eta^* \, .$$

If $\lambda > 0$, then the set of all solutions is equal to

$$\{T_{\Theta}[\sigma] : \ \sigma \in \mathcal{S}(\ell_{\mathcal{N}}^2; \ell_{\mathcal{M}}^2)\} \, ,$$

where Θ may be prescribed by (4.17)–(4.20). The requisite M_{12} and M_{22} exist by Theorem 4.3. In fact

$$M_{12} = \mathrm{diag}(\mu_j) \quad \text{with} \quad \mu_j = \mu$$
$$M_{22} = \mathrm{diag}(\nu_j) \quad \text{with} \quad \nu_j = \nu$$

for $j = 0, \pm 1, \ldots$, where $\mu \in \mathbb{C}^{k \times p}$ and $\nu \in \mathbb{C}^{p \times p}$ are chosen to make the matrix

$$\begin{bmatrix} \lambda^{\frac{1}{2}} \omega^* & \mu \\ x & \nu \end{bmatrix} \begin{bmatrix} (\lambda + y^* y)^{-\frac{1}{2}} & 0 \\ 0 & I_p \end{bmatrix}$$

unitary. Accordingly, it now follows from (4.17)–(4.20), that

$$\Theta_{11} = \sum_{j=0}^{\infty} Z^j A_{[j]} \, ,$$

with

$$A_{[0]} = \mathrm{diag}(\nu) \, ,$$
$$A_{[j]} = \mathrm{diag}(x \omega^{*j-1} \lambda^{-\frac{1}{2}} \mu) \, , \qquad j = 1, 2, \ldots \ ;$$

$$\Theta_{12} = \sum_{j=0}^{\infty} Z^j B_{[j]} \, ,$$

with

$$B_{[j]} = \mathrm{diag}\left(x \omega^{*j} \lambda^{-1} y^* \{I_q + y \lambda^{-1} y^*\}^{-\frac{1}{2}} \right) \, , \qquad j = 0, 1, \ldots \ ;$$

$$\Theta_{21} = \sum_{j=1}^{\infty} Z^j C_{[j]}$$

with

$$C_{[j]} = \mathrm{diag}\left(y \omega^{*j-1} \lambda^{-\frac{1}{2}} \mu \right) \, , \qquad j = 1, 2, \ldots \ ;$$

and

$$\Theta_{22} = \sum_{j=0}^{\infty} Z^j D_{[j]}$$

with

$$D_{[0]} = \mathrm{diag}(I_q + y \lambda^{-1} y^*)^{\frac{1}{2}}$$

and

$$D_{[j]} = \mathrm{diag}\left(y \omega^{*j} \lambda^{-1} y^* \{I_q + y \lambda^{-1} y^*\}^{-\frac{1}{2}} \right) \, , \qquad j = 1, 2, \ldots \ .$$

In particular, each of the operators Θ_{ij}, $i, j = 1, 2$, is block Toeplitz and $T_\Theta[\sigma]$ will again be block Toeplitz for every $\sigma \in \mathcal{S}$ which is block Toeplitz.

The operators A, B, C and D are completely determined by their generating functions:

$$A(\zeta) = \sum_{j=0}^{\infty} A_{0j}\zeta^{j}$$

$$= \nu + \zeta x(I_k - \zeta\omega^*)^{-1}\lambda^{-\frac{1}{2}}\mu \ ,$$

$$B(\zeta) = \sum_{j=0}^{\infty} B_{0j}\zeta^{j}$$

$$= x(I_k - \zeta\omega^*)^{-1}\lambda^{-1}y^*\{I_q + y\lambda^{-1}y^*\}^{-\frac{1}{2}} \ ,$$

$$C(\zeta) = \sum_{j=0}^{\infty} C_{0j}\zeta^{j}$$

$$= y(I_k - \zeta\omega^*)^{-1}\lambda^{-\frac{1}{2}}\mu$$

and

$$D(\zeta) = \sum_{j=0}^{\infty} D_{0j}\zeta^{j}$$

$$= \{I_q + y\lambda^{-1}y^* + y\zeta\omega^*(I_k - \zeta\omega^*)^{-1}\lambda^{-1}y^*\}\{I_q + y\lambda^{-1}y^*\}^{-\frac{1}{2}}$$

$$= \{I_q + y(I_k - \zeta\omega^*)^{-1}\lambda^{-1}y^*\}\{I_q + y\lambda^{-1}y^*\}^{-\frac{1}{2}} \ .$$

For further developments, turn to Section 11.

EXAMPLE 2 (BLOCK TOEPLITZ).

Let $\mathcal{B} = \mathbb{C}^k$, $\mathcal{M} = \mathbb{C}^p$, $\mathcal{N} = \mathbb{C}^q$,

$$V = \begin{bmatrix} V_1 & 0 \\ 0 & V_2 \end{bmatrix}$$

with

$$V_j = \begin{bmatrix} \ddots & & & & \\ & \omega_j & & & \\ & & \boxed{\omega_j} & & \\ & & & \omega_j & \\ & & & & \ddots \end{bmatrix} \quad \text{and} \quad \omega_j \in \mathbb{C}^{k\times k} \ ;$$

$$\alpha^* = [\xi_1 \ \ \xi_2]$$

with

$$\xi_j = \begin{bmatrix} \ddots & & & \\ & x_j & & \\ & & \boxed{x_j} & \\ & & & x_j \\ & & & & \ddots \end{bmatrix} \quad \text{and} \quad x_j \in \mathbb{C}^{p \times k} \; ;$$

$$\beta^* = [\eta_1 \quad \eta_2]$$

with

$$\eta_j = \begin{bmatrix} \ddots & & & \\ & y_j & & \\ & & \boxed{y_j} & \\ & & & y_j \\ & & & & \ddots \end{bmatrix} \quad \text{and} \quad y_j \in \mathbb{C}^{q \times k} \; .$$

Clearly,

$$\ell_V = \max\{r_{sp}(\omega_1) \, , \; r_{sp}(\omega_2)\} \; .$$

We shall assume that

$$\ell_V < 1 \, ,$$

as usual. Then

$$\Lambda = [\Lambda_{ij}] \, , \qquad i,j = 1,2 \; ,$$

where

$$\Lambda_{ij} = \sum_{t=0}^{\infty} V_i^t (\xi_i^* \xi_j - \eta_i^* \eta_j) V_j^{t*}$$

$$= \begin{bmatrix} \ddots & & & \\ & \lambda_{ij} & & \\ & & \boxed{\lambda_{ij}} & \\ & & & \lambda_{ij} \\ & & & & \ddots \end{bmatrix}$$

and

$$\lambda_{ij} = \sum_{t=0}^{\infty} \omega_i^t (x_i^* x_j - y_i^* y_j) \omega_j^{t*} \; .$$

Again, if $k = 1$, then

$$\lambda_{ij} = \frac{x_i^* x_j - y_i^* y_j}{1 - \omega_i \omega_j^*} \, ,$$

which is a familiar entry in the Pick matrix for the tangential NP problem with two points (see e.g., formula (5.2) of [D1]). For $k > 1$, the recipe is more complicated. Nevertheless, Θ may be calculated from formulas (4.17)–(4.20) when the $2k \times 2k$ matrix

$$\lambda = [\lambda_{ij}] \, , \qquad i,j = 1,2 \; ,$$

is positive definite. Once again it will turn out that Θ_{ij} are Toeplitz operators and that, if $\lambda > 0$, then

$$\{S = T_\Theta[\sigma] : \ \sigma \in \mathcal{S}(\ell^2_\mathcal{N}; \ell^2_\mathcal{M})\}$$

is a complete description of the set of all solutions to the interpolation problem

$$(\xi_j^* S)^\wedge(V_j) = \eta_j^* , \qquad j = 1, 2 \ .$$

Moreover, if σ is block Toeplitz, then so is S. More details are provided in Section 11.

EXAMPLE 3.

Let $\mathcal{M} = \mathcal{N} = \mathcal{B} = \ell^2_+$, the space of one-sided scalar sequences $x = (x_0, x_1, \ldots)$ with $\|x\|^2 = \sum_{j=0}^\infty |x_j|^2 < \infty$. Let

$$V = \mathrm{diag}(V_{jj}) , \qquad j = 0, \pm 1, \ldots \ ,$$

be the block Toeplitz operator with

$$V_{jj} = c\tau ,$$

where $c \in \mathbb{R}$,

$$\tau \ = \ \begin{bmatrix} 0 & 1 & 0 & 0 & \\ 0 & 0 & 1 & 0 & \\ 0 & 0 & 0 & 1 & \\ & & & & \ddots & \ddots \end{bmatrix}$$

denotes the backwards shift on $\mathcal{B} = \ell^2_+$:

$$\tau : \ (x_0, x_1, \ldots) \ \longrightarrow \ (x_1, x_2, \ldots) \ ,$$

and let

$$\mathbf{V} \ = \ \begin{bmatrix} V & 0 \\ I & V \end{bmatrix} ,$$

$$\alpha \ = \ \begin{bmatrix} I \\ 0 \end{bmatrix}$$

and (to make the calculations simple)

$$\beta \ = \ b \begin{bmatrix} 0 \\ I \end{bmatrix} ,$$

with $b \in \mathbb{R}$.

Since

$$r_{sp}(\mathbf{Z}\,\mathbf{V}^*) = r_{sp}(ZV) = |c| \ ,$$

we shall take $|c| < 1$. Then the sum

$$\Lambda = \sum_{t=0}^{\infty} V^{[t]}(\alpha\alpha^* - \beta\beta^*)^{(t)} V^{[t]*}$$

$$= \sum_{t=0}^{\infty} \begin{bmatrix} V^t & 0 \\ tV^{t-1} & V^t \end{bmatrix} \begin{bmatrix} I & 0 \\ 0 & b^2 I \end{bmatrix} \begin{bmatrix} V^{*t} & tV^{*t-1} \\ 0 & V^{*t} \end{bmatrix}$$

converges. Upon multiplying out and taking advantage of the fact that $VV^* = c^2 I$, the sum can be evaluated explicitly to yield

$$\Lambda = \frac{1}{1-c^2} \begin{bmatrix} I & \frac{V}{1-c^2} \\ \frac{V^*}{1-c^2} & \frac{k+c^2}{(1-c^2)^2} I \end{bmatrix} \ ,$$

where

$$k = 1 + b^2(1 - c^2)^2 \ .$$

Since the Schur complement of the top entry in the matrix of the last formula is equal to

$$(1 - c^2)^{-2}\{kI + c^2 I - V^*V\} \geq (1 - c^2)^{-2}kI \ ,$$

it follows readily that Λ is uniformly positive on $(\ell_B^2)^2$ for any choice of $b \in \mathbb{R}$. It is also readily checked that

$$\Lambda^{-1} = (1 - c^2) \begin{bmatrix} \frac{k+c^2}{k} I & \frac{(c^2-1)}{k} V \\ \frac{(c^2-1)}{k} V^* & E \end{bmatrix}$$

with

$$E = \frac{(1 - c^2)^2}{k + c^2} I + \frac{(1 - c^2)^2}{k(k + c^2)} V^*V \ ,$$

for short.

By the analysis in Case 4 of Section 5, the kernel condition II is automatically satisfied and the associated LCS operator Θ can be calculated by formulas (4.17)–(4.20), where

$$M_{12} = (\Lambda^{-\frac{1}{2}})^{(1)} YQ = \Lambda^{-\frac{1}{2}} YQ \ ,$$

$$M_{22} = V^2 Q \ ,$$

and (for the present block Toeplitz case)

$$Q = \left\{ [-V^* \ I]\Lambda^{-1} \begin{bmatrix} -V \\ I \end{bmatrix} + V^{*2}V^2 \right\}^{-\frac{1}{2}} \ .$$

The term in curly brackets in the formula for Q is equal to

$$\frac{(1-c^2)^3}{k+c^2}\left\{I + \frac{(k+1)^2}{k(1-c^2)^2}V^*V + \frac{k+c^2}{(1-c^2)^3}V^{*2}V^2\right\} \, ,$$

which, apart from the constant multiplier on the far left, is of the form

$$I + xP_1 + yP_2 \, ,$$

where

$$P_1 = W^*W \, , \qquad P_2 = W^{*2}W^2$$

and

$$W = c^{-1}V$$

is a coisometry: $WW^* = I$. Because of this, P_1 and P_2 are, as the notation suggests, orthogonal projectors. Moreover,

$$\mathrm{ran}\ P_2 \subset \mathrm{ran}\ P_1 \, ,$$

as follows from the calculation

$$P_1P_2 = P_2P_1 = P_2 \, .$$

Thus

$$I + xP_1 + yP_2 = I - P_1 + (1+x)(P_1 - P_2) + (1+x+y)P_2$$

admits the orthogonal decomposition exhibited on the right of the last equality. From this in turn, it is readily checked that

$$(I + xP_1 + yP_2)^{-\frac{1}{2}} = I - P_1 + (1+x)^{-\frac{1}{2}}(P_1 - P_2) + (1+x+y)^{-\frac{1}{2}}P_2 \, ,$$

providing of course that both

$$1 + x > 0 \qquad \text{and} \qquad 1 + x + y > 0 \, .$$

In the case at hand these two conditions are clearly met, since

$$x = \frac{(k+1)^2c^2}{k(1-c^2)^2}$$

and

$$y = \frac{(k+c^2)c^4}{(1-c^2)^3}$$

are both already clearly nonnegative.

It is now routine to solve explicitly first for Q, secondly for M_{12} and M_{22} and then finally for the associated LCS operator Θ via (4.17)–(4.20). The details are left to the hardy.

EXAMPLE 4.

Let $\mathcal{B} = \mathbb{C}^k$, $\mathcal{M} = \mathbb{C}^p$, $\mathcal{N} = \mathbb{C}^q$ and let

$$V = \begin{bmatrix} \ddots & & & & & & \\ & 0 & & & & & \\ & & \boxed{0} & & & & \\ & & & \omega_1 & & & \\ & & & & \omega_2 & & \\ & & & & & 0 & \\ & & & & & & \ddots \end{bmatrix}$$

and
$$\alpha = \xi^* = \mathrm{diag}(x_j^*) \quad \text{with} \quad x_j \in \mathbb{C}^{p \times k}\,,$$
$$\beta = \eta^* = \mathrm{diag}(y_j^*) \quad \text{with} \quad y_j \in \mathbb{C}^{q \times k}\,,$$

for $j = 0, \pm 1, \ldots$.

Since $V^{[n]} = 0$ for $n \geq 3$,

$$\ell_V = r_{sp}(ZV^*) = 0$$

for any choice of the $k \times k$ matrices ω_1 and ω_2.

For this choice of the data operators,

$$(\xi^* S)^{\wedge}(V) = (\xi^* S)_{[0]} + V(\xi^* S)_{[1]} + VV^{(1)}(\xi^* S)_{[2]}$$
$$= (\xi^* S)_{[0]} + \mathrm{diag}(\gamma_j)\,,$$

where
$$\gamma_1 = \omega_1 x_0^* S_{01}$$
$$\gamma_2 = \omega_2 x_1^* S_{12} + \omega_2 \omega_1 x_0^* S_{02}$$
$$\gamma_j = 0 \quad \text{for} \quad j \leq 0 \quad \text{and} \quad j \geq 3\,.$$

Thus the NP interpolation problem reduces to finding an $S \in \mathcal{S}(\ell^2_{\mathcal{N}}; \ell^2_{\mathcal{M}})$ with

$$x_j^* S_{jj} = y_j^* \quad \text{for} \quad j \leq 0 \quad \text{and} \quad j \geq 3$$

$$x_1^* S_{11} + \omega_1 x_0^* S_{01} = y_1^*$$

$$x_2^* S_{22} + \omega_2 x_1^* S_{12} + \omega_2 \omega_1 x_0^* S_{02} = y_2^* .$$

Moreover,

$$\Lambda = \text{diag}(\lambda_j)$$

with

$$\lambda_j = x_j^* x_j - y_j^* y_j \quad \text{for} \quad j \leq 0 \quad \text{and} \quad j \geq 3$$

$$\lambda_1 = x_1^* x_1 - y_1^* y_1 + \omega_1 (x_0^* x_0 - y_0^* y_0) \omega_1^*$$

$$\lambda_2 = x_2^* x_2 - y_2^* y_2 + \omega_2 (x_1^* x_1 - y_1^* y_1) \omega_2^* + \omega_2 \omega_1 (x_0^* x_0 - y_0^* y_0) \omega_1^* \omega_2^* .$$

Thus Λ is uniformly positive on $\ell^2_{\mathcal{B}}$ if and only if there exists an $\epsilon > 0$ such that

$$\lambda_j \geq \epsilon I_k$$

for $j = 0, \pm 1, \dots$.

The next step is to find the operators

$$M_{12} = \text{diag}(\mu_j)$$

$$M_{22} = \text{diag}(\nu_j) ,$$

$j = 0, \pm 1, \dots$, by finding $\mu_j \in \mathbb{C}^{k \times p}$ and $\nu_j \in \mathbb{C}^{p \times p}$ such that for each j, the $(k + p) \times (k \mid p)$ matrix

$$\begin{bmatrix} \lambda_{j-1}^{\frac{1}{2}} \omega_j^* (\lambda_j + y_j^* y_j)^{-\frac{1}{2}} & \mu_j \\ x_j (\lambda_j + y_j^* y_j)^{-\frac{1}{2}} & \nu_j \end{bmatrix}$$

is unitary, where (because of the chosen V)

$$\omega_j = 0 \quad \text{for} \quad j \leq 0 \quad \text{and} \quad j \geq 3 .$$

This is a less formidable task than it might seem because for $j \leq 0$ and $j \geq 3$

$$(\lambda_j + y_j^* y_j)^{-\frac{1}{2}} = (x_j^* x_j)^{-\frac{1}{2}}$$

and the matrix

$$\begin{bmatrix} 0 & (x_j^* x_j)^{-\frac{1}{2}} x_j^* \\ x_j (x_j^* x_j)^{-\frac{1}{2}} & I_p - x_j (x_j^* x_j)^{-1} x_j^* \end{bmatrix}$$

is readily seen to be unitary. Thus we may choose

$$\mu_j = (x_j^* x_j)^{-\frac{1}{2}} x_j^*$$

and

$$\nu_j = I_p - x_j(x_j^* x_j)^{-1} x_j^*$$

for every j apart from $j = 1$ and $j = 2$.

The associated LCS operator may now be calculated from (4.17)–(4.20).

To this end it is advisable to first calculate ρ_V^{-1}. Because of the special choice of V this is just a sum of three terms:

$$\rho_V^{-1} = I + ZV^* + Z^2 V^{(1)*} V^*$$

$$= I + \begin{bmatrix} \ddots & & & & & \\ & 0 & & & & \\ & & \boxed{0} & \omega_1^* & \omega_1^* \omega_2^* & \\ & & 0 & 0 & \omega_2^* & \\ & & 0 & 0 & 0 & \\ & & & & & 0 \\ & & & & & & \ddots \end{bmatrix} .$$

In particular there are only three nonzero entries in the second matrix on the right in the last line of the equality. Nevertheless, it is convenient to write

$$V = \operatorname{diag}(\omega_j)$$

with the understanding that $\omega_j = 0$ for $j \leq 0$ and $j \geq 3$. Now substituting the above calculations into (4.17)–(4.20) we obtain:

$$\Theta_{11} = \sum_{t=0}^{3} Z^t A_{[t]} ,$$

where

$$(A_{[0]})_{jj} = \nu_j$$

$$(A_{[1]})_{jj} = x_{j-1} \lambda_{j-1}^{-\frac{1}{2}} \mu_j$$

$$(A_{[2]})_{jj} = x_{j-2} \omega_{j-1}^* \lambda_{j-1}^{-\frac{1}{2}} \mu_j$$

$$(A_{[3]})_{jj} = x_{j-3} \omega_{j-2}^* \omega_{j-1}^* \lambda_{j-1}^{-\frac{1}{2}} \mu_j ,$$

for $j = 0, \pm 1, \ldots$. In particular, $(A_{[2]})_{jj} = 0$ unless $j = 2$ or $j = 3$, while $(A_{[3]})_{jj} = 0$ unless $j = 2$.

$$\Theta_{12} = \sum_{t=0}^{2} Z^t B_{[t]} \, ,$$

where

$$(B_{[0]})_{jj} = x_j \lambda_j^{-1} y_j^* (I_q + y_j \lambda_j^{-1} y_j^*)^{-\frac{1}{2}}$$

$$(B_{[1]})_{jj} = x_{j-1} \omega_j^* \lambda_j^{-1} y_j^* (I_q + y_j \lambda_j^{-1} y_j^*)^{-\frac{1}{2}}$$

$$(B_{[2]})_{jj} = x_{j-2} \omega_{j-1}^* \omega_j^* \lambda_j^{-1} y_j^* (I_q + y_j \lambda_j^{-1} y_j^*)^{-\frac{1}{2}} \, ,$$

for $j = 0, \pm 1, \ldots$. Notice that because of the special choice of V, $(B_{[1]})_{jj} = 0$ unless $j = 1$ or $j = 2$ and $(B_{[2]})_{jj} = 0$ for $j \neq 2$.

$$\Theta_{21} = \sum_{t=1}^{3} Z^t C_{[t]} \, ,$$

where

$$(C_{[1]})_{jj} = y_{j-1} \lambda_{j-1}^{-\frac{1}{2}} \mu_j$$

$$(C_{[2]})_{jj} = y_{j-2} \omega_{j-1}^* \lambda_{j-1}^{-\frac{1}{2}} \mu_j$$

$$(C_{[3]})_{jj} = y_{j-3} \omega_{j-2}^* \omega_{j-1}^* \lambda_{j-1}^{-\frac{1}{2}} \mu_j \, ;$$

for $j = 0, \pm 1, \ldots$. Because of the choice of V, $(C_{[2]})_{jj} = 0$ except for $j = 2$ or 3, while $(C_{[3]})_{jj} = 0$ except for $j = 3$.

Finally,

$$\Theta_{22} = \sum_{t=0}^{2} Z^t D_{[t]} \, ,$$

where

$$(D_{[0]})_{jj} = (I_q + y_j \lambda_j^{-1} y_j^*)^{\frac{1}{2}}$$

$$(D_{[1]})_{jj} = y_{j-1} \omega_j^* \lambda_j^{-1} y_j^* (I_q + y_j \lambda_j^{-1} y_j^*)^{-\frac{1}{2}}$$

$$(D_{[2]})_{jj} = y_{j-2} \omega_{j-1}^* \omega_j^* \lambda_j^{-1} y_j^* (I_q + y_j \lambda_j^{-1} y_j^*)^{-\frac{1}{2}} \, ,$$

for $j = 0, \pm 1, \ldots$. Again, because of the special choice of V, $[D_{[1]})_{jj} = 0$ except for $j = 1$ or 2, while $[D_{[2]})_{jj} = 0$ except for $j = 2$.

EXAMPLE 5 (Specialization of Example 4).

Let $k = p = q = 1$, $x_j = 1$ and $\omega_2 = 0$ in Example 4. Then the formulas reduce further:

$$(\xi^* S)^{\wedge}(V) = S_{[0]} + V S_{[1]}$$

and hence

$$\{(\xi^* S)^{\wedge}(V)\}_{jj} = \begin{cases} S_{jj} & \text{for} \quad j \neq 1 \\ S_{11} + \omega_1 S_{01} & \text{for} \quad j = 1 \ . \end{cases}$$

The NP problem reduces to finding an $S \in \mathcal{S}(\ell^2_{\mathcal{N}}; \ell^2_{\mathcal{M}})$ with

$$S_{jj} = y_j^* \quad \text{for} \quad j \neq 1$$

and

$$S_{11} + \omega_1 S_{01} = y_1^* \ .$$

Moreover,

$$\lambda_j = \begin{cases} 1 - |y_j|^2 & \text{for} \quad j \neq 1 \\ 1 - |y_1|^2 + |\omega_1|^2(1 - |y_0|^2) & \text{for} \quad j = 1 \ , \end{cases}$$

$$\mu_j = \begin{cases} 1 & \text{for} \quad j \neq 1 \\ (\lambda_1 + |y_1|^2)^{-\frac{1}{2}} & \text{for} \quad j = 1 \ , \end{cases}$$

and

$$\nu_j = \begin{cases} 0 & \text{for} \quad j \neq 1 \\ -\lambda_0^{\frac{1}{2}} \omega_1 (\lambda_1 + |y_1|^2)^{-\frac{1}{2}} & \text{for} \quad j = 1 \ . \end{cases}$$

Thus Λ will be uniformly positive on $\ell^2_{\mathcal{B}}$ if and only if there exists an $\epsilon > 0$ such that

$$1 - |y_j|^2 > \epsilon \quad \text{for} \quad j \neq 1$$

and

$$1 - |y_1|^2 + |\omega_1|^2(1 - |y_0|^2) > 0 \ .$$

If this condition is in force, then the formulas for Θ_{ij} now drop out easily by suitably specializing the formulas in Example 5.

In particular we find that

$$\theta_{11} = \begin{bmatrix} \ddots & & \ddots & & \ddots & \\ & 0 & a_{-2} & 0 & 0 & \\ & 0 & \boxed{0} & a_{-1} & a_{+} & \\ & & 0 & a_{-} & a_{0} & \ddots \\ & & 0 & 0 & 0 & a_{1} \ddots \\ & & & \ddots & & \ddots \end{bmatrix}$$

is a tridiagonal, upper triangular infinite matrix with

$$a_j = \lambda_j^{-\frac{1}{2}}\mu_{j+1} = \begin{cases} \lambda_j^{-\frac{1}{2}}, & j \neq 0 \\[2mm] \lambda_0^{-\frac{1}{2}}\mu_1, & j = 0 \end{cases}$$

$$a_- = \nu_1 = -\lambda_0^{\frac{1}{2}}\omega_1(\lambda_1 + |y_1|^2)^{-\frac{1}{2}}$$

$$a_+ = \omega_1^* \lambda_1^{-\frac{1}{2}} ;$$

$$\theta_{12} = \begin{bmatrix} \ddots & \ddots & & & \\ & b_{-1} & 0 & & \\ & & \boxed{b_0} & b_+ & \\ & & & b_1 & 0 \\ & & & & \ddots & \ddots \end{bmatrix}$$

is a bidiagonal upper triangular matrix with

$$b_j = \begin{cases} y_j^*(1 - |y_j|^2)^{-\frac{1}{2}}, & j \neq 1 \\[2mm] \lambda_1^{-\frac{1}{2}}y_1^*(\lambda_1 + |y_1|^2)^{-\frac{1}{2}}, & j = 1, \end{cases}$$

$$b_+ = \omega_1^* \lambda_1 b_1 ;$$

$$\theta_{21} = \begin{bmatrix} \ddots & \ddots & & & \\ \ddots & c_0 & 0 & & \\ & \boxed{0} & c_1 & c_+ & \\ & & 0 & c_2 & 0 \\ & & & \ddots & \ddots & \ddots \end{bmatrix}$$

is a bidiagonal strictly upper triangular matrix with

$$c_j = y_{j-1}\lambda_{j-1}^{-\frac{1}{2}}\mu_j$$

$$c_+ = y_0\omega_1^*\lambda_1^{-\frac{1}{2}} \; ;$$

$$\theta_{22} = \begin{bmatrix} \ddots & \ddots & & & \\ & d_{-1} & \mathbf{O} & & \\ & & \boxed{d_0} & d_+ & \\ & & & d_1 & \mathbf{O} \\ & & & & \ddots & \ddots \end{bmatrix}$$

is a bidiagonal upper triangular matrix with

$$d_j = (\lambda_j + |y_j|^2)^{\frac{1}{2}}\lambda_j^{-\frac{1}{2}} = \begin{cases} \lambda_j^{-\frac{1}{2}} , & j \neq 1 \\ (\lambda_1 + |y_1|^2)^{\frac{1}{2}}\lambda_1^{-\frac{1}{2}} , & j = 1 , \end{cases}$$

$$d_+ = y_0\omega_1^*\lambda_1^{-\frac{1}{2}}y_1^*(\lambda_1 + |y_1|^2)^{-\frac{1}{2}} .$$

EXAMPLE 6 (A Matrix Interpolation Problem).

We shall once again specialize the setting of Example 4. This time we choose $p = k$, $x_j = I_p$ for all j, and $y_j = 0$ for $j < 0$ and $j \geq 3$. Thus the basic interpolation problem is to find an $S \in \mathcal{S}(\ell^2_{\mathcal{N}}; \ell^2_{\mathcal{M}})$ with

$$S_{jj} = 0 \quad \text{for} \quad j < 0 \quad \text{and} \quad j \geq 3$$

$$S_{00} = y_0^*$$

$$S_{11} + \omega_1 S_{01} = y_1^*$$

$$S_{22} + \omega_2 S_{12} + \omega_2\omega_1 S_{02} = y_2^* .$$

As already noted, for such a choice of V,

$$r_{sp}(ZV^*) = 0$$

for any choice of the $k \times k$ matrices ω_1 and ω_2. It further follows from the analysis of Example 4 that

$$\lambda_j = I_k \quad \text{for} \quad j < 0 \quad \text{and} \quad j \geq 3$$

$$\lambda_0 = I_k - y_0^* y_0$$

$$\lambda_1 = I_k - y_1^* y_1 + \omega_1(I_k - y_0^* y_0)\omega_1^*$$

$$\lambda_2 = I_k - y_2^* y_2 + \omega_2(I_k - y_1^* y_1)\omega_2^* + \omega_2\omega_1(I_k - y_0^* y_0)\omega_1^*\omega_2^* \ .$$

Thus Λ will be uniformly positive on $\ell_{\mathcal{B}}^2$ if and only if the last three matrices are positive definite. If this condition is met, then, by Theorem 7.3, the NP problem under present consideration is solvable and the set of all solutions is given by

$$\left\{ T_\Theta[\sigma] : \ \sigma \in \mathcal{S}(\ell_{\mathcal{N}}^2; \ell_{\mathcal{M}}^2) \right\} \ .$$

In this instance, since $\alpha = I$, Θ can be calculated from formula (5.4) as well as by specializing the general formulas developed in Example 4. We shall use the former for the sake of change. It also avoids the need to calculate μ_j and ν_j.

It is convenient to write the normalizing factors on the far right of (5.4) as

$$\left\{ \Lambda^{(1)} + \Lambda^{(1)} V^* V \Lambda^{(1)} \right\}^{-\frac{1}{2}} = \gamma = \mathrm{diag}(\gamma_j)$$

and

$$(I + \beta^* \Lambda^{-1} \beta)^{-\frac{1}{2}} = \epsilon = \mathrm{diag}(\epsilon_j) \ .$$

It is then readily checked that

$$\gamma_j = \left(\lambda_{j-1} + \lambda_{j-1} \omega_j^* \omega_j \lambda_{j-1} \right)^{-\frac{1}{2}}$$

for $j = 0, \pm 1, \dots$. But, for the present choice of the data operators, this simplifies to

$$\gamma_j = I_k \quad \text{for} \quad j \leq 0 \quad \text{and} \quad j > 3$$

and

$$\gamma_3 = \lambda_2^{-\frac{1}{2}} \ .$$

Similarly,

$$\epsilon_j = (I_q + y_j \lambda_j^{-1} y_j^*)^{-\frac{1}{2}}$$

for $j = 0, \pm 1, \dots$, which simplifies to

$$\epsilon_j = I_q \quad \text{for} \quad j < 0 \quad \text{and} \quad j \geq 3 \ .$$

Next, since

$$(I - ZV^*)^{-1} = I + ZV^* + Z^2 V^{[2]*} \, ,$$

it follows from (5.4) that:

$$\Theta_{11} = \sum_{j=0}^{3} Z^j A_{[j]}$$

with

$$(A_{[0]})_{jj} = \begin{cases} -\omega_j \lambda_{j-1} \gamma_j \, , & j = 1, 2 \\ 0 & \text{otherwise} \end{cases}$$

$$(A_{[1]})_{jj} = \gamma_j \quad \text{for all } j$$

$$(A_{[2]})_{jj} = \begin{cases} \omega_{j-1}^* \gamma_j \, , & j = 2, 3 \\ 0 & \text{otherwise} \end{cases}$$

$$(A_{[3]})_{jj} = \begin{cases} \omega_{j-2}^* \omega_{j-1}^* \gamma_j \, , & j = 3 \\ 0 & \text{otherwise} \, ; \end{cases}$$

$$\Theta_{12} = \sum_{j=0}^{2} Z^j B_{[j]}$$

with

$$(B_{[0]})_{jj} = \begin{cases} \lambda_j^{-1} y_j^* \epsilon_j \, , & j = 0, 1, 2 \\ 0 & j < 0 \text{ and } j \geq 3 \end{cases}$$

$$(B_{[1]})_{jj} = \begin{cases} \omega_j^* \lambda_j^{-1} y_j^* \epsilon_j \, , & j = 1, 2 \\ 0 & j \leq 0 \text{ and } j \geq 3 \end{cases}$$

$$(B_{[2]})_{jj} = \begin{cases} \omega_{j-1}^* \omega_j^* \lambda_j^{-1} y_j^* \epsilon_j \, , & j = 2 \\ 0 & j \neq 2 \, ; \end{cases}$$

$$\Theta_{21} = \sum_{j=1}^{3} Z^j C_{[j]}$$

with

$$(C_{[1]})_{jj} = \begin{cases} y_{j-1} \gamma_j \, , & j = 1, 2, 3 \\ 0 & j \leq 0 \text{ and } j \geq 4 \end{cases}$$

$$(C_{[2]})_{jj} = \begin{cases} y_{j-2}\omega_{j-1}^*\gamma_j \,, & j = 2,3 \\ 0 & \text{otherwise} \end{cases}$$

$$(C_{[3]})_{jj} = \begin{cases} y_{j-3}\omega_{j-2}^*\omega_{j-1}^*\gamma_j \,, & j = 3 \\ 0 & \text{otherwise} \,; \end{cases}$$

$$\Theta_{22} = \sum_{j=0}^{2} Z^j D_{[j]}$$

with

$$(D_{[0]})_{jj} = \epsilon_j^{-1} \quad \text{all } j$$

$$(D_{[1]})_{jj} = \begin{cases} y_{j-1}\omega_j^*\lambda_j^{-1}y_j^*\epsilon_j \,, & j = 1,2 \\ 0 & \text{otherwise} \end{cases}$$

$$(D_{[2]})_{jj} = \begin{cases} y_{j-2}\omega_{j-1}^*\omega_j^*\lambda_j^{-1}y_j^*\epsilon_j \,, & j = 2,3 \\ 0 & \text{otherwise} \,. \end{cases}$$

After some manipulation, the preceding formulas for the Θ_{ij} can be reexpressed in the following more concise form:

$$\theta_{11} = \begin{bmatrix} \ddots & I_k & \\ & A & \\ & & I_k \ddots \end{bmatrix} Z$$

with

$$A = \begin{bmatrix} \boxed{\gamma_1} & \omega_1^*\gamma_2 & \omega_1^*\omega_2^*\gamma_3 \\ -\omega_1\lambda_0\gamma_1 & \gamma_2 & \omega_2^*\gamma_3 \\ 0 & -\omega_2\lambda_1\gamma_2 & \gamma_3 \end{bmatrix}$$

$$\theta_{12} = \begin{bmatrix} \ddots & 0 & \\ & B & \\ & & 0 \ddots \end{bmatrix}$$

with

$$B = \begin{bmatrix} \boxed{b_0} & \omega_1^* b_1 & \omega_1^* \omega_2^* b_2 \\ 0 & b_1 & \omega_2^* b_2 \\ 0 & 0 & b_2 \end{bmatrix}$$

and

$$b_j = \lambda_j^{-1} y_j^* \epsilon_j \ , \qquad j = 0,1,2, \ \ ;$$

$$\theta_{21} = \begin{bmatrix} \ddots & & \raisebox{0.5em}{O} \\ & C & \\ \raisebox{-0.5em}{O} & & \ddots \end{bmatrix} Z$$

with

$$C = \begin{bmatrix} \boxed{y_0 \gamma_1} & y_0 \omega_1^* \gamma_2 & y_0 \omega_1^* \omega_2^* \gamma_3 \\ 0 & y_1 \gamma_2 & y_1 \omega_2^* \gamma_3 \\ 0 & 0 & y_2 \gamma_3 \end{bmatrix}$$

and finally,

$$\theta_{22} = \begin{bmatrix} \ddots & & \\ & I_q & \\ & & D \\ & & & I_q \\ & & & & \ddots \end{bmatrix}$$

with

$$D = \begin{bmatrix} \boxed{\epsilon_0^{-1}} & y_0 d_1 & y_0 \omega_1^* d_2 \\ 0 & \epsilon_1^{-1} & y_1 d_2 \\ 0 & 0 & \epsilon_2^{-1} \end{bmatrix}$$

and

$$d_j = \omega_j^* \lambda_j^{-1} y_j^* \epsilon_j \ , \qquad j = 1,2 \ .$$

Thus all the "action" is concentrated in the four matrices A, B, C and D.

To keep the solutions $T_\Theta[\sigma]$ in terms of upper triangular block matrices, we choose

$$\sigma = \begin{bmatrix} \ddots & & & & \\ & \boxed{0} & 0 & 0 & \\ & 0 & \sigma_{01} & \sigma_{02} & \\ & 0 & 0 & \sigma_{12} & \\ & & & 0 & \\ & & & & \ddots \end{bmatrix}$$

with at most the three itemized entries nonzero. Then

$$Z\sigma = \begin{bmatrix} \ddots & & & \\ & \boxed{0} & \sigma_{01} & \sigma_{02} \\ & 0 & 0 & \sigma_{12} \\ & & & 0 \\ & & & \ddots \end{bmatrix}$$

and so the general solution $S = T_\Theta[\sigma]$ is given by the formula

$$\begin{bmatrix} S_{00} & S_{01} & S_{02} \\ 0 & S_{11} & S_{12} \\ 0 & 0 & S_{22} \end{bmatrix} = \left\{ A \begin{bmatrix} 0 & \sigma_{01} & \sigma_{02} \\ 0 & 0 & \sigma_{12} \\ 0 & 0 & 0 \end{bmatrix} + B \right\} \left\{ C \begin{bmatrix} 0 & \sigma_{01} & \sigma_{02} \\ 0 & 0 & \sigma_{12} \\ 0 & 0 & 0 \end{bmatrix} + D \right\}^{-1},$$

when Λ is uniformly positive. The indicated inverse is easily calculated since it is an upper triangular matrix and so one can obtain explicit formulas for the S_{ij}; the details are left to the reader.

Finally, we remark that the formulas for A, B, C and D in this example can also be obtained by suitably specializing the corresponding formulas in Example 4. In so doing, it is useful to bear in mind that, for $x_j = I_p$,

$$\mu_j \mu_j^* = I_p - \lambda_{j-1}^{\frac{1}{2}} \omega_j^* (\lambda_j + y_j^* y_j)^{-1} \omega_j \lambda_{j-1}^{\frac{1}{2}}$$

$$= I_p - \lambda_{j-1}^{\frac{1}{2}} \omega_j^* (I_p + \omega_j \lambda_{j-1} \omega_j^*)^{-1} \omega_j \lambda_{j-1}^{\frac{1}{2}}$$

$$= \lambda_{j-1}^{\frac{1}{2}} (\lambda_{j-1} + \lambda_{j-1} \omega_j^* \omega_j \lambda_{j-1})^{-1} \lambda_{j-1}^{\frac{1}{2}} .$$

11. BLOCK TOEPLITZ AND SOME IMPLICATIONS

In this section we briefly review the block Toeplitz case and then explain how to use the already established results to solve a matrix interpolation problem of independent interest.

Recall that the data operators, in the basic setting of Theorem 4.4, are said to be block Toeplitz if

$$Z^*VZ = V, \quad Z^*\alpha Z = \alpha, \quad \text{and} \quad Z^*\beta Z = \beta$$

(for appropriately sized shift operators), or equivalently, in terms of components,

$$Z^*V_{ij}Z = V_{ij}, \quad Z^*\alpha_i Z = \alpha_i \text{ and } Z^*\beta_i Z = \beta_i,$$

for $i,j = 1,\ldots,n$. This in turn means that

$$V_{ij} = \begin{bmatrix} \ddots & & & & \\ & \omega_{ij} & & & \\ & & \boxed{\omega_{ij}} & & \\ & & & \omega_{ij} & \\ & & & & \ddots \end{bmatrix}, \quad i,j = 1,\ldots,n,$$

$$\alpha_j^* = \xi_j = \begin{bmatrix} \ddots & & & & \\ & x_j & & & \\ & & \boxed{x_j} & & \\ & & & x_j & \\ & & & & \ddots \end{bmatrix}, \quad j = 1,\ldots,n,$$

$$\beta_j^* = \eta_j = \begin{bmatrix} \ddots & & & & \\ & y_j & & & \\ & & \boxed{y_j} & & \\ & & & y_j & \\ & & & & \ddots \end{bmatrix}, \quad j = 1,\ldots,n,$$

where the ω_{ij}, x_j and y_j are bounded linear operators from $\mathcal{B}$ to $\mathcal{B}$, $\mathcal{B}$ to $\mathcal{M}$ and $\mathcal{B}$ to $\mathcal{N}$, respectively.

It is convenient to let

$$\Omega = \begin{bmatrix} \omega_{11} & \cdots & \omega_{1n} \\ \vdots & & \vdots \\ \omega_{n1} & \cdots & \omega_{nn} \end{bmatrix},$$

$$X = [x_1 \ \cdots \ x_n],$$

and

$$Y = [y_1 \ \cdots \ y_n].$$

It is then readily checked that

$$\ell_V = r_{sp}(\boldsymbol{Z}\boldsymbol{V}^*) \;=\; r_{sp}(\Omega)$$

and hence, if $\ell_V < 1$, that the formulas for Δ and Λ reduce to

$$\Delta = \sum_{t=0}^{\infty} \boldsymbol{V}^t \alpha\alpha^* \boldsymbol{V}^{t*}$$

and

$$\Lambda = \sum_{t=0}^{\infty} \boldsymbol{V}^t(\alpha\alpha^* - \beta\beta^*)\boldsymbol{V}^{t*}\ ,$$

respectively. Since the product of two block Toeplitz operators is again block Toeplitz, these formulas exhibit both Δ and Λ as block Toeplitz operators. More precisely,

$$\Lambda_{ij} \;=\; \begin{bmatrix} \ddots & & & & \\ & \lambda_{ij} & & & \\ & & \boxed{\lambda_{ij}} & & \\ & & & \lambda_{ij} & \\ & & & & \ddots \end{bmatrix} , \qquad i,j = 1,\ldots,n\ ,$$

where

$$\lambda = \begin{bmatrix} \lambda_{11} & \cdots & \lambda_{1n} \\ \vdots & & \vdots \\ \omega_{n1} & \cdots & \omega_{nn} \end{bmatrix} ,$$

$$= \sum_{t=0}^{\infty} \Omega^t(X^*X - Y^*Y)\Omega^{*t}$$

and a similar decomposition prevails for Δ in terms of

$$\delta = \begin{bmatrix} \delta_{11} & \cdots & \delta_{1n} \\ \vdots & & \vdots \\ \delta_{n1} & \cdots & \delta_{nn} \end{bmatrix}$$

$$= \sum_{t=0}^{\infty} \Omega^t X^*X\Omega^{*t}\ .$$

It is then readily checked that the basic conditions I, II and III are each equivalent to their corresponding primed version, which is stated below:

I' λ *is uniformly positive on* $\mathcal{B}^n$.

II′   *There exist a pair of bounded linear operators* $\mu : \mathcal{M} \to \mathcal{B}^n$ *and* $\nu : \mathcal{M} \to \mathcal{M}$ *such that the operator matrix*

$$\begin{bmatrix} \lambda^{\frac{1}{2}} \Omega^* & \mu \\ X & \nu \end{bmatrix} \begin{bmatrix} (\lambda + Y^* Y)^{-\frac{1}{2}} & 0 \\ 0 & I \end{bmatrix}$$

is unitary on $\mathcal{B}^n \oplus \mathcal{M}$.

III′ δ *is uniformly positive on* $\mathcal{B}^n$.

Now, for any block Toeplitz operator

$$S = \begin{bmatrix} \ddots & & & & & \\ & s_0 & s_1 & s_2 & & \cdots \\ & \boxed{s_0} & s_1 & s_2 & & \cdots \\ & 0 & s_0 & s_1 & s_2 & \cdots \\ & & & \ddots & \ddots & \ddots \end{bmatrix}$$

in $U(\ell^2_{\mathcal{N}}; \ell^2_{\mathcal{M}})$, the sum

$$s(\zeta) = \sum_{t=0}^{\infty} \zeta^t s_t \tag{11.1}$$

is an operator valued analytic function of the complex variable ζ in the open unit disc $\mathbb{D} = \{\zeta \in \mathbb{C} : |\zeta| < 1\}$. Moreover, as is well known (see e.g., p.193 of [FF]) for the general idea),

$$\|S\| \leq 1 \iff \|s(\zeta)\| \leq 1 \quad \text{for every} \ \zeta \in \mathbb{D} \,.$$

An operator valued function of the form (11.1) which is both contractive and analytic in $\mathbb{D}$ will be referred to as a Schur function. The interpolation theorems established in Sections 7, 8 and 9, all translate into statements about operator valued Schur functions with I′, II′ and III′ in place of I, II and III, respectively. Correspondingly, if I and II are in force, then the LCS operator Θ associated with a given set of block Toeplitz data corresponds to the operator valued function

$$\Psi(\zeta) \ = \ \begin{bmatrix} A(\zeta) & B(\zeta) \\ C(\zeta) & D(\zeta) \end{bmatrix} \,,$$

with

$$A(\zeta) = \nu + \zeta X (I - \zeta \Omega^*)^{-1} \lambda^{-\frac{1}{2}} \mu \,, \tag{11.1}$$

$$C(\zeta) = \zeta Y (I - \zeta \Omega^*)^{-1} \lambda^{-\frac{1}{2}} \mu \,, \tag{11.2}$$

$$B(\zeta) = X(I - \zeta \Omega^*)^{-1} \lambda^{-1} Y^* (I + Y \lambda^{-1} Y^*)^{-\frac{1}{2}} \,, \tag{(11.3)}$$

and

$$D(\zeta) = \{I + Y(I - \zeta \Omega^*)^{-1} \lambda^{-1} Y^*\}(I + Y \lambda^{-1} Y^*)^{-\frac{1}{2}} \,, \tag{11.4}$$

providing of course that $r_{sp}(\Omega) < 1$ and $|\zeta| \leq 1$.

The solutions to the interpolation problem under consideration are then given by the set of linear fractional transformations

$$T_\Psi[s_L](\zeta) = \{A(\zeta)s_L(\zeta) + B(\zeta)\}\{C(\zeta)s_L(\zeta) + D(\zeta)\}^{-1} \,, \tag{11.5}$$

of the set of all Schur functions s_L.

It is reassuring to provide a direct proof that the exhibited operator valued function $\Psi(\zeta)$ is doubly J contractive in $\mathbb{D}$ and J unitary on the boundary. This is immediate from the factorization formulas

$$\frac{J - \Psi(\zeta)^* J \Psi(\gamma)}{1 - \zeta^* \gamma} = G(\zeta)^* \lambda^{-1} G(\gamma) \tag{11.6}$$

and

$$\frac{J - \Psi(\zeta) J \Psi(\gamma)^*}{1 - \zeta \gamma^*} = F(\zeta) \lambda^{-1} F(\gamma)^* \,, \tag{11.7}$$

where

$$G(\zeta) = \left[\lambda(I - \zeta \Omega^*)^{-1} \lambda^{-\frac{1}{2}} \mu \quad \lambda \Omega^*(I - \zeta \Omega^*)^{-1} \lambda^{-1} Y^*(I + Y \lambda^{-1} Y^*)^{-\frac{1}{2}}\right] \,, \tag{11.8}$$

$$F(\zeta) = \begin{bmatrix} X \\ Y \end{bmatrix} (I - \zeta \Omega^*)^{-1} \,, \tag{11.9}$$

and both ζ and γ are restricted to the closed unit disc $\overline{\mathbb{D}}$. Both of these formulas are established by lengthy but straightforward calculations which depend heavily upon II'. To help clarify the issues, we shall now spell out the implications of the preceding analysis in a little more detail for the two basic interpolation problems under study: the NP problem and the CF problem. To ease the exposition, we shall assume that $B = \mathbb{C}^k$, $\mathcal{M} = \mathbb{C}^p$ and $\mathcal{N} = \mathbb{C}^q$.

THE NP PROBLEM: The data in the present setting is a given set of matrices

$$\omega_j \in \mathbb{C}^{k \times k} \,, \quad x_j \in \mathbb{C}^{p \times k} \,, \quad \text{and} \quad y_j \in \mathbb{C}^{q \times k} \,,$$

for $j = 1, \ldots, n$. Let

$$X = [x_1 \ \cdots \ x_n] \quad \text{and} \quad Y = [y_1 \ \cdots \ y_n]$$

as above, let

$$\Omega = \begin{bmatrix} \omega_1 & & \\ & \ddots & \\ & & \omega_n \end{bmatrix}$$

and suppose that

$$r_{sp}(\Omega) < 1$$

and that the $n \times n$ matrix

$$\delta = \sum_{t=0}^{\infty} \Omega^t X^* X \Omega^{*t}$$

is positive definite.

THEOREM 11.1. *There exists a $p \times q$ matrix valued Schur function*

$$s(\zeta) = \sum_{t=0}^{\infty} \zeta^t s_t$$

such that

$$\sum_{t=0}^{\infty} \omega_j^t x_j^* s_t = y_j^*$$

for $j = 1, \ldots, n$, if and only if the $n \times n$ matrix

$$\lambda = \sum_{t=0}^{\infty} \Omega^t (X^* X - Y^* Y) \Omega^{*t}$$

is positive semidefinite.

If λ is positive definite, then the set of linear fractional transformations

$$\{T_\Psi[\sigma] : \quad \sigma \ \ is \ a \ p \times q \ matrix \ Schur \ function\}$$

is a complete description of the set of all solutions.

PROOF. In view of the preceding discussion, this is immediate from Theorem 7.3. ∎

THE CF PROBLEM: The data in the present setting is a given set of matrices

$$\omega \in \mathbb{C}^{k \times k}, \quad x \in \mathbb{C}^{p \times k}, \quad \text{and} \quad y_j \in \mathbb{C}^{q \times k},$$

for $j = 0, \ldots, n$. Let

$$X = [x \ 0 \ \cdots \ 0], \qquad Y = [y_0 \ \cdots \ y_n]$$

and

$$\Omega = \begin{bmatrix} \omega & 0 & & 0 & 0 \\ 1 & \omega & & 0 & 0 \\ 0 & 1 & \ddots & \vdots & \vdots \\ \vdots & \vdots & \ddots & 0 & 0 \\ 0 & 0 & & 1 & \omega \end{bmatrix},$$

and suppose that

$$r_{sp}(\Omega) = r_{sp}(\omega) < 1$$

and that the $(n+1) \times (n+1)$ matrix

$$\delta = \sum_{t=0}^{\infty} \Omega^t X^* X \Omega^{*t}$$

is positive definite.

THEOREM 11.2. *There exists a $p \times q$ matrix valued Schur function $s(\zeta)$ such that*

$$(x^* s)(\zeta) = \sum_{t=0}^{\infty} (\zeta I_k - \Omega)^t x^* s_t$$

with

$$x^* s_t = y_t^*$$

for $t = 0, \ldots, n$ if and only if λ is positive semidefinite.

If λ is positive definite, then the set of linear fractional transformations

$$\{T_\Psi[\sigma] : \quad \sigma \quad is\ a\ p \times q\ matrix\ Schur\ function\}$$

is a complete description of the set of all solutions.

PROOF. In view of the preceding discussion, this is immediate from Theorem 8.4.   ∎

Both of these theorems reduce to well known results in the special case $k = 1$; see e.g., Chapters 5 and 6 of [D1].

12. VARYING COORDINATE SPACES

In the preceding analysis the spaces $\ell_{\mathcal{B}}^2$, $\ell_{\mathcal{M}}^2$ and $\ell_{\mathcal{N}}^2$ were all chosen with identical coordinate spaces: $\mathcal{B}_j = \mathcal{B}$, $\mathcal{M}_j = \mathcal{M}$ and $\mathcal{N}_j = \mathcal{N}$ for $j = 0, \pm 1, \ldots$. However, with proper adaptation of the notation everything goes through even if the coordinate spaces $\mathcal{B}_j$, $\mathcal{M}_j$ and $\mathcal{N}_j$ are allowed to vary with j. Such a general framework was introduced by van der Veen and Dewilde in [VD] for other purposes, but is is also particularly appropriate for matrix interpolation problems, as we shall illustrate by example.

From now on, $\ell_{\mathcal{B}}^2$ will denote the set of sequences

$$u = (\ldots, u_{-1}, u_0, u_1, \ldots)$$

with $u_j \in \mathcal{B}_j$ and

$$\|u\|^2 = \sum_{j=-\infty}^{\infty} \|u_j\|_{\mathcal{B}_j}^2 \, .$$

Since the $\mathcal{B}_j$ may now be all different, some care has to be taken with the definition of the shift operator Z. To this end it is useful to introduce the notation $\ell_{\mathcal{B}(k)}^2$ for the direct sum

$$\cdots \oplus \mathcal{B}_{k-1} \oplus \boxed{\mathcal{B}_k} \oplus \mathcal{B}_{k+1} \oplus \cdots$$

where the square denotes the zeroth coordinate space. Thus

$$Z \; : \; \begin{bmatrix} \vdots \\ u_{-1} \\ \boxed{u_0} \\ u_1 \\ \vdots \end{bmatrix} \in \ell_{\mathcal{B}}^2 \; \longrightarrow \; \begin{bmatrix} \vdots \\ u_0 \\ \boxed{u_1} \\ u_2 \\ \vdots \end{bmatrix} \in \ell_{\mathcal{B}(1)}^2$$

To keep the notation reasonable, we shall use the same symbol Z [resp. $Z^{-1} = Z^*$] to denote the "upwards" [resp. "downwards"] shift by one index from any space $\ell_{\mathcal{B}(k)}^2 \to \ell_{\mathcal{B}(k+1)}^2$ [resp. $\ell_{\mathcal{B}(k-1)}^2$]. With this convention in mind, it is readily checked that the product

$$Z^j \; : \; \begin{bmatrix} \vdots \\ u_{k-1} \\ \boxed{u_k} \\ u_{k+1} \\ \vdots \end{bmatrix} \in \ell_{\mathcal{B}(k)}^2 \; \longrightarrow \; \begin{bmatrix} \vdots \\ u_{k-1+j} \\ \boxed{u_{k+j}} \\ u_{k+1+j} \\ \vdots \end{bmatrix} \in \ell_{\mathcal{B}(k+j)}^2$$

for every integer j: positive, negative or zero. Correspondingly, if $A \in \mathcal{X}(\ell_{\mathcal{N}}^2, \ell_{\mathcal{B}}^2)$, then

$$Z^j A \in \mathcal{X}(\ell_{\mathcal{N}}^2; \ell_{\mathcal{B}(j)}^2) \, .$$

Thus, if $A \in \mathcal{U}(\ell_{\mathcal{N}}^2; \ell_{\mathcal{B}}^2)$ is expressed in terms of the proper analogue of the diagonal decomposition used earlier:

$$A = \sum_{j=0}^{\infty} Z^j A_{[j]} \, ,$$

then we choose the operators

$$A_{[j]} \in \mathcal{D}(\ell^2_{\mathcal{N}}; \ell^2_{\mathcal{B}(-j)})$$

so that

$$Z^j A_{[j]} \in \mathcal{U}(\ell^2_{\mathcal{N}}; \ell^2_{\mathcal{B}}) \ .$$

We next define the diagonal transform

$$\hat{A}(V) = \sum_{t=0}^{\infty} V^{[t]} A_{[t]}$$

much as before with

$$V^{[0]} = I \ ,$$
$$V^{[t]} = V V^{(1)} \cdots V^{(t-1)} \ , \quad \text{for} \quad t \geq 1 \ ,$$

except that now

$$V \in \mathcal{D}(\ell^2_{\mathcal{B}(-1)}; \ell^2_{\mathcal{B}})$$

and

$$V^{(t)} = Z^{*t} V Z^t \in \mathcal{D}(\ell^2_{\mathcal{B}(-t-1)}; \ell^2_{\mathcal{B}(-t)}) \ .$$

To get a better feeling for these formulas, it is helpful to write

$$(V)_{jj} = T_{j,j-1} \ ,$$

wherein T_{ij} designates an operator from $\mathcal{B}_j$ to $\mathcal{B}_i$. Then

$$V = \begin{bmatrix} \ddots & & & & \\ & T_{-1,-2} & & & \\ & & \boxed{T_{0,-1}} & & \\ & & & T_{1,0} & \\ & & & & \ddots \end{bmatrix}$$

$$V^{(1)} = \begin{bmatrix} \ddots & & & & \\ & T_{-2,-3} & & & \\ & & \boxed{T_{-1,-2}} & & \\ & & & T_{0,-1} & \\ & & & & \ddots \end{bmatrix}$$

$$V^{(2)} = \begin{bmatrix} \ddots & & & & \\ & T_{-3,-4} & & & \\ & & \boxed{T_{-2,-3}} & & \\ & & & T_{-1,-2} & \\ & & & & \ddots \end{bmatrix}$$

and so forth. These pictures make it pretty clear that the product $VV^{(1)} \cdots V^{(t-1)}$ makes sense for every $t \geq 2$, and that

$$V^{(t)} \in \mathcal{D}(\ell^2_{\mathcal{B}(-t-1)}; \ell^2_{\mathcal{B}(-t)}) \,,$$

as noted earlier; as usual the entries in $V^{(t)}$ are obtained by shifting those in V by t blocks in the southeast direction. Moreover,

$$V^{[t]} \in \mathcal{D}(\ell^2_{\mathcal{B}(-t)}; \ell^2_{\mathcal{B}})$$

and

$$V^{[t]} A_{[t]} \in \mathcal{D}(\ell^2_{\mathcal{N}}; \ell^2_{\mathcal{B}}) \,,$$

for $t = 0, 1, \ldots$, as needed.

Since Z is unitary,

$$V^{[t]} = (VZ^*)^t Z^t$$

and so we can just as well reexpress the diagonal transform as

$$\hat{A}(V) = \sum_{t=0}^{\infty} (VZ^*)^t Z^t A_{[t]} \,, \tag{12.1}$$

whenever the sum converges. The advantage of this formulation is that $VZ^* \in \mathcal{L}(\ell^2_{\mathcal{B}}; \ell^2_{\mathcal{B}})$ and hence there is less trouble in keeping track of the underlying spaces, as will become clearer upon consideration of the example given below. This formulation also makes it clear that the sum will converge for $A \in \mathcal{U}(\ell^2_{\mathcal{N}}; \ell^2_{\mathcal{B}})$ if

$$r_{sp}(VZ^*) = r_{sp}(ZV^*) < 1 \,.$$

The basic NP problem in this setting is (once again) to obtain conditions for the existence of an $S \in \mathcal{S}(\ell^2_{\mathcal{N}}; \ell^2_{\mathcal{M}})$ such that

$$(\xi_j^* S)^{\wedge}(V_j) = \eta_j^*$$

for $j = 1, \ldots, n$ and a given set of data operators

$$V_j \in \mathcal{D}(\ell^2_{\mathcal{B}(-1)}; \ell^2_{\mathcal{B}}) \,, \quad \xi_j \in \mathcal{D}(\ell^2_{\mathcal{B}}; \ell^2_{\mathcal{M}}) \ \text{ and } \ \eta_j \in \mathcal{D}(\ell^2_{\mathcal{B}}; \ell^2_{\mathcal{N}})$$

which, just as before, are subject to two sets of constraints: that

$$r_{sp}(ZV_j^*) < 1 \,, \quad j = 1, \ldots, n \,,$$

and that the operator

$$\Delta = \sum_{j=0}^{\infty} V^{[j]}(\alpha\alpha^*)^{(j)} V^{[j]}$$

is uniformly positive on $(\ell^2_{\mathcal{B}})^n$. Here, V, α and β are defined just as in (1.8). However, in light of the preceding discussion, it is best to reexpress Δ as

$$\Delta = \sum_{j=0}^{\infty} (VZ^*)^j \alpha\alpha^* (ZV^*)^j \,, \tag{12.2}$$

and correspondingly

$$\Lambda = \sum_{j=0}^{\infty} (VZ^*)^j (\alpha\alpha^* - \beta\beta^*)(ZV^*)^j \,. \tag{12.3}$$

The upshot is that all the analysis in the early sections of this paper goes through and all the basic interpolation theorems derived in Sections 7, 8 and 9 continue to hold for this more general setting, with some minor adaptations in notation in the formulas for Θ. The added flexibility provided by this broader setting is particularly useful for dealing with finite block matrix interpolation problems wherein the blocks in the matrices have different sizes, including "zero".

To illustrate this new setting, we consider the one point NP interpolation problem of obtaining conditions for the existence of $S \in \mathcal{S}(\ell^2_{\mathcal{N}}; \ell^2_{\mathcal{M}})$ such that

$$(\xi^* S)^{\wedge}(V) = \eta^*$$

when the spaces $\mathcal{B}_j$, $\mathcal{M}_j$ and $\mathcal{N}_j$ all vanish for $j < 0$ and $j > 3$. To keep things relatively simple, we shall further assume that

$$\mathcal{M}_j = \mathbb{C}^{p_j} = \mathcal{B}_j = \mathbb{C}^{k_j} \,, \quad j = 0, 1, 2 \,,$$

$$\mathcal{N}_j = \mathbb{C}^{q_j} \,, \quad j = 0, 1, 2 \,,$$

$$\alpha = \xi^* = \begin{bmatrix} I_{p_0} & 0 & 0 \\ 0 & I_{p_1} & 0 \\ 0 & 0 & I_{p_2} \end{bmatrix}$$

$$\beta = \eta^* = \begin{bmatrix} y_0^* & 0 & 0 \\ 0 & y_1^* & 0 \\ 0 & 0 & y_2^* \end{bmatrix} , \quad \text{with} \quad y_j \in \mathbb{C}^{q_j \times k_j} ,$$

$$(V)_{11} = \omega_1 \in \mathbb{C}^{k_1 \times k_0}$$

$$(V)_{22} = \omega_2 \in \mathbb{C}^{k_2 \times k_1}$$

and

$$(V)_{jj} = 0 \quad \text{(by default) for} \quad j \leq 0 \quad \text{and} \quad j \geq 3 .$$

Thus

$$\ell_{\mathcal{B}}^2 = \ell_{\mathcal{M}}^2 = \mathbb{C}^{p_0} \oplus \mathbb{C}^{p_1} \oplus \mathbb{C}^{p_2} ,$$

$$\ell_{\mathcal{N}}^2 = \mathbb{C}^{q_0} \oplus \mathbb{C}^{q_1} \oplus \mathbb{C}^{q_2} ,$$

and the problem reduces to finding a contractive

$$S = \begin{bmatrix} S_{00} & S_{01} & S_{02} \\ 0 & S_{11} & S_{12} \\ 0 & 0 & S_{22} \end{bmatrix}$$

with entries $S_{ij} \in \mathbb{C}^{p_i \times q_j}$ such that

$$\hat{S}(V) = \begin{bmatrix} y_0^* & 0 & 0 \\ 0 & y_1^* & 0 \\ 0 & 0 & y_2^* \end{bmatrix} .$$

The diagonal transform is based on the expansion

$$S = S_{[0]} + Z S_{[1]} + Z^2 S_{[2]}$$

$$= \begin{bmatrix} S_{00} & 0 & 0 \\ 0 & S_{11} & 0 \\ 0 & 0 & S_{22} \end{bmatrix} + \begin{bmatrix} \square & I_{p_0} & 0 \\ \square & 0 & I_{p_1} \\ \square & 0 & 0 \end{bmatrix} \begin{bmatrix} \square & \square & \square \\ 0 & S_{01} & 0 \\ 0 & 0 & S_{02} \end{bmatrix}$$

$$+ \begin{bmatrix} \square & \square & I_{p_0} \\ \square & \square & 0 \\ \square & \square & 0 \end{bmatrix} \begin{bmatrix} \square & \square & \square \\ \square & \square & \square \\ 0 & 0 & S_{02} \end{bmatrix}$$

of

$$S = \begin{bmatrix} S_{00} & 0 & 0 \\ 0 & S_{11} & 0 \\ 0 & 0 & S_{22} \end{bmatrix} + \begin{bmatrix} 0 & S_{01} & 0 \\ 0 & 0 & S_{12} \\ 0 & 0 & 0 \end{bmatrix} + \begin{bmatrix} 0 & 0 & S_{02} \\ 0 & 0 & 0 \\ 0 & 0 & 0 \end{bmatrix}$$

must be treated with care because the rows and columns containing "square zeros" should
be deleted: they correspond to deleted spaces. Since

$$VZ^* = \begin{bmatrix} \square & 0 & 0 \\ \square & \omega_1 & 0 \\ \square & 0 & \omega_2 \end{bmatrix} \begin{bmatrix} \square & \square & \square \\ I_{p_0} & 0 & 0 \\ 0 & I_{p_1} & 0 \end{bmatrix} = \begin{bmatrix} 0 & 0 & 0 \\ \omega_1 & 0 & 0 \\ 0 & \omega_2 & 0 \end{bmatrix}$$

is a block matrix with square diagonals, it is better to write

$$\hat{S}(V) = S_{[0]} + (VZ^*)(ZS_{[1]}) + (VZ^*)^2 Z^2 S_{[2]}$$

$$= \begin{bmatrix} S_{00} & 0 & 0 \\ 0 & S_{11} & 0 \\ 0 & 0 & S_{22} \end{bmatrix} + \begin{bmatrix} 0 & 0 & 0 \\ \omega_1 & 0 & 0 \\ 0 & \omega_2 & 0 \end{bmatrix} \begin{bmatrix} 0 & S_{01} & 0 \\ 0 & 0 & S_{12} \\ 0 & 0 & 0 \end{bmatrix}$$

$$+ \begin{bmatrix} 0 & 0 & 0 \\ \omega_1 & 0 & 0 \\ 0 & \omega_2 & 0 \end{bmatrix}^2 \begin{bmatrix} 0 & 0 & S_{02} \\ 0 & 0 & 0 \\ 0 & 0 & 0 \end{bmatrix} .$$

Thus the interpolation problem reduces to finding a contractive 3×3 upper triangular
block matrix of the indicated form such that

$$S_{00} = y_0^*$$

$$S_{11} + \omega_1 S_{01} = y_1^*$$

$$S_{22} + \omega_2 S_{12} + \omega_2 \omega_1 S_{02} = y_2^* .$$

By Theorem 7.3, this problem is solvable if and only if the 3×3 block matrix

$$\Lambda = \sum_{t=0}^{\infty} (VZ^*)^t (\alpha\alpha^* - \beta\beta^*)(ZV^*)^t$$

is positive semidefinite. For the present choice of the data operators,

$$\Lambda = \sum_{t=0}^{2} \begin{bmatrix} 0 & 0 & 0 \\ \omega_1 & 0 & 0 \\ 0 & \omega_2 & 0 \end{bmatrix}^t \begin{bmatrix} L_0 & 0 & 0 \\ 0 & L_1 & 0 \\ 0 & 0 & L_2 \end{bmatrix} \begin{bmatrix} 0 & \omega_1^* & 0 \\ 0 & 0 & \omega_2^* \\ 0 & 0 & 0 \end{bmatrix}^t ,$$

where

$$L_j = I_{k_j} - y_j^* y_j , \qquad j = 0, 1, 2 ,$$

for short.

Now if Λ is positive definite, then, by Theorem 7.3, the set of all solution-
s to the interpolation problem is equal to the set of linear fractional transformations

$\{T_\Theta[\sigma] : \sigma \in \mathcal{S}(\ell^2_\mathcal{N}; \ell^2_{\mathcal{M}(-1)})\}$. Since $\alpha = I$, Θ may be calculated by (5.4), properly adapted to the present setting. In particular, we now write

$$\Theta_{11} = (\rho_V^{-1} - VZ^*\Lambda)(\Lambda + \Lambda ZV^*VZ^*\Lambda)^{-\frac{1}{2}} Z \tag{12.4}$$

and

$$\Theta_{21} = \beta^* \rho_V^{-1}(\Lambda + \Lambda ZV^*VZ^*\Lambda)^{-\frac{1}{2}} Z \tag{12.5}$$

so that only 3×3 block matrices intervene in the terms to the left of the rightmost Z; Θ_{12} and Θ_{22} can be used directly as written. It is now readily checked by simply using the already established formulas for ZV^*, ρ_V^{-1} and

$$\Lambda = \begin{bmatrix} \lambda_0 & 0 & 0 \\ 0 & \lambda_1 & 0 \\ 0 & 0 & \lambda_2 \end{bmatrix} ,$$

that

$$\Theta_{11} = AZ , \quad \Theta_{12} = B , \quad \Theta_{21} = CZ , \quad \Theta_{22} = D ,$$

where A, B, C and D are the 3×3 block matrices which are computed towards the end of Example 6 in Section 10, with appropriate adjustments of size. The main difference is that now all calculations are carried out in the setting of 3×3 block matrices.

Finally, for $\sigma \in \mathcal{S}(\ell^2_\mathcal{N}; \ell^2_{\mathcal{M}(-1)})$ we have

$$\sigma = \begin{array}{c} \quad \overbrace{q_0} \quad \overbrace{q_1} \quad \overbrace{q_2} \quad \overbrace{q_3} \\ \begin{bmatrix} \square & \square & \square & \square \\ 0 & \sigma_{01} & \sigma_{02} & \square \\ 0 & 0 & \sigma_{12} & \square \\ 0 & 0 & 0 & \square \end{bmatrix} \begin{array}{l} \} \, p_{-1} \\ \} \, p_0 \\ \} \, p_1 \\ \} \, p_2 \end{array} \end{array}$$

$$Z\sigma = \begin{array}{c} \quad \overbrace{q_0} \quad \overbrace{q_1} \quad \overbrace{q_2} \\ \begin{bmatrix} 0 & \sigma_{01} & \sigma_{02} \\ 0 & 0 & \sigma_{12} \\ 0 & 0 & 0 \end{bmatrix} \begin{array}{l} \} \, p_0 \\ \} \, p_1 \\ \} \, p_2 \end{array} \end{array}$$

and hence, when Λ is positive definite, the set of all solutions to the stated interpolation problem is given by the set of matrices

$$\left\{ A \begin{bmatrix} 0 & \sigma_{01} & \sigma_{02} \\ 0 & 0 & \sigma_{12} \\ 0 & 0 & 0 \end{bmatrix} + B \right\} \left\{ C \begin{bmatrix} 0 & \sigma_{01} & \sigma_{02} \\ 0 & 0 & \sigma_{12} \\ 0 & 0 & 0 \end{bmatrix} + D \right\}^{-1} , \tag{12.6}$$

over all contractions of the form

$$
\begin{array}{ccc}
\overbrace{q_0} & \overbrace{q_1} & \overbrace{q_2} \\
0 & \sigma_{01} & \sigma_{02} & \}\, p_0 \\
0 & 0 & \sigma_{12} & \}\, p_1 \\
0 & 0 & 0 & \}\, p_2
\end{array} \;.
$$

The linear fractional representation (12.6) leads readily to explicit formulas for the entries S_{ij} in the solution matrix to the stated interpolation problem. The details are left to the reader.

With the preceding discussion and example as a guide, it should now be clear that all the interpolation theorems presented earlier can be adapted to the present general setup with variable coordinate spaces. To minimize grief in calculations, it is advisable to work with formulas (12.1)–(12.5) instead of their predecessors, and, in the same spirit, to bear in mind that in (4.17) and (4.18) $Z\,M_{12}$ maps $\ell^2_{\mathcal{M}}$ into $(\ell^2_{\mathcal{B}})^n$.

13. REFERENCES

[AG]   N.I. Akhiezer and I.M. Glazman, *Theory of Linear Operators in Hilbert Space*, Vols. I and II (Second Edition), Pitman, London, 1981.

[AD]   D. Alpay and H. Dym, *On applications of reproducing kernel spaces to the Schur algorithm and rational J unitary factorization*, in: I. Schur Methods in Operator Theory and Signal Processing (I. Gohberg, ed.), Oper. Theory: Adv. Appl. **OT18**, Birkhäuser Verlag, Basel, 1986, pp. 89-159.

[ADD]   D. Alpay, P. Dewilde and H. Dym, *Lossless inverse scattering and reproducing kernels for upper triangular operators*, in: Extension and Interpolation of Linear Operators and Matrix Functions (I. Gohberg, ed.), Oper. Theory: Adv. Appl. **OT47**, Birkhäuser Verlag, Basel, 1990, pp. 61-135.

[Be]   B. Beauzamy, *Introduction to Operator Theory and Invariant Subspaces*, North Holland, Amsterdam, 1988.

[BGK1]   J. A. Ball, I. Gohberg and M.A. Kaashoek, *Nevanlinna Pick interpolation for time-varying input-output maps: the discrete time case*, this volume.

[BGK2]   J. A. Ball, I. Gohberg and M.A. Kaashoek, *Time-varying systems: Nevanlinna-Pick interpolation and sensitivity minimization*, preprint, 1991.

[Br]   M.S. Brodskii, *Unitary operator colligations and their characteristic function-s*, Uspekhi Mat. Nauk, **33:4** (1978), 141–168 (Eng. transl.: Russian Math. Surveys, **33:4** (1978), 159–191).

[De]   P. Dewilde, *A course on the algebraic Schur and Nevanlinna-Pick interpolation problems*, in: Algorithms and Parallel VLSI Architectures, Volume A: Tutorials (E.F. Deprettere and A-J. van der Veen, eds.), Elsevier, Amsterdam, 1991, pp. 13–69.

[D1]   H. Dym, *J Contractive Matrix Functions, Reproducing Kernel Hilbert Spaces and Interpolation*, Regional Conference Series in Mathematics, **71**, Amer. Math. Soc., Providence, R.I., 1989.

[D2]   ——, *On reproducing kernel spaces, J unitary matrix functions, interpolation and displacement rank*, in: The Gohberg Anniversary Collection (H. Dym, S. Goldberg, M.A. Kaashoek and P. Lancaster, eds.), Oper. Theory: Adv. Appl. **OT41**, Birkhäuser Verlag, Bazel, 1989, pp. 173–239.

[FF]   C. Foias and A.E. Frazho, *The Commutant Lifting Approach to Interpolation Problems*, Birkhäuser Verlag, Basel, 1990.

[GKW]   I. Gohberg, M.A.Kaashoek and H.J. Woerdeman, *The band method for positive and strictly contractive extension problems: An alternative version and new applications*, Integral Equations Operator Theory, **12** (1989), 343–382.

[RR]   M. Rosenblum and J. Rovnyak, *Hardy Classes and Operator Theory*, Oxford University Press, Oxford, 1985.

[Ru]   W. Rudin, *Real and Complex Analysis*, McGraw Hill, New York, 1966.

[VD]   A-J. van der Veen and P. Dewilde, *Time-varying system theory for computational networks*, in: Algorithms and Parallel VLSI Architectures, Volume II (Y. Robert and P. Quinton, eds.), Elsevier, Amsterdam, in press.

Patrick Dewilde   Harry Dym
Department of Electrical Engineering   Department of Theoretical Mathematics
Delft University of Technology   The Weizmann Institute of Science
2600GA Delft, The Netherlands   Rehovot 76100, Israel

Operator Theory:
Advances and Applications, Vol. 56
© 1992 Birkhäuser Verlag Basel

MINIMALITY AND REALIZATION OF DISCRETE TIME-VARYING SYSTEMS

I. Gohberg, M.A. Kaashoek and L. Lerer*

The minimality and realization theory is developed for discrete time-varying finite dimensional linear systems with time-varying state spaces. The results appear as a natural generalization of the corresponding theory for the time-independent case. Special attention is paid to periodical systems. The case when the state space dimensions do not change in time is re-examined.

INTRODUCTION

Let T be a lower triangular double infinite block Toeplitz matrix, $T = \left[t_{i-j}\right]^{\infty}_{i,j=-\infty}$, where t_k are $m \times p$ matrices and $t_k = 0$ for $k < 0$. A discrete time-invariant finite dimensional system

$$(0.1) \qquad \sum \begin{cases} x_{k+1} = Ax_k + Bu_k \, , \\ y_k = Cx_k + Du_k \, , \end{cases}$$

is called a *realization* of T if $t_0 = D$ and $t_k = CA^{k-1}B$ for $k \geq 1$; in other words, if T is the input-output map of (0.1). Here $A: X \to X$ is a linear transformation acting on the finite dimensional *state space* X. In what follows we write $\Sigma = (A,B,C,D)$ instead of (0.1).

In connection with the notion of realization, four problems appear in a natural way. The first is the existence problem and concerns the question as to when the given lower triangular Toeplitz matrix admits a realization. The second is the minimality problem, i.e., the problem of finding criteria for the state space dimension to be as small as possible. The third is the reduction problem, namely, how to reduce a given realization to a minimal one. The fourth is the uniqueness problem for minimal realizations, and asks to what extent a realization with minimal state space dimension is unique.

---

* The work of this author was supported by the Netherlands organization
 for scientific research (NWO) and by the fund for promotion of research
 at the Technion.

In the classical theory of time invariant causal systems (see [8,14] and the books [7,9,11] these four problems are completely solved in the following way. Let H_T be the Hankel matrix

$$(0.2) \qquad H_T = \begin{pmatrix} t_1 & t_2 & t_3 & \cdots \\ t_2 & t_3 & & \\ t_3 & & & \\ \vdots & & & \end{pmatrix}$$

considered as a linear transformation of the space ℓ_0^p of all $\mathbb{C}^p$-valued sequences with a finite number of nonzero entries into the space ℓ^m of all $\mathbb{C}^m$-valued sequences. The double infinite lower triangular block Toeplitz matrix $T = \left[t_{i-j}\right]_{k,j=-\infty}^{\infty}$ admits a realization if and only if rank $H_T < \infty$, and in that case a realization $\tilde{\Sigma} = (\tilde{A},\tilde{B},\tilde{C},\tilde{D})$ of smallest possible state space dimension is obtained by taking $\mathrm{Im}\ H_T$ to be the state space and

$$\tilde{A}:\ \mathrm{Im}\ H_T \to \mathrm{Im}\ H_T, \qquad \tilde{A} = V\big|_{\mathrm{Im}\ H_T},$$

$$\tilde{B} = \begin{pmatrix} t_1 \\ t_2 \\ t_3 \\ \vdots \end{pmatrix} :\ \mathbb{C}^p \to \mathrm{Im}\ H_T,$$

$$\tilde{C} = (I\ O\ O\ \ldots):\ \mathrm{Im}\ H_T \to \mathbb{C}^m, \qquad \tilde{D} = t_0,$$

where V is the backward shift on the sequence space ℓ^m. In particular, rank H_T is the smallest possible state space dimension of a realization of T. Furthermore, a system $\Sigma = (A,B,C,D)$ is a minimal realization of T, i.e., has smallest possible state space dimension among all realizations of T, if and only if

$$(0.3) \qquad \mathcal{N}(\Sigma) := \mathrm{Ker} \begin{pmatrix} C \\ CA \\ \vdots \\ CA^{n-1} \end{pmatrix} = (0),$$

$$(0.4) \qquad \mathcal{R}(\Sigma) := \mathrm{Im}\begin{bmatrix} B & AB & \ldots & A^{n-1}B \end{bmatrix} = X,$$

where $n = \dim X$. If the conditions (0.3) and (0.4) are not both fulfilled, then the realization Σ may be reduced to a minimal one in the following way. Put $X_1 = \mathcal{N}(\Sigma)$, let X_0 be a direct complement of $X_1 \cap \mathcal{R}(\Sigma)$ in $\mathcal{R}(\Sigma)$, and choose X_2 such that $X = X_1 \oplus X_0 \oplus X_2$. Then relative to the decomposition $X = X_1 \oplus X_0 \oplus X_2$ the operators A, B and C partition as follows:

$$A = \begin{pmatrix} * & * & * \\ 0 & A_0 & * \\ 0 & 0 & * \end{pmatrix}, \quad B = \begin{pmatrix} * \\ B_0 \\ 0 \end{pmatrix}, \quad C = \begin{bmatrix} 0 & C_0 & * \end{bmatrix},$$

and $\left(A_0, B_0, C_0, D\right)$ is a minimal realization of T.   Finally, we mention that a minimal realization $\Sigma = \left(A, B, C, D\right)$ of a given T is unique up to similarity, which means that any other minimal realization $\Sigma_1 = \left(A_1, B_1, C_1, D_1\right)$ of T is of the form $A_1 = SAS^{-1}$, $B_1 = SB$, $C_1 = CS^{-1}$, $D_1 = D$, where S is some invertible operator.

The four problems referred to above appear in an analogous way for non-Toeplitz lower triangular block matrices, $T = \left[t_{jk}\right]_{j,k=-\infty}^{\infty}$ with $t_{jk} = 0$ for $k > j$, and the realizations being provided by discrete time-varying linear systems:

$$(0.5) \qquad \begin{cases} x_{k+1} = A_k x_k + B_k u_k \ , \\ y_k = C_k x_k + D_k u_k \ . \end{cases}$$

Here $A_k: X_k \to X_{k+1}$, $B_k: \mathbb{C}^{p_k} \to X_{k+1}$, $C_k: X_k \to \mathbb{C}^{m_k}$ and $D_k: \mathbb{C}^{p_k} \to \mathbb{C}^{m_k}$ are linear transformations and X_k is a finite dimensional space for each k.   Furthermore, we allow the dimension of the kth state space X_k to vary in time.

For time-variant systems (0.5) with a time-constant state space dimension the theory of structural properties such as controllability and structurability, reachability and observability is well-developed (see [13], [6], [3], [5], [2] and references therein).   For the minimality and realization theory, the situation is different and in this area only partial results are available (see [13], [6]).   The main aim of the present paper is to extend the minimality and realization theory for time-invariant systems as summarized above to the case of non-Toeplitz lower triangular block matrices and time-varying systems, allowing the state space dimensions to change in time.   We also re-examine the theory for the case when the state space dimensions are constant in time;   here our results clarify the additional complication that may appear as a consequence of this extra restriction.

In a number of problems (see, e.g., [4], [1]) it is natural to require the input-output map to be a bounded operator and to consider appropriate restrictions on the corresponding realizations and the related state space transformations.   This leads to a different minimality and realization theory which will be a topic for a future publication.   Results in this direction can be found in [1], [12].

The paper consists of seven sections (not counting the present introduction).   The first section has a preliminary character.   In Section 2 we consider for the system (0.5) with time-varying state space dimensions the notions of observability and reachability.   Section 3 contains the main minimality theorems; the proofs are given in the fourth section.   The realization theory is presented in Section 5.   The sixth section re-examines the theory of time-variant systems with time-constant state space dimensions.   In the seventh section the minimality and realization theory is illustrated by and developed further for periodical systems. Throughout the paper we also analyse how the relevant properties of the realization change when the sizes of the blocks in the lower triangular matrix T are enlarged by regrouping the entries without changing the action of the operator.

1. PRELIMINARIES

Consider a discrete time-varying system Σ

(1.1)
$$\sum \begin{cases} x(j+1) = A(j)x(j) + B(j)u(j) , \\ y(j) = C(j)x(j) + D(j)u(j), \ j \in \mathbb{Z}, \end{cases}$$

where the *state space operators* $A(j)$: $X(j) \to X(j+1)$ are linear operators acting between complex vector spaces $X(j)$, called *state spaces* of Σ, the *input operators* $B(j)$ are linear operators acting from the *input spaces* $\mathbb{C}^{p(j)}$ into $X(j+1)$, the *output operators* $C(j)$ are linear operators acting from $X(j)$ into the *output spaces* $\mathbb{C}^{m(j)}$ and $D(j)$ are $m(j) \times p(j)$ matrices $(j \in \mathbb{Z})$.   Note that the systems under consideration may have time-dependent state spaces as well as time dependent input and output spaces.   As a rule we consider systems (1.1) with state spaces $X(j)$ $(j \in \mathbb{Z})$ of finite dimensions and call such systems *finite-dimensional*.   In this case we sometimes identify $X(j)$ with $\mathbb{C}^{n(j)}$, $n(j) := \dim X(j)$, and the operators $A(j)$, $B(j)$ and $C(j)$ with their matrix representations with respect to the standard bases in the spaces $\mathbb{C}^{n(j)}$, $\mathbb{C}^{p(j)}$ and $\mathbb{C}^{m(j)}$.   The class of all systems (1.1) with input spaces $\mathbb{C}^{p(j)}$ and output spaces $\mathbb{C}^{m(j)}$ will be denoted by $\mathscr{S}(p(\cdot), m(\cdot))$.   The notation $\Sigma = \big(A(\cdot), B(\cdot), C(\cdot), D(\cdot); X(\cdot)\big)$ will be used throughout to denote the system (1.1).

The linear operator $S(k, \ell)$: $X(k) \to X(\ell)$ defined by

(1.2)
$$S(k, \ell) := \begin{cases} A(k-1)A(k-2) \cdots A(\ell), & \text{if } k > \ell \\ I & , \ \text{if } k = \ell \\ 0 & , \ \text{if } k < \ell \end{cases}$$

is called the *state-transition operator* of Σ.   It is clear that

(1.3)
$$S(k, \ell) = S(k, r)S(r, \ell) \quad (k \geq r \geq \ell) .$$

The double infinite lower triangular block matrix

(1.4)
$$T_\Sigma := \left[t_{ki}\right]_{k,i=-\infty}^{\infty}$$

with $m(k)\times p(i)$ matrix entries

(1.5)
$$t_{ki} = t_{ki}(\Sigma) := \begin{cases} C(k)S(k,i+1)B(i), & \text{if } i < k \\ D(i), & \text{if } i = k \\ 0, & \text{if } i > k \end{cases}$$

will be referred to as the *input/output map* of the system Σ. It is clear that the output $y(k)$ of the system (1.1) which starts operating at time $j = k_0$ with the initial state $x_0 = x(k_0)$ is given by the formula

(1.6)
$$y(k) = C(k)S(k,k_0)x_0 + \sum_{i=k_0}^{k} t_{ki}u(i) \quad (k > k_0) \ .$$

Two systems $\Sigma = \left(\left(A(\cdot),B(\cdot),C(\cdot),D(\cdot);X(\cdot)\right)\right)$ and $\tilde{\Sigma} = \left(\left(\tilde{A}(\cdot),\tilde{B}(\cdot),\tilde{C}(\cdot),\tilde{D}(\cdot);\tilde{X}(\cdot)\right)\right)$ in $\mathcal{S}\left(p(\cdot),m(\cdot)\right)$ are called *similar* if $D(\cdot) = \tilde{D}(\cdot)$ and there is a sequence of invertible operators $K(j) : X(j) \to \tilde{X}(j)$ $(j \in \mathbb{Z})$ such that

(1.7) $\tilde{A}(j) = K(j+1)A(j)K^{-1}(j),$ $\tilde{B}(j) = K(j+1)B(j),$ $\tilde{C}(j) = C(j)K^{-1}(j)$ $(j \in \mathbb{Z})$

Clearly, the input/output map of similar systems coincide: $T_\Sigma = T_{\tilde{\Sigma}}$.

The system (1.1) is called N-*periodical* if $m(j+N) = m(j)$, $p(j+N) = p(j)$, $X(j+N) = X(j)$ $(j \in \mathbb{Z})$ and

(1.8) $A(j+N) = A(j), B(j+N) = B(j), C(j+N) = C(j), D(j+N) = D(j)$ $(j \in \mathbb{Z})$.

The entries of the input/output map of a N–periodical system have the following periodicity property:

(1.9)
$$t_{k+N,i+N} = t_{k,i} \quad (k,i \in \mathbb{Z}) \ .$$

We denote by $\mathcal{S}_N\left(p(\cdot),m(\cdot)\right)$ the collection of all N-periodical systems with inputs $u_j \in \mathbb{C}^{p(j)}$ and outputs $y_j \in \mathbb{C}^{m(j)}$. Two systems Σ and $\tilde{\Sigma}$ in $\mathcal{S}_N\left(p(\cdot),m(\cdot)\right)$ are called N-*periodically similar* if the sequence of invertible operators $K(j)$ in (1.7) is N-periodical $K(j+N) = K(j)$.

Sometimes it is useful to partition the time axis $\mathbb{Z}$ by means of a partitioning $\alpha = \left\{\alpha_j\right\}_{j=-\infty}^{\infty}$ $(\alpha_{j+1}-\alpha_j \geq 1)$ into time intervals $\tau_j = \left[\alpha_j,\alpha_j+1,\ldots\alpha_{j+1}-1\right]$ and to view the interval τ_j as the jth instant of time. To be more precise, introduce "enlarged" input and output vectors $U^{(\alpha)}(k) = \text{col}\left(u(\alpha_k+i-1)\right)_{i=1}^{\alpha_{k+1}-\alpha_k}$, $Y^{(\alpha)}(k) = \text{col}\left(y(\alpha_k+i-1)\right)_{i=1}^{\alpha_{k+1}-\alpha_k}$ and consider the system

$$(1.10) \qquad \begin{cases} \rho(k+1) = A^{(\alpha)}(k)\rho(k) + B^{(\alpha)}(k)U^{(\alpha)}(k) & , \\[2mm] Y^{(\alpha)}(k) = C^{(\alpha)}(k)\rho(k) + D^{(\alpha)}(k)\rho(k), \ k \in \mathbb{Z} , \end{cases}$$

of which the state spaces are $X^{(\alpha)}(j) = X(\alpha_j)$ $(j \in \mathbb{Z})$ and the operators $A^{(\alpha)}(\cdot)$, $B^{(\alpha)}(\cdot)$, $C^{(\alpha)}(\cdot)$ and $D^{(\alpha)}(\cdot)$ are defined as follows:

$$A^{(\alpha)}(j) := S\big(\alpha_{j+1}, \alpha_j\big),$$

$$B^{(\alpha)}(j) := \mathrm{row}\Big(S(\alpha_{j+1},\alpha_j+i)B(\alpha_j+i-1)\Big)_{i=1}^{\alpha_{j+1}-\alpha_j},$$

$$(1.11) \qquad C^{(\alpha)}(j) := \mathrm{col}\Big(C(\alpha_j+i-1)S(\alpha_j+i-1,\alpha_j)\Big)_{i=1}^{\alpha_{j+1}-\alpha_j},$$

$$D^{(\alpha)}(j) = \Big[D_{ik}^{(\alpha)}(j)\Big]_{i,k=1}^{\alpha_{j+1}-\alpha_j}$$

$$= \Big[\delta_{ik}D(\alpha_j+i-1) + C(\alpha_j+i-1)S(\alpha_j+i-1,\alpha_j+k)B(\alpha_j+k-1)\Big]_{i,k=1}^{\alpha_{j+1}-\alpha_j}.$$

Note that in fact $D_{ik}^{(\alpha)}(j) = 0$ for $k > i$ and $D_{ii}^{(\alpha)}(j) = D(\alpha_j+i-1)$ $(i = 1,2,\ldots,$ $\alpha_{j+1}-\alpha_j)$. Here and elsewhere if $M_1,M_2,\ldots,M_\omega$ are matrices of appropriate sizes, we use the notation $\mathrm{col}\big(M_i\big)_{i=1}^{\omega}$ for the block column matrix

$$\begin{bmatrix} M_1 \\ M_2 \\ \vdots \\ M_\omega \end{bmatrix},$$

while the notation $\mathrm{row}\big(M_i\big)_{i=1}^{\omega}$ stands for the block row matrix $\big[M_1 M_2 \ldots M_\omega\big]$.

The "enlarged" system (1.10) will be denoted by $\Sigma^{E(\alpha)} := \big(A^{(\alpha)}(\cdot),B^{(\alpha)}(\cdot),C^{(\alpha)}(\cdot),D^{(\alpha)}(\cdot);X^{(\alpha)}(\cdot)\big)$ and will be referred to as the *system associated with* Σ *and corresponding to the partitioning* α. The system $\Sigma^{E(\alpha)}$ belongs to the class $\mathscr{S}\big(p^{(\alpha)}(\cdot),m^{(\alpha)}(\cdot)\big)$, where $p^{(\alpha)}(j) := \displaystyle\sum_{i=\alpha_j}^{\alpha_{j+1}-1} p(i)$, $m^{(\alpha)}(j) := \displaystyle\sum_{i=\alpha_j}^{\alpha_{j+1}-1} m(i)$, and the entries $t_{ki}^{(\alpha)}$ of its input/output map $T_\Sigma E(\alpha) = \big[t_{rs}^{(\alpha)}\big]_{r,s=-\infty}^{\infty}$ are of size $m^{(\alpha)}(k) \times p^{(\alpha)}(i)$. It is clear that the matrix $T_\Sigma E(\alpha)$ is obtained from the input/output map $T_\Sigma = \big[t_{ki}\big]_{k,i=-\infty}^{\infty}$ of Σ just by an appropriate grouping of the entries t_{ki} into new block entries $t_{rs}^{(\alpha)}$, and in this sense the matrices T_Σ and $T_\Sigma E(\alpha)$ coincide. One may, however, expect that the matrix $\big[t_{rs}^{(\alpha)}\big]_{r,s=-\infty}^{\infty}$ is more nicely structured than the matrix $\big[t_{ki}\big]_{k,i=-\infty}^{\infty}$. Of course, the system $\Sigma^{E(\alpha)}$ gives a more rough model for the input/output behaviour of the system Σ, in particular, regarding such properties as reachability, observability, stability, etc. In some interesting cases the input/output behaviour of the system Σ can be fully described by a finite family of associated systems $\Sigma^{E(\,\cdot\,)}$. As we shall see in the sequel this happens if one takes the family of associated systems $\Sigma^{E(\,\cdot\,)}$ corresponding to the partitionings

$$(1.12) \qquad \alpha(k) = \left\{\alpha_j^k\right\}_{j=-\infty}^{\infty}, \quad \alpha_j^k = k+jN \ (j\in\mathbb{Z}), \ k = 0,1,\ldots,N-1 \ ,$$

where N is a fixed positive integer. To simplify the notations we shall write in this case $\Sigma^{(k)}$ instead of $\Sigma^{E(\alpha(k))}$.

The passage to the associated systems $\Sigma^{(k)}$ ($k = 0,1,\ldots,N-1$) is especially useful if the system Σ is N-periodical. One easily sees that in this case the systems $\Sigma^{(k)}$ are time-invariant (cf. [10]). In fact, $\Sigma^{(k)}$ is the system

$$(1.13) \qquad \begin{cases} \rho(i+1) = A^{(k)}\rho(i) + B^{(k)}U(i) \ , \\ Y(i) = C^{(k)}\rho(i) + D^{(k)}U(i) \ , \quad i \in \mathbb{Z} \ , \end{cases}$$

of which the state space is $X(k)$, and the input and output vectors are given by

$$(1.14) \qquad U(i) = \mathrm{col}\left(u(iN+k+\ell)\right)_{\ell=0}^{N-1}, \quad Y(i) = \mathrm{col}\left(y(iN+k+\ell)\right)_{\ell=0}^{N-1} \ , \quad (i \in \mathbb{Z})$$

and the operators $A^{(k)}$, $B^{(k)}$, $C^{(k)}$ and $D^{(k)}$ are defined as follows:

$$(1.15) \qquad \begin{aligned} A^{(k)} &:= \ S(k+N,k), \\ B^{(k)} &:= \ \mathrm{row}\left(S(k+N,k+\ell)B(k+\ell-1)\right)_{\ell=1}^{N} \ , \\ C^{(k)} &:= \ \mathrm{col}\left(C(k+\ell-1)S(k+\ell-1,k)\right)_{\ell-1}^{N} \ , \\ D^{(k)} &= \ \left[D_{\ell q}^{(k)}\right]_{\ell,q=1}^{N} \\ &= \ \left[\delta_{\ell q}D(k+q-1)+C(k+\ell-1)S(k+\ell-1,k+q)B(k+q-1)\right]_{\ell,q=1}^{N} \ . \end{aligned}$$

In this N-periodic case we call $\Sigma^{(k)}$ the kth *linear time-invariant system* (abbreviated LTI-system) *associated with* Σ. Thus, with any N-periodical system $\Sigma = \left(A(\,\cdot\,),B(\,\cdot\,),C(\,\cdot\,),D(\,\cdot\,);X(\,\cdot\,)\right) \in \mathcal{S}_N\left(p(\,\cdot\,),m(\,\cdot\,)\right)$ one associates N LTI-systems $\Sigma^{(k)} = \left(A^{(k)},B^{(k)},C^{(k)},D^{(k)};X(k)\right)$ ($k = 0,1,\ldots N-1$) given by (1.13)–(1.15).

Throughout the rest of this paper the following notations will be used. If $\left(\Gamma_k\right)_{k=\omega}^{\delta}$ $(-\infty \leq \omega < \delta \leq \infty)$ is a sequence of vector spaces, then $\ell\left((\Gamma_k)_{k=\omega}^{\delta}\right)$ will denote the vector space of all sequences $\left(x_k\right)_{k=\omega}^{\delta}$ with $x_k \in \Gamma_k$. If $\omega = -\infty$ and/or $\delta = \infty$, then by $\ell_0\left((\Gamma_k)_{k=\omega}^{\delta}\right)$ we denote the space of all sequences $(x_k)_{k=\omega}^{\delta} \in \ell\left((\Gamma_k)_{k=\omega}^{\delta}\right)$ with finite support, i.e., $x_j = 0$ for all $j \notin \left[\omega_1,\delta_1\right]$ where $\omega_1(\geq \omega)$ and $\delta_1(\leq \delta)$ are finite numbers depending on the sequence $(x_k)_{k=\omega}^{\delta}$. Next, $\ell_+\left((\Gamma_k)_{k=\omega}^{\delta}\right)$ will denote the space of all sequences $\left(x_k\right)_{k=\omega}^{\delta} \in \ell\left(\left(\Gamma_k\right)_{k=\omega}^{\delta}\right)$ with finite "negative" support, i.e., $x_k = 0$ for all $\omega \le k \le k_0$ for some integer k_0 depending on the sequence $\left(x_k\right)_{k=\omega}^{\delta}$.

2. OBSERVABILITY AND REACHABILITY

In this section we extend the well known notions and results concerning observability and reachability of discrete time-varying systems with a constant state space (and constant input and output spaces) to systems with time-varying state spaces $X(j)$ (and time-varying input and output spaces).

A vector $x \in X(j)$ is called an *unobservable state* of the system $\Sigma = \left(A(\cdot),B(\cdot),C(\cdot),D(\cdot);X(\cdot)\right)$ given by (1.1) if for any time $k \ge j$ the free motion of the system Σ starting at time j with the initial value $x(j) = x$ ends up with a zero output, i.e.,

$$y(k) = C(k)S(k,j)x = 0 \quad (k \ge j) ,$$

where $S(k,j)$ is the state-transition operator defined by (1.2). The set of all unobservable states in $X(j)$ is a subspace of $X(j)$, called the *unobservable subspace at time j*. The system Σ is said to be *observable at time* j if $x = 0$ is the only unobservable state in $X(j)$. In other words, the system Σ given by (1.1) is observable at time j if and only if the equations

$$C(k)S(k,j)x = 0 \quad (k = j,j+1,\ldots; \ x \in X(j))$$

imply $x = 0$. The system Σ is called *completely observable* if it is observable at all time $j \in \mathbf{Z}$.

To state a rank test for observability consider the infinite block column matrix

$$(2.1) \qquad N_j(\Sigma) := \operatorname{col}\left(C(i+j-1)S(i+j-1,j)\right)_{i=1}^{\infty} = \begin{bmatrix} C(j) \\ C(j+1)S(j+1,j) \\ \vdots \\ C(j+\ell)S(j+\ell,j) \\ \vdots \end{bmatrix}$$

as a linear operator acting from $X(j)$ into the space $\ell\left(\left(\mathbb{C}^{m(\nu+j-1)}\right)_{\nu=1}^{\infty}\right)$. It is clear that the system Σ is observable at time j if and only if the operator $N_j(\Sigma)$ is injective. Thus, we have the following observability test.

PROPOSITION 2.1. *The system* $\Sigma = \left(A(\cdot),B(\cdot),C(\cdot),D(\cdot);X(\cdot)\right)$ *is observable at time j if and only if*

$$(2.2) \qquad \operatorname{rank} N_j(\Sigma) = \operatorname{rank} \operatorname{col}\left(C(i+j-1)S(i+j-1,j)\right)_{i=1}^{\infty} = \dim X(j) ,$$

and Σ is completely observable if and only if (2.2) holds true for any $j \in \mathbb{Z}$.

Now we consider the connections between the observability of a system Σ and the observability of the enlarged systems $\Sigma^{E(\cdot)}$.

PROPOSITION 2.2. *Let $\alpha = \left\{\alpha_j\right\}_{j=-\infty}^{\infty}$ be a partitioning of $\mathbb{Z}$. Then the system Σ is observable at all times α_j $(j \in \mathbb{Z})$ if and only if the corresponding associated system $\Sigma^{E(\alpha)}$ is completely observable.*

Proof. We first apply (1.3) and (1.11) to compute $S^{(\alpha)}(k,\ell)$, the state-transition operator of $\Sigma^{E(\alpha)}$, for $k > \ell$:

$$S^{(\alpha)}(k,\ell) = A^{(\alpha)}(k-1)A^{(\alpha)}(k-2)\ldots A^{(\alpha)}(\ell)$$
$$= S(\alpha_k,\alpha_{k-1})S(\alpha_{k-1},\alpha_{k-2})\ldots S(\alpha_{\ell+1},\alpha_\ell)$$
$$= S(\alpha_k,\alpha_\ell) \ .$$

Then we have for each $i = 0,1,\ldots$ and $j \in \mathbb{Z}$:

$$C^{(\alpha)}(i+j)S^{(\alpha)}(i+j,j)$$
$$= \mathrm{col}\left(C(\alpha_{i+j}+\ell-1)S(\alpha_{i+j}+\ell-1,\alpha_{i+j})\right)_{\ell=1}^{\alpha_{i+j+1}-\alpha_{i+j}}S(\alpha_{i+j},\alpha_j)$$
$$= \mathrm{col}\left(C(\alpha_{i+j}+\ell-1)S(\alpha_{i+j}+\ell-1,\alpha_j)\right)_{\ell=1}^{\alpha_{i+j+1}-\alpha_{i+j}} \ .$$

Now one easily sees that the matrix

$$N_{\alpha_j}(\Sigma) = \mathrm{col}\left(C(\alpha_j+i)S(\alpha_j+i,\alpha_j)\right)_{i=0}^{\infty}$$

can be represented in the form

$$(2.3) \qquad N_{\alpha_j}(\Sigma) = \mathrm{col}\left[\mathrm{col}\left(C(\alpha_{i+j}+\ell-1)S(\alpha_{i+j}+\ell-1,\alpha_j)\right)_{\ell=1}^{\alpha_{i+j+1}-\alpha_{i+j}}\right]_{i=0}^{\infty} = N_j\left(\Sigma^{E(\alpha)}\right),$$

and hence rank $N_j\left(\Sigma^{(\alpha)}\right) = $ rank $N_{\alpha_j}(\Sigma)$. In view of Proposition 2.1 this completes the proof. $\square$

Now fix a positive integer N and consider the enlarged systems $\Sigma^{(k)}$ corresponding to the partitionings (1.12)). The preceding proposition shows that the system Σ is observable at times $k + jN$ $(j \in \mathbb{Z})$ if and only if the system $\Sigma^{(k)}$ is completely observable. Thus, the following result follows.

PROPOSITION 2.3. *The system Σ is completely observable if and only if all systems $\Sigma^{(k)}$ $(k = 0,1,\ldots,N-1)$ are completely observable.*

Regarding N-periodical systems the last two propositions can be reformulated in the following way:

PROPOSITION 2.4. *An N-periodical system Σ is observable at times $k + jN$ $(j \in \mathbb{Z})$ if and only if the kth associated with Σ LTI-system $\Sigma^{(k)}$ is observable.*

Furthermore, Σ is completely observable if and only if each LTI-system $\Sigma^{(k)}$ $(k = 0,1,...,N-1)$ is observable.

We note that in the case of an N-periodical system $\Sigma = \left(A(\cdot),B(\cdot),C(\cdot),D(\cdot);X(\cdot)\right)$ the equality (2.3) can be rewritten as follows:

$$(2.4) \qquad N_k(\Sigma) = \mathrm{col}\left(C^{(k)}(A^{(k)})^{i-1}\right)_{i=1}^{\infty},$$

and since $A^{(k)}$ acts on a finite dimensional space X(k), we have

$$(2.5) \qquad \mathrm{Ker}\ N_k(\Sigma) = \mathrm{Ker}\ \mathrm{col}\left(C^{(k)}(A^{(k)})^{i-1}\right)_{i=1}^{\dim X(k)}.$$

We pass now to the notion of reachability of a system $\Sigma = \left(A(\cdot),B(\cdot),C(\cdot),D(\cdot);X(\cdot)\right)$. A vector $x \in X(j)$ is called a *reachable state* of the system Σ if there is a positive integer $0 < j_0 < j$ and a control function $u(\cdot)$ defined over the interval $\left[j-j_0,j-1\right]$ which carries the state $x(j-j_0) = 0$ into the state $x(j) = x$, i.e.,

$$x = \sum_{i=j-j_0}^{j-1} S(j,i+1)B(i)u(i).$$

The set of all reachable states in X(j) forms a linear subspace of X(j) called the *reachable subspace of* Σ *at time* j. If all the vectors in X(j) are reachable states the system (1.1) is said to be *reachable at time* j. The system Σ is called *completely reachable* if it is reachable at all times $j \in \mathbb{Z}$.

To state a rank test for reachability we consider

$$(2.6) \qquad R_j(\Sigma):= \mathrm{row}\ \left(S(j,j-k+1)B(j-k)\right)_{k=1}^{\infty} := \left[S(j,j)B(j-1)\ S(j,j-1)B(j-2)...\right]$$

as a linear operator acting from the space $\ell_0\left(\left(\mathbb{C}^{p(j-\nu)}\right)_{\nu=1}^{\infty}\right)$ into X(j). Now it is clear that the system Σ is reachable at time j if and only if the operator $R_j(\Sigma)$ is surjective. Thus, we have the following test for reachability.

PROPOSITION 2.5. *The system Σ defined by (1.1) is reachable at time j if and only if*

$$(2.7) \qquad \mathrm{rank}\ \mathrm{row}\ \left(S(j,j-k+1)B(j-k)\right)_{k=1}^{\infty} = \dim X(j),$$

and Σ is completely reachable if and only if (2.7) holds true for all $j \in \mathbb{Z}$.

Now consider the connections between the reachability of a system Σ and reachability of the associated systems $\Sigma^{E(\cdot)}$. Arguing as in the proof of Proposition 2.2 one easily sees that if $\alpha = \left\{\alpha_j\right\}_{j=-\infty}^{\infty}$ is a partitioning of $\mathbb{Z}$, then

$$(2.8) \qquad \mathrm{Im}\ R_{\alpha_j}(\Sigma) = \mathrm{Im}\ R_j\left(\Sigma^{E(\alpha)}\right),$$

and therefore the following result holds true.

PROPOSITION 2.6.   *Let* $\alpha = \left\{\alpha_j\right\}_{j=-\infty}^{\infty}$ *be a partitioning of* $\mathbb{Z}$. *Then the system* Σ *is reachable at all times* α_j *($j \in \mathbb{Z}$) if and only if the corresponding associated system* $\Sigma^{E(\alpha)}$ *is completely reachable.*

Regarding the enlarged systems $\Sigma^{(k)}$ corresponding to the partitionings (1.12) we obtain the following result.

PROPOSITION 2.7.   *The system* Σ *is completely reachable if and only if all systems* $\Sigma^{(k)}$ *($k = 0,1,\ldots,N-1$) are completely reachable.*

For N-periodical systems the last two propositions can be reformulated in the following way.

PROPOSITION 2.8.   *An N-periodical system* Σ *is reachable at times* $k + jN$ *($j \in \mathbb{Z}$) if and only if the kth associated with* Σ *LTI-system* $\Sigma^{(k)}$ *is reachable. Furthermore,* Σ *is completely reachable if and only if each LTI-system* $\Sigma^{(k)}$ *($k = 0,1,\ldots,N-1$) is reachable.*

Note that in the case of an N-periodical system $\Sigma = \left(A(\cdot),B(\cdot),C(\cdot),D(\cdot);X(\cdot)\right)$ the equality (2.8) can be written as follows:

$$(2.9) \qquad \operatorname{Im} R_k(\Sigma) = \operatorname{Im} \operatorname{row}\left(B^{(k)}(A^{(k)})^{i-1}\right)_{i=1}^{\dim X(k)}.$$

3. MINIMALITY FOR TIME-VARYING SYSTEMS

In this section we solve the minimality problem for time-varying systems of the type (1.1) with time-varying state spaces (and time-varying input and output spaces).

A system $\Sigma = \left(A(\cdot),B(\cdot),C(\cdot),D(\cdot);X(\cdot)\right)$ in $\mathscr{S}\left(p(\cdot),m(\cdot)\right)$ is called *minimal* if the dimensions of its state spaces $X(j)$ are as small as possible among all systems in $\mathscr{S}\left(p(\cdot),m(\cdot)\right)$ with the same input/output map as Σ. In other words, Σ is minimal if for any other system $\tilde{\Sigma} = \left(\tilde{A}(\cdot),\tilde{B}(\cdot),\tilde{C}(\cdot),\tilde{D}(\cdot);\tilde{X}(\cdot)\right) \in \mathscr{S}\left(p(\cdot),m(\cdot)\right)$ such that $T_{\tilde{\Sigma}} = T_{\Sigma}$ we have $\dim X(j) \le \dim \tilde{X}(j)$ ($j \in \mathbb{Z}$). The existence of minimal systems will follow from the theorems below.

We shall state now four theorems which provide a complete solution to the minimality problem for time-varying discrete time systems.   First we characterize minimal systems in terms of reachability and observability.

THEOREM 3.1.   A *system* $\Sigma = \left(A(\cdot),B(\cdot),C(\cdot),D(\cdot);X(\cdot)\right)$ *is minimal if and only if it is completely reachable and completely observable.*

Now we are going to describe how to get from a given system $\Sigma \in \mathscr{S}\left(p(\cdot),m(\cdot)\right)$ a minimal system $\Sigma_0 \in \mathscr{S}\left(p(\cdot),m(\cdot)\right)$ with the same input/output map as Σ: $T_{\Sigma_0} = T_{\Sigma}$. To this end we need the following notions of reduction and dilation. We say that a system $\Sigma_0 = \left(A^0(\cdot),B^0(\cdot),C^0(\cdot),D^0(\cdot);X^0(\cdot)\right) \in \mathcal{S}\left(p(\cdot),m(\cdot)\right)$ is a *reduction* of the system $\Sigma = \left(A(\cdot),B(\cdot),C(\cdot),D(\cdot);X(\cdot)\right)$ if the spaces $X(j)$ admit a direct sum decomposition $X(j) = X^1(j) \oplus X^0(j) \oplus X^2(j)$ $(j \in \mathbb{Z})$ such that relative to this decomposition the following partitionings hold true:

$$(3.1) \qquad A(j) = \begin{pmatrix} * & * & * \\ 0 & A^0(j) & * \\ 0 & 0 & * \end{pmatrix}: X^1(j) \oplus X^0(j) \oplus X^2(j) \to X^1(j{+}1) \oplus X^0(j{+}1) \oplus X^2(j{+}1),$$

$$(3.2) \qquad B(j) = \begin{bmatrix} * \\ B^0(j) \\ 0 \end{bmatrix}: \mathbb{C}^{p(j)} \to X^1(j{+}1) \oplus X^0(j{+}1) \oplus X^2(j{+}1),$$

$$(3.3) \qquad C(j) = \begin{bmatrix} 0 & C^0(j) & * \end{bmatrix}: X^1(j) \oplus X^0(j) \oplus X^2(j) \to \mathbb{C}^{m(j)},$$

where $*$ stands for unspecified entries. In this case we say that Σ is a *dilation* of Σ_0. Note that if the system Σ_0 is a reduction of the system Σ, then the input/output maps of the two systems coincide.

Given a system $\Sigma = \left(A(\cdot),B(\cdot),C(\cdot),D(\cdot);X(\cdot)\right)$ the following subspaces in $X(j)$ will be used in our constructions:

$$(3.4) \qquad \mathcal{N}_j(\Sigma) := \operatorname{Ker} N_j(\Sigma), \quad \mathcal{R}_j(\Sigma) := \operatorname{Im} R_j(\Sigma) \quad (j \in \mathbb{Z}),$$

where $N_j(\Sigma)$ and $R_j(\Sigma)$ are the operators defined by (2.1) and (2.5), respectively.

THEOREM 3.2. *Any system $\Sigma = \left(A(\cdot),B(\cdot),C(\cdot),D(\cdot);X(\cdot)\right)$ is a dilation of a minimal system. Namely, for each $j \in \mathbb{Z}$, put $X^1(j) = \mathcal{N}_j(\Sigma)$, let $X^0(j)$ be a direct complement of $X^1(j) \cap \mathcal{R}_j(\Sigma)$ in $\mathcal{R}_j(\Sigma)$ and choose a direct complement $X^2(j)$ of $\mathcal{R}_j(\Sigma) + \mathcal{N}_j(\Sigma)$ in $X(j)$. Then relative to the decompositions*

$$X(j) = X^1(j) \oplus X^0(j) \oplus X^2(j) \quad (j \in \mathbb{Z})$$

the partitionings (3.1)–(3.3) hold true and the system $\Sigma_0 = \left(A^0(\cdot),B^0(\cdot),C^0(\cdot),D^0(\cdot);X^0(\cdot)\right)$ is minimal.

If Σ and Σ_0 are related as in the previous theorem, we call Σ_0 a *minimal reduction* of Σ. Our next result shows that any minimal system $\tilde{\Sigma}$ with the same input/output map as Σ can be obtained by the reduction procedure described in Theorem 3.2.

THEOREM 3.3. *Let $\Sigma, \tilde{\Sigma} \in \mathcal{S}\left(p(\cdot),m(\cdot)\right)$ be minimal systems. Then the systems Σ and $\tilde{\Sigma}$ are similar if and only if they have the same input/output map: $T_{\Sigma} = T_{\tilde{\Sigma}}$.*

In order to compute the state–space dimensions of a minimal system we define the following notion of a degree sequence. Given an infinite lower triangular matrix $T = \left[t_{ki}\right]_{k,i=-\infty}^{\infty}$ ($t_{ki} = 0$, if $k < i$) with $m(k) \times p(i)$ matrix entries the semi–infinite matrices $H_T^{(j)} := \left[t_{j+i-1,j-k}\right]_{i,k=1}^{\infty}$ are viewed as linear operators acting from $\ell_0\left((\mathbb{C}^{p(j-\nu)})_{\nu=1}^{\infty}\right)$ into $\ell\left((\mathbb{C}^{m(\nu+j)})_{j=0}^{\infty}\right.$ Denote $\delta_j(T) = \operatorname{rank} H_T^{(j)}$. The sequence $\left\{\delta_j(T)\right\}_{j=-\infty}^{\infty}$ will be referred to as the *degree sequence* of the matrix T. Note that if T is a block–Toeplitz matrix: $t_{k,i} = h_{k-i}$ ($k,i \in \mathbb{Z}$, $k \geq i$), then all the matrices $H_T^{(j)}$ coincide with the familiar block–Hankel matrix $H = \left[h_{i+k-1}\right]_{i,k=1}^{\infty}$, and hence the integers $\delta_j(T)$ ($j \in \mathbb{Z}$) are equal to the (McMillan) degree of the matrix H.

THEOREM 3.4. *A system* $\Sigma = \left(A(\cdot),B(\cdot),C(\cdot),D(\cdot);X(\cdot)\right)$ *is minimal if and only if the sequence of its state space dimensions* $\left\{\dim X(j)\right\}_{j=-\infty}^{\infty}$ *coincides with the degree sequence* $\left\{\delta_j(T_\Sigma)\right\}_{j=-\infty}^{\infty}$ *of its input/output map* T_Σ.

Theorems 3.1–3.4 will be proved in the next section. Here we proceed with the analysis of the connections between the minimality of a system Σ and minimality of the enlarged systems $\Sigma^{E(\cdot)}$.

First, using Propositions 2.2 and 2.6 one immediately deduces from Theorem 3.1 the following result.

THEOREM 3.5. *Given a partitioning* $\alpha = \left\{\alpha_j\right\}_{j=-\infty}^{\infty}$ *of the time axis* $\mathbb{Z}$, *a system* $\Sigma \in \mathcal{S}\left(p(\cdot),m(\cdot)\right)$ *is reachable and observable at times* α_j ($j \in \mathbb{Z}$) *if and only if the system* $\Sigma^{E(\alpha)}$ *is minimal.*

Next, using Theorem 3.6 we see that a minimal reduction of $\Sigma^{E(\alpha)}$ can be obtained from the minimal reduction Σ_0 of Σ.

THEOREM 3.6. *Let* $\alpha = \left\{\alpha_j\right\}_{j=-\infty}^{\infty}$ *be a partitioning of* $\mathbb{Z}$, *and* $\Sigma \in \mathcal{S}\left(p(\cdot),m(\cdot)\right)$. *Then the system* $\Sigma^{E(\alpha)}$ *is a dilation of a minimal system* $\tilde{\Sigma} \in \mathcal{S}\left(p^{(\alpha)}(\cdot),m^{\alpha}(\cdot)\right)$ *which can be represented as* $\tilde{\Sigma} = \Sigma_0^{E(\alpha)}$, *where* $\Sigma_0 \in \mathcal{S}\left(p(\cdot),m(\cdot)\right)$ *is a minimal reduction of* Σ.

Note that, generally speaking, a minimal reduction of $\Sigma^{E(\alpha)}$ need not be of the form $\Sigma_0^{E(\alpha)}$, where Σ_0 is a minimal reduction of Σ.

Concerning the degree sequence of $T_\Sigma E(\cdot)$ it is clear that it is a subsequence of the degree sequence of T_Σ. To be more precise,

$$\delta_j\bigl(T_\Sigma E(\alpha)\bigr) = \delta_{\alpha_j}(T_\Sigma) \quad (j \in \mathbb{Z}).$$

Thus, a straightforward application of Theorem 3.4 leads to the following result.

THEOREM 3.7. *Let* $\alpha = \bigl\{\alpha_j\bigr\}_{j=-\infty}^{\infty}$ *be a partitioning of the time axis* $\mathbb{Z}$ *and let* $\Sigma = \bigl(A(\cdot),B(\cdot),C(\cdot),D(\cdot);X(\cdot)\bigr) \in \mathcal{S}\bigl(p(\cdot),m(\cdot)\bigr)$. *If*

$$\text{(3.5)} \qquad\qquad \dim X(\alpha_j) = \delta_{\alpha_j}(T_\Sigma) \quad (j \in \mathbb{Z}) ,$$

then the system $\Sigma^{E(\alpha)}$ *is minimal. Conversely, if* $\Sigma^{E(\alpha)}$ *is a minimal system, then* *(3.5) holds true.*

In the case of enlarged systems $\Sigma^{(k)}$ corresponding to the partitionings (1.12) Propositions 2.3 and 2.7 and Theorem 3.1 lead to the following result.

THEOREM 3.8. *The system* Σ *is minimal if and only if all the associated systems* $\Sigma^{(k)}$ *(k = 0,1,...,n-1) are minimal.*

4. PROOFS OF THE MINIMALITY THEOREMS

In this section we prove Theorems 3.1–3.4 from the preceding section. For the sake of convenience, we present the proofs of Theorems 3.1–3.4 as a chain of steps which yield the desired result.

Proof of Theorems 3.1–3.4

Step 1. Let $\Sigma = \bigl(A(\cdot),B(\cdot),C(\cdot),D(\cdot);X(\cdot)\bigr)$, and let $\mathcal{N}_j(\Sigma)$ and $R_j(\Sigma)$ $(j \in \mathbb{Z})$ be the subspaces defined by (3.4). In this step we show that

$$\text{(4.1)} \qquad\qquad A(j)\mathcal{N}_j(\Sigma) \subset \mathcal{N}_{j+1}(\Sigma) \quad (j \in \mathbb{Z}) ,$$

$$\text{(4.2)} \qquad\qquad A(j)\mathcal{R}_j(\Sigma) \subset \mathcal{R}_{j+1}(\Sigma) \quad (j \in \mathbb{Z}) .$$

Indeed, let V_ℓ denote the left shift operator acting from $\ell\bigl((\mathbb{C}^{m(\nu+j-1)})_{\nu=1}^{\infty}\bigr)$ into $\ell\bigl((\mathbb{C}^{m(\nu+j)})_{\nu=1}^{\infty}\bigr)$:

$$V_\ell\bigl(f_i\bigr)_{i=1}^{\infty} = \bigl(f_{i+1}\bigr)_{i=1}^{\infty}, \quad f_\nu \in \mathbb{C}^{m(\nu+j)} \quad (\nu=1,2,\ldots).$$

Then one easily sees that

$$\text{(4.3)} \qquad\qquad V_\ell N_j(\Sigma) = N_{j+1}(\Sigma)A(j),$$

and hence

$$A(j)\mathrm{Ker}N_j(\Sigma) \subset A(j)\mathrm{Ker}V_\ell N_j(\Sigma) \subset \mathrm{Ker}N_{j+1}(\Sigma),$$

which proves (4.1).

To prove (4.2) denote by V_r the right shift operator acting from $\ell_0\bigl((\mathbb{C}^{p(j-\nu)})_{\nu=1}^{\infty}\bigr)$ into $\ell_0\bigl((\mathbb{C}^{p(j+1-\nu)})_{\nu=1}^{\infty}\bigr)$:

$$V_r(g_i)_{i=1}^{\infty} = (h_i)_{i=1}^{\infty}, \quad h_i = g_{i-1} \ (i = 2,3,\ldots), \quad h_1 = 0 \in \mathbb{C}^{p(j)}.$$

A simple verification shows that

(4.4)
$$R_{j+1}(\Sigma)V_r = A(j)R_j(\Sigma) \ ,$$

and therefore

$$\mathrm{Im}A(j)R_j(\Sigma) = \mathrm{Im}R_{j+1}(\Sigma)V_r \subset \mathrm{Im}R_{j+1}(\Sigma),$$

which proves (4.2).

Step 2. Let $\Sigma = \big(A(\cdot),B(\cdot),C(\cdot),D(\cdot);X(\cdot)\big)$, and make the decompositions $X(j) = X^1(j) \oplus X^0(j) \oplus X^2(j) \ (j \in \mathbb{Z})$ as in Theorem 3.2. In this step we shall show that relative to these decompositions the partitionings (3.1)–(3.3) hold true and that the reduction $\Sigma_0 = \big(A^0(\cdot),B^0(\cdot),C^0(\cdot),D(\cdot);X^0(\cdot)\big)$ is completely reachable and completely observable.

Indeed, from the definitions of the subspaces $X^1(j)$ and $X^0(j)$ and from the inclusions (4.1),(4.2) we see that

$$A(j)X^1(j) \subset X^1(j+1), \quad A(j)\big(X^1(j) \oplus X^0(j)\big) \subset X^1(j+1) \oplus X^0(j+1) \ (j \in \mathbb{Z}),$$

which implies the partitionings (3.1) of the operators $A(j)$.

Since
$$\mathrm{Im}B(j) \subset \mathcal{R}_{j+1}(\Sigma) \subset X^1(j) \oplus X^0(j) \quad (j \in \mathbb{Z}),$$
and
$$\mathrm{Ker}C(j) \supset \mathcal{N}_j(\Sigma) = X^1(j) \quad (j \in \mathbb{Z}),$$

the partitionings (3.2) and (3.3) are also clear.

To prove that the system Σ_0 is completely observable, assume that $x_0 \in X(j_0)$ is an unobservable state of Σ_0 at time j_0, i.e., $x_0 \in \mathrm{Ker}N_{j_0}(\Sigma_0)$. From the partitionings (3.1) and (3.3) we see that the operator $N_{j_0}(\Sigma)$ admits the partitioning

$$N_{j_0}(\Sigma) = \big[0 \ N_{j_0}(\Sigma_0) \ *\big]: X^1(j_0) \oplus X^0(j_0) \oplus X^2(j_0) \to \ell\big((\mathbb{C}^{m(\nu+j-1)})_{\nu=1}^{\infty}\big),$$

and therefore the vector $\tilde{x}_0 = \begin{bmatrix} 0 \\ x_0 \\ 0 \end{bmatrix} \in X^0(j_0)$ is also in $\mathrm{Ker}N_{j_0}(\Sigma) = X^1(j_0)$. But then $\tilde{x}_0 = 0$, and hence $x_0 = 0$. This proves the complete observability of Σ_0.

Now use the partitioning (3.1),(3.2) to see that the operators $R_j(\Sigma)$ admit the partitionings

$$R_j(\Sigma) = \begin{bmatrix} * \\ R_j(\Sigma_0) \\ 0 \end{bmatrix}: \ell_0\big((\mathbb{C}^{(j-\nu)})_{\nu=1}^{\infty}\big) \to X^1(j) \oplus X^0(j) \oplus X^2(j) \quad (j \in \mathbb{Z}).$$

Since $\mathcal{R}_j(\Sigma) = \mathrm{Im}R_j(\Sigma) \supset X^0(j)$ and $\mathrm{Im}R_j(\Sigma_0) \subset X^0(j)$, we infer that $X^0(j) = \mathrm{Im}R_j(\Sigma_0)$ $(j \in \mathbb{Z})$, i.e., Σ is completely reachable.

Remark that this step also shows that in order to prove Theorem 3.2 it is enough to prove Theorem 3.1.

Step 3. The previous step shows that a minimal system is necessarily completely observable and completely reachable. Indeed, if $\Sigma = \big(A(\cdot),B(\cdot),C(\cdot),D(\cdot);X(\cdot)\big)$ is a minimal system and for some j_0 the operator $R_{j_0}(\Sigma)$ is not surjective and/or the operator $N_{j_0}(\Sigma)$ is not injective, then using Step 2 we can construct a reduction $\Sigma_0 = \big(A^0(\cdot),B^0(\cdot),C^0(\cdot),D(\cdot);X^0(\cdot)\big)$ of Σ such that $\dim X^0(j) \leq \dim X(j)$ $(j \in \mathbb{Z})$ and $\dim X^0(j_0) < \dim X(j_0)$. This contradicts the minimality of Σ, because, clearly, Σ and Σ_0 have the same input/output map.

Step 4. In this step we shall prove that if the systems $\Sigma = \big(A(\cdot),B(\cdot),C(\cdot),D(\cdot);X(\cdot)\big) \in \mathscr{S}\big(p(\cdot),m(\cdot)\big)$ and $\tilde{\Sigma} = \big(\tilde{A}(\cdot),\tilde{B}(\cdot),\tilde{C}(\cdot),\tilde{D}(\cdot);\tilde{X}(\cdot)\big) \in \mathscr{S}\big(p(\cdot),m(\cdot)\big)$ are completely reachable and completely observable and if the input/output maps of Σ and $\tilde{\Sigma}$ coincide, then Σ and $\tilde{\Sigma}$ are similar.

Consider the matrix $N_j(\Sigma)R_j(\Sigma) = \big[\gamma_{ik}(\Sigma)\big]_{i,k=1}^{\infty}$ as a linear operator acting $\ell_0\big((\mathbb{C}^{p(j-\nu)})_{\nu=1}^{\infty}\big) \to \ell\big((\mathbb{C}^{m(j+\nu-1)})_{\nu=1}^{\infty}\big)$. Using (1.3) one computes the entries $\gamma_{ik}(\Sigma)$:

$$\gamma_{ik}(\Sigma) = C(i+j-1)S(i+j-1,j)S(j,j-k+1)B(j-k)$$
$$= C(i+j-1)S(i+j-1,j-k+1)B(j-k) = t_{i+j-1,\,j-k}(\Sigma) \quad (i,k=1),$$

where $t_{\alpha\beta}$ are defined by (1.5).

Thus we have proved that

(4.5) $$H_{T_\Sigma}^{(j)} = N_j(\Sigma)R_j(\Sigma) \quad (j \in \mathbb{Z}) .$$

Since $T_{\tilde{\Sigma}} = T_\Sigma$ we infer that

(4.6) $$N_j(\Sigma)R_j(\Sigma) = N_j(\tilde{\Sigma})R_j(\tilde{\Sigma}) \quad (j \in \mathbb{Z}) .$$

As Σ and $\tilde{\Sigma}$ are completely observable, the operators $N(\Sigma)$ and $N_j(\tilde{\Sigma})$ have left inverses $\big(N_j(\Sigma)\big)^{(-1)}$ and $\big(N_j(\tilde{\Sigma})\big)^{(-1)}$. Similarly, the complete reachability of Σ and $\tilde{\Sigma}$ implies that the operators $R_j(\Sigma)$ and $R_j(\tilde{\Sigma})$ have right inverses $\big(R_j(\Sigma)\big)^{(-1)}$ and $\big(R_j(\tilde{\Sigma})\big)^{(-1)}$. Set $K(j) := (N_j(\tilde{\Sigma}))^{(-1)}N_j(\Sigma)$ $(j \in \mathbb{Z})$. Then, from (4.6) we see that $K(j) = R_j(\tilde{\Sigma})\big(R_j(\Sigma)\big)^{(-1)}$. The operators $K(j)$ act from $X(j)$ into $\tilde{X}(j)$ and have inverses $K^{-1}(j) = \big(N_j(\Sigma)\big)^{(-1)}N_j(\tilde{\Sigma}) = R_j(\Sigma)\big(R_j(\tilde{\Sigma})\big)^{(-1)}$. Indeed,

$$\big(N_j(\Sigma)\big)^{(-1)}N_j(\tilde{\Sigma})K(j) = \big(N_j(\Sigma)\big)^{(-1)}N_j(\tilde{\Sigma})R_j(\tilde{\Sigma})\big(R_j(\Sigma)\big)^{(-1)}$$
$$= \big(N_j(\Sigma)\big)^{(-1)}N_j(\Sigma)R_j(\Sigma)\big(R_j(\Sigma)\big)^{(-1)} = I_{X(j)} \quad (j \in \mathbb{Z}),$$

and

$$K(j)R_j(\Sigma)\left(R_j(\tilde{\Sigma})\right)^{(-1)} = \left(N_j(\tilde{\Sigma})\right)^{(-1)}N_j(\Sigma)R_j(\Sigma)\left(R_j(\tilde{\Sigma})\right)^{(-1)}$$

$$= \left(N_j(\tilde{\Sigma})\right)^{(-1)}N_j(\tilde{\Sigma})R_j(\tilde{\Sigma})\left(R_j(\tilde{\Sigma})\right)^{(-1)} = I_{\tilde{X}(j)} \quad (j \in \mathbb{Z}).$$

Next, in view of (4.3) and (4.6)

$$(4.7) \qquad N_{j+1}(\Sigma)A(j)R_j(\Sigma) = V_\ell N_j(\Sigma)R_j(\Sigma) = V_\ell N_j(\tilde{\Sigma})R_j(\tilde{\Sigma})$$

$$= N_{j+1}(\tilde{\Sigma})\tilde{A}(j)R_j(\tilde{\Sigma}) \quad (j \in \mathbb{Z}).$$

These equalities yield

$$(4.8) \qquad K(j+1)A(j)K^{-1}(j) = \tilde{A}(j) \quad (j \in \mathbb{Z}).$$

From (4.6) we see also that

$$N_j(\tilde{\Sigma}) = N_j(\Sigma)K^{-1}(j)$$

and

$$R_{j+1}(\tilde{\Sigma}) = K(j+1)R_{j+1}(\Sigma),$$

and therefore

$$(4.9) \qquad \tilde{C}(j) = C(j)K^{-1}(j), \quad \tilde{B}(j) = K(j+1)B(j), \quad (j \in \mathbb{Z}),$$

which concludes Step 4.

Step 5. In this step we conclude the proofs of Theorems 3.1–3.4. Let $\Sigma = \left(A(\cdot),B(\cdot),C(\cdot),D(\cdot);X(\cdot)\right) \in \mathcal{S}\left(p(\cdot),m(\cdot)\right)$ be a completely observable and completely reachable system, and let $\tilde{\Sigma} = \left(\tilde{A}(\cdot),\tilde{B}(\cdot),\tilde{C}(\cdot),D(\cdot);\tilde{X}(\cdot)\right) \in \mathcal{S}\left(p(\cdot),m(\cdot)\right)$ be another system with the same input/output map as Σ: $T_{\tilde{\Sigma}} = T_{\Sigma}$. To prove Theorem 3.1 we have to show that $\dim X(j) \le \dim\tilde{X}(j)$ $(j \in \mathbb{Z})$. Applying Step 2 we can construct a reduction $\tilde{\Sigma}^0 = \left(\tilde{A}^0(\cdot),\tilde{B}^0(\cdot),\tilde{C}^0(\cdot),D(\cdot);\tilde{X}^0(\cdot)\right)$ of $\tilde{\Sigma}$ which is completely reachable and completely observable. Since $T_{\tilde{\Sigma}^0} = T_{\tilde{\Sigma}} = T_{\Sigma}$, Step 4 implies that Σ and $\tilde{\Sigma}^0$ are similar. In particular, $\dim X(j) = \dim\tilde{X}^0(j)$ $(j \in \mathbb{Z})$. But $\dim\tilde{X}^0(j) \le \dim\tilde{X}(j)$ $(j \in \mathbb{Z})$ and Theorem 3.1 is proved.

Now Theorem 3.3 follows from Theorem 3.1 and Step 4. As already mentioned, Theorem 3.2 also follows from Theorem 3.1 and Step 2. It remains to prove Theorem 3.4. To this end assume that $\Sigma = \left(A(\cdot),B(\cdot),C(\cdot),D(\cdot);X(\cdot)\right)$ is a minimal system. Then in view of Theorem 3.1 and Propositions 2.1 and 2.5 the operators $N_j(\Sigma): X(j) \to \ell\left((\mathbb{C}^{m(j+\nu-1)})_{\nu=1}^\infty\right)$ $(j \in \mathbb{Z})$ are injective, while the operators $R_j(\Sigma): \ell_0\left((\mathbb{C}^{p(j-\nu)})_{\nu=1}^\infty\right) \to X(j)$ $(j \in \mathbb{Z})$ are surjective. It follows, therefore, from (4.5) that

$$(4.10) \qquad \text{rank}H_{T_\Sigma}^{(j)} = \delta_j(T_\Sigma) = \dim X(j) \quad (j \in \mathbb{Z}).$$

Conversely, if (4.10) holds true, then the operators $N_j(\Sigma)$ $(j \in \mathbb{Z})$ have to be injective and the operators $R_j(\Sigma)$ $(j \in \mathbb{Z})$ have to be surjective, i.e., by Propositions 2.1 and 2.5 the system Σ is completely observable and completely reachable. But then Theorem 3.1 implies that Σ is minimal. $\qquad\square$

5. REALIZATIONS OF INFINITE LOWER TRIANGULAR MATRICES

In this section we solve the following problem. Given a double-infinite lower triangular block-matrix $T = \left[T_{ki}\right]_{k,i=-\infty}^{\infty}$ with $m(k) \times p(i)$ matrix entries t_{ki}, find a discrete time-varying system $\Sigma = \left(A(\cdot),B(\cdot),C(\cdot),D(\cdot);X(\cdot)\right) \in \mathscr{S}\left(p(\cdot),m(\cdot)\right)$ of which the input/output map coincides with T, i.e., $T_\Sigma = T$. A system Σ with this property will be called a *realization* of the matrix T.

The given matrix T defines in a natural way a linear transformation (also denoted by T) from $\ell_+\left(\left(\mathbb{C}^{p(\nu)}\right)_{\nu=-\infty}^{\infty}\right)$ into $\ell_+\left(\left(\mathbb{C}^{m(\nu)}\right)_{\nu=-\infty}^{\infty}\right)$, namely, via the rule

$$T\left(u_\nu\right)_{\nu=-\infty}^{\infty} = \left(y_\nu\right)_{\nu=-\infty}^{\infty}, \quad y_\nu = \sum_{i=-\infty}^{\infty} t_{\nu i} u_i \quad (\nu = 0,\pm 1,\dots) \ .$$

In fact, the infinite sum above reduces to a finite one in view of the structure of the matrix T and the definition of the spaces $\ell_+\left(\left(\Gamma_\nu\right)_{\nu=-\infty}^{\infty}\right)$. Thus, the operator T is well defined.

Furthermore, for a fixed $j \in \mathbb{Z}$ the semi-infinite matrix

$$H_T^{(j)} = \left[t_{j+i-1,\,j-k}\right]_{i,k=1}^{\infty}$$

defines a linear operator (also denoted by $H_T^{(j)}$) acting from $\ell_0\left(\left(\mathbb{C}^{p(j-\nu)}\right)_{\nu=1}^{\infty}\right)$ into $\ell\left(\left(\mathbb{C}^{m(j+\nu-1)}\right)_{\nu=1}^{\infty}\right)$ according to the rule

$$H_T^{(j)}\left(u_\nu\right)_{\nu=1}^{\infty} = \left(y_\nu\right)_{\nu=1}^{\infty}, \quad y_\nu = \sum_{k=1}^{\infty} t_{j+\nu-1,\,j-k} u_k \ .$$

Again, the infinite sum above is in fact a finite sum, due to the definition of the spaces $\ell_0\left(\left(\Gamma_\nu\right)_{\nu=1}^{\infty}\right)$, and hence the action of $H_T^{(j)}$ is well defined. As in Section 3 the sequence $\left\{\delta_j(T)\right\}_{j=-\infty}^{\infty}$, where $\delta_j(T) := \operatorname{rank} H_T^{(j)}$ will be referred to as the *degree sequence* of the matrix T. Modifying slightly the notations in Section 4 we denote by $V_\ell^{(j)}$ the left shift operator acting from $\ell\left(\left(\mathbb{C}^{m(\nu+j-1)}\right)_{\nu=1}^{\infty}\right)$ into $\ell\left(\left(\mathbb{C}^{m(\nu+j)}\right)_{\nu=1}^{\infty}\right)$:

$$V_\ell^{(j)}\left(f_i\right)_{i=1}^{\infty} = \left(f_{i+1}\right)_{i=1}^{\infty}, \quad f_\nu \in \mathbb{C}^{m(\nu+j-1)} \quad (\nu=1,2,\dots) \ ,$$

and by $V_r^{(j)}$ the right shift operator acting from $\ell_0\left(\left(\mathbb{C}^{p(j-\nu)}\right)_{\nu=1}^{\infty}\right)$ into $\ell_0\left(\left(\mathbb{C}^{p(j+1-\nu)}\right)_{\nu=1}^{\infty}\right)$:

$$V_r\left(g_i\right)_{i=1}^{\infty} = \left(h_i\right)_{i=1}^{\infty}, \quad h_i = g_{i-1} \ (i = 2,3,\dots), \quad h_1 = 0 \in \mathbb{C}^{p(j)} \ .$$

A simple verification shows that

$$(5.1) \qquad\qquad V_\ell^{(j)} H_T^{(j)} = H_T^{(j+1)} V_r^{(j)} \quad (j \in \mathbb{Z}).$$

This equality implies that

$$(5.2) \qquad V_{\ell}^{(j)} \operatorname{Im} H_T^{(j)} \subset \operatorname{Im} H_T^{(j+1)} .$$

THEOREM 5.1. *Let* $T = \left[t_{ki}\right]_{k,i=-\infty}^{\infty}$ *be a lower triangular block matrix with* $m(k) \times p(i)$ *matrix entries* t_{ki}. *Then* T *is an input/output map of a finite-dimensional system* Σ *if and only if* rank $H_T^{(j)}$ *is finite for all* $j \in \mathbb{Z}$. *If this condition is satisfied, then a system* $\Sigma = \left(A(\cdot),B(\cdot),C(\cdot),D(\cdot);X(\cdot)\right)$ *such that* $T = T_{\Sigma}$ *is constructed in the following way. Denote* $X(j) = \operatorname{Im} H_T^{(j)}$ *and for* $j \in \mathbb{Z}$ *put*

$$(5.3) \qquad A(j) \colon \operatorname{Im} H_T^{(j)} \to \operatorname{Im} H_T^{(j+1)}, \quad A(j) = V_{\ell}^{(j)} \Big|_{\operatorname{Im} H_T^{(j)}},$$

$$(5.4) \qquad B(j) \colon \mathbb{C}^{p(j)} \to \operatorname{Im} H_T^{(j+1)}, \quad B(j)u = \left(t_{j+\nu,j} u\right)_{\nu=1}^{\infty},$$

$$(5.5) \qquad C(j) \colon \operatorname{Im} H_T^{(j)} \to \mathbb{C}^{m(j)}, \quad C(j)\left(y_{\nu}\right)_{\nu=1}^{\infty} = y_1 ,$$

$$(5.6) \qquad D(j) = t_{jj}.$$

Moreover, the system Σ *is minimal.*

Before proving Theorem 5.1 let us introduce the following notations which will be convenient in the sequel. By $\pi_k^{(j)}$ we denote the standard projection of the space $\ell\left((\mathbb{C}^{m(j+\nu-1)})_{\nu=1}^{\infty}\right)$ onto the space $\ell\left((\mathbb{C}^{m(j+\nu-1)})_{\nu=1}^{k}\right)$:

$$(5.7) \qquad \pi_k^{(j)}\left(f_{\nu}\right)_{\nu=1}^{\infty} = \left(f_{\nu}\right)_{\nu=1}^{k} ,$$

and let $\varepsilon_k^{(j)}$ stand for the natural embedding of the space $\ell\left((\mathbb{C}^{p(j-\nu)})_{\nu=1}^{k}\right)$ into the space $\ell_0\left((\mathbb{C}^{p(j-\nu)})_{\nu=1}^{\infty}\right)$:

$$(5.8) \qquad \varepsilon_k^{(j)}\left(f_{\nu}\right)_{\nu=1}^{k} = \left(h_{\nu}\right)_{\nu=1}^{\infty}, \quad h_i = f_i \ (i = 1,\ldots,k), \quad h_i = 0 \in \mathbb{C}^{p(j-i)} \ (i > k).$$

With these notations formulas (5.4) and (5.5) become

$$(5.4') \qquad B(j) = H_T^{(j+1)} \varepsilon_1^{(j+1)} \colon \mathbb{C}^{p(j)} \to \operatorname{Im} H_T^{(j+1)},$$

$$(5.5') \qquad C(j) = \pi_1^{(j)} \Big|_{\operatorname{Im} H_T^{(j)}} \colon \operatorname{Im} H_T^{(j)} \to \mathbb{C}^{m(j)},$$

respectively.

Proof of Theorem 5.1. Let $\Sigma = \left(A(\cdot),B(\cdot),C(\cdot),D(\cdot);X(\cdot)\right)$ be the system given by (5.3)–(5.6). To prove the "if" part of the theorem we have to show that the matrices

$$\gamma_{ki} := C(k)A(k-1)A(k-2)\ldots A(i+1)B(i)$$

coincide with the matrices t_{ki} for all $k > i$. To this end assume first that $k > i+1$ and substitute the formulas for $A(i+1)$ and $B(i)$ from (5.3) and (5.4$'$) and use (5.1) to obtain

$$\gamma_{ki} = C(k)A(k-1)A(k-2)\ldots A(i+2)V_\ell^{(i+1)}H_T^{(i+1)}\varepsilon_1^{(i+1)}$$

$$= C(k)A(k-1)A(k-2)\ldots A(i+2)H_T^{(i+2)}V_r^{(i+1)}\varepsilon_1^{(i+1)} \; .$$

Then substitute $A(i+2) = V_\ell^{(i+2)}\Big|_{\mathrm{Im}H_T^{(i+2)}}: \mathrm{Im}H_T^{(i+2)} \to \mathrm{Im}H_T^{(i+3)}$ and use again (5.1) to

obtain $\gamma_{ki} = C(k)A(k-1)A(k-2)\ldots A(i+3)H_T^{(i+3)}V_r^{(i+2)}V_r^{(i+1)}\varepsilon_1^{(i+1)}$.   Proceeding in this way one gets the equality

$$\gamma_{ki} = C(k)H_T^{(k)}V_r^{(k-1)}V_r^{(k-2)}\ldots V_r^{(i+1)}\varepsilon_1^{(i+1)} \; .$$

It is clear that

$$V_r^{(k-1)}V_r^{(k-2)}\ldots V_r^{(i+1)}\varepsilon_1^{(i+1)}u = \left(u_\nu\right)_{\nu=1}^\infty,$$

$$u_{k-i} = u \in \mathbb{C}^{p(i)}, \quad u_\nu = 0 \in \mathbb{C}^{p(k-\nu)} \quad (\nu \neq k-i) \; ,$$

and therefore substituting the formula (5.5$'$) for $C(k)$ we see that γ_{ki} coincides with t_{ki}. Also, for $i = k-1$ we see that

$$\gamma_{k,k-1} = C(k)B(k-1) = \pi_1^{(k)}H_T^{(k)}\varepsilon_1^{(k)} = t_{k,k-1}.$$

Thus, comparing with (1.5) we infer that the given matrix T coincides with T_Σ, the input/output map of the system Σ given by (5.3)–(5.6). The minimality of Σ follows from Theorem 3.4, since $\dim X(j) = \mathrm{rank}\, H_T^{(j)} = \delta_j(T_\Sigma)$ $(j \in \mathbb{Z})$.

The "only if" part of the theorem follows from Theorems 3.2 and 3.4. Indeed, if $T = T_\Sigma$ for some $\Sigma \in \mathcal{S}\big(p(\cdot),m(\cdot)\big)$, then in view of Theorem 3.2 $T = T_{\Sigma_0}$, where $\Sigma_0 = \big(A^0(\cdot),B^0(\cdot),C^0(\cdot),D^0(\cdot);X^0(\cdot)\big)$ is a minimal reduction of Σ and Theorem 3.4 implies that $\mathrm{rank}H_T^{(j)} = \delta_j\big(T_{\Sigma_0}\big) = \dim X^0(j)$.   □

We remark that the system Σ defined in the theorem is a realization of T also in the case when some of the numbers $\delta_j(T) = \mathrm{rank}H_T^{(j)}$ are equal to infinity. Of course, in such a case we cannot claim the finite-dimensionality of Σ and its minimality.

Note also that if the given matrix T has the periodicity property: $t_{k+N,i+N} = t_{k,i}$ $(i,k \in \mathbb{Z})$, then the realization Σ in Theorem 5.1 is N-periodical.

The realization Σ constructed in Theorem 5.1 is the time-varying analog of the well known "shift realization" of lower triangular block-Toeplitz matrices. We now present the time-varying analog of the "Hankel realization" of such matrices.

THEOREM 5.2. *Let* $T = \left[t_{ki}\right]_{k,i=-\infty}^{\infty}$ *be a lower triangular block matrix with* $m(k)\times p(i)$ *matrix entries* t_{ki}, *and assume that all numbers* rank $H_T^{(j)}$ $(j \in \mathbb{Z})$ *are finite. For each* $j \in \mathbb{Z}$ *write a minimal rank decomposition of* $H_T^{(j)}$:

(5.9)
$$H_T^{(j)} = \Phi_j \Psi_j \,,$$
$$\Psi_j \colon \ell_0\bigl((\mathbb{C}^{p(j-\nu)})_{\nu=1}^{\infty}\bigr) \to \tilde{X}(j), \quad \Phi_j \colon \tilde{X}(j) \to \ell\bigl((\mathbb{C}^{m(j+\nu-1)})_{\nu=1}^{\infty}\bigr) \,,$$

where $\tilde{X}(j)$ *is a linear space of dimension* $\delta_j(T)(= \text{rank } T^{(j)})$. *Denote* $\tilde{H}_T^{(j)} = \left[t_{j+i,j-k}\right]_{i,k=1}^{\infty}$ *and for each* $j \in \mathbb{Z}$ *set*

(5.10)
$$\tilde{A}(j) = \Phi_{j+1}^{(-1)}\tilde{H}_T^{(j)}\Psi_j^{(-1)} \,, \qquad\qquad \tilde{B}(j) = \Psi_{j+1}\varepsilon_1^{(j+1)},$$
$$\tilde{C}(j) = \pi_1^{(j)}\Phi_j, \qquad\qquad\qquad \tilde{D}(j) = t_{jj} \,.$$

Then the system $\tilde{\Sigma} = \bigl(\tilde{A}(\cdot),\tilde{B}(\cdot),\tilde{C}(\cdot),\tilde{D}(\cdot);\tilde{X}(\cdot)\bigr)$ *is a minimal realization of the matrix* T.

We recall that the notation $M^{(-1)}$ is used to denote a one-sided inverse of the operator M provided it exists. The operators Φ_j and Ψ_j in (5.9) are left invertible and right invertible, respectively.

Proof of Theorem 5.2. We shall prove the theorem by showing that the system $\tilde{\Sigma}$ is similar to the system Σ constructed in Theorem 5.1. To this end first note that Φ_j is injective and $\Phi_j\tilde{X}(j) = \text{Im } H_T^{(j)} = X(j)$. Hence, for any left inverse $\Phi_j^{(-1)}$ of Φ_j we have $\Phi_j^{(-1)}X(j) = \tilde{X}(j)$. It follows that the operators

$$K(j):= \Phi_j^{(-1)}\Big|_{X(j)} : X(j) \to \tilde{X}(j) \quad (j \in \mathbb{Z})$$

are invertible with inverses

$$K^{-1}(j) = \Phi_j \colon \tilde{X}(j) \to X(j) \quad (j \in \mathbb{Z}).$$

We claim that the operators $K(j)$ provide the similarity relations (1.7) between the system $\Sigma = \bigl(A(\cdot),B(\cdot),C(\cdot),D(\cdot);X(\cdot)\bigr)$ defined in Theorem 5.1 and the system $\tilde{\Sigma} = \bigl(\tilde{A}(\cdot),\tilde{B}(\cdot),\tilde{C}(\cdot),\tilde{D}(\cdot);\tilde{X}(\cdot)\bigr)$ defined by (5.10). Indeed, using (5.9) and right invertibility of Ψ_j, we have

$$K(j+1)A(j)K(j)^{-1} = K(j+1)V_\ell^{(j)}\Big|_{\text{Im}H_T^{(j)}}\Phi_j\Psi_j\Psi_j^{(-1)} = K(j+1)V_\ell^{(j)}H_T^{(j)}\Psi_j^{(-1)}.$$

But $V_\ell^{(j)}H_T^{(j)} = \tilde{H}_T^{(j)}$, and in view of (5.1) Im $\tilde{H}_T^{(j)} \subset \text{Im } H_T^{(j+1)} = X(j+1)$. Hence

$$K(j+1)V_\ell^{(j)}H_T^{(j)} = \Phi_{j+1}^{(-1)}\tilde{H}_T^{(j)} \,,$$

and we see that

$$K(j+1)A(j)K(j)^{-1} = \tilde{A}(j) \quad (j \in \mathbb{Z}).$$

Next, in view of (5.9), we have

$$
\begin{aligned}
K(j+1)B(j) &= K(j+1)H_T^{(j+1)}\varepsilon_1^{(j+1)} \\
&= \Phi_{j+1}^{(-1)}\Phi_{j+1}\Psi_{j+1}\varepsilon_1^{(j+1)} = \Psi_{j+1}\varepsilon_1^{(j+1)} = \tilde{B}(j) \quad (j \in \mathbb{Z}).
\end{aligned}
$$

Also,

$$C(j)K(j)^{-1} = \pi_1^{(j)}\Big|_{X(j)}\Phi_j = \pi_1^{(j)}\Phi_j = \tilde{C}(j) ,$$

which completes the proof of the theorem. $\qquad\qquad\qquad\qquad\qquad\square$

Now we present a realization procedure which involves finite submatrices of T only.

THEOREM 5.3. *Let* $T = \left[t_{ki}\right]_{k,i=-\infty}^{\infty}$ *be a lower triangular block matrix with* $m(k)\times p(i)$ *entries* t_{ki}, *and assume that the numbers* $\delta_j := \operatorname{rank} H_T^{(j)}$ $(j \in \mathbb{Z})$ *are finite. For each* $j \in \mathbb{Z}$ *fix positive integers* μ_j *and* ν_j *such that the finite submatrix* $T_{\mu_j,\nu_j}^{(j)} = \left[t_{j+i-1,j-k}\right]_{i,k=1}^{\mu_j,\nu_j}$ *of* $H_T^{(j)}$ *is of rank* δ_j, *and write a minimal rank decomposition of* $T_{\mu_j,\nu_j}^{(j)}$:

$$
\text{(5.11)} \qquad T_{\mu_j,\nu_j}^{(j)} = \varphi_j\psi_j ,
$$

$$
\varphi_j = \operatorname{col}\left(\varphi_{ji}\right)_{i=1}^{\mu_j}, \quad \varphi_{ji} \in \mathbb{C}^{m(j+i-1)\times\delta_j}; \quad \psi_j = \operatorname{row}\left(\psi_{jk}\right)_{k=1}^{\nu_j}, \quad \psi_{jk} \in \mathbb{C}^{\delta_j\times p(j-k)} .
$$

Denote $\hat{T}_{\mu_{j+1},\nu_j}^{(j)} = \left[t_{j+i,j-k}\right]_{i,k=1}^{\mu_{j+1},\nu_j}$ and for each $j \in \mathbb{Z}$ set

$$
\text{(5.12)} \qquad \hat{X}(j) = \mathbb{C}^{\delta_j}, \quad \hat{A}(j) = \varphi_{j+1}^{(-1)}\hat{T}_{\mu_{j+1},\nu_j}^{(j)}\psi_j^{(-1)},
$$

$$
\hat{B}(j) = \psi_{j+1,1}, \quad \hat{C}(j) = \varphi_{j1}, \quad \hat{D}(j) = t_{jj} .
$$

Then the system $\hat{\Sigma} = \left(\hat{A}(\cdot),\hat{B}(\cdot),\hat{C}(\cdot),\hat{D}(\cdot);\hat{X}(\cdot)\right)$ *is a minimal realization of the matrix T.*

Proof. Let a minimal rank decomposition (5.11) of $T_{\mu_j,\nu_j}^{(j)}$ be given. We shall first find minimal rank decompositions (5.9) of $H_T^{(j)}$ $(j \in \mathbb{Z})$ such that

$$
\text{(5.13)} \qquad \varphi_j = \pi_{\mu_j}^{(j)}\Phi_j, \quad \psi_j = \Psi_j\varepsilon_{\nu_j}^{(j)} \quad (j \in \mathbb{Z}) ,
$$

where $\pi_k^{(j)}$ and $\varepsilon_k^{(j)}$ are defined by (5.7) and (5.8), respectively. To this end take a minimal rank decomposition $H_T^{(j)} = \tilde{\Phi}_j\tilde{\Psi}_j$ if $H_T^{(j)}$ with $X(j) = \mathbb{C}^{\delta_j}$ as the intermediate space. Then

(5.14)
$$T^{(j)}_{\mu_j,\nu_j} = \pi^{(j)}_{\mu_j} H^{(j)}_T \varepsilon^{(j)}_{\nu_j} = \tilde{\varphi}_j \tilde{\psi}_j \;,$$

where $\tilde{\varphi}_j := \pi^{(j)}_{\mu_j} \tilde{\Phi}_j$, $\tilde{\psi}_j := \tilde{\Psi}_j \varepsilon^{(j)}_{\nu_j}$. Obviously, rank $\tilde{\varphi}_j \leq \delta_j$, rank $\tilde{\psi}_j \leq \delta_j$, and since rank $T^{(j)}_{\mu_j,\nu_j} = \delta_j$, we see that rank $\tilde{\varphi}_j$ = rank $\tilde{\psi}_j = \delta_j$, i.e., (5.14) is a minimal rank decomposition of $T^{(j)}_{\mu_j,\nu_j}$. But then there is an invertible operator S_j on $\mathbb{C}^{\delta_j}$ such that $\varphi_j = \tilde{\varphi}_j S_j$, $\psi_j = S_j^{-1} \tilde{\psi}_j$, where φ_j and ψ_j are the factors in the given minimal rank decomposition (5.11) of $T^{(j)}_{\mu_j,\nu_j}$. Now set $\Phi_j = \tilde{\Phi}_j S_j$, $\Psi_j = S_j^{-1} \tilde{\Psi}_j$. Then $H^{(j)}_T = \Phi_j \Psi_j$ is a minimal rank decomposition and

$$\varphi_j = \tilde{\varphi}_j S_j = \pi^{(j)}_{\mu_j} \tilde{\Phi}_j S_j = \pi^{(j)}_{\mu_j} \Phi_j, \quad \psi_j = S_j^{-1} \tilde{\psi}_j = S_j^{-1} \tilde{\Psi}_j \varepsilon^{(j)}_{\nu_j} = \Psi_j \varepsilon^{(j)}_{\nu_j} \;,$$

i.e., (5.13) is satisfied.

We note that if the minimal rank decompositions (5.9) of $H^{(j)}_T$ are such that (5.13) holds true, and if $\varphi_j^{(-1)}$ and $\psi_j^{(-1)}$ denote some one-sided inverses of φ_j and ψ_j, respectively, then the operators

(5.15)
$$\Phi_j^{(-1)} := \varphi_j^{(-1)} \pi^{(j)}_{\mu_j}, \quad \Psi_j^{(-1)} = \varepsilon^{(j)}_{\nu_j} \psi_j^{(-1)}$$

are one-sided inverses of Φ_j and Ψ_j, respectively. Indeed, in view of (5.13) we have

$$\Phi_j^{(-1)} \Phi_j = \varphi_j^{(-1)} \pi^{(j)}_{\mu_j} \Phi_j = \varphi_j^{(-1)} \varphi_j = I_{\delta_j} \;,$$

and

$$\Psi_j \Psi_j^{(-1)} = \Psi_j \varepsilon^{(j)}_{\nu_j} \psi_j^{(-1)} = \psi_j \psi_j^{(-1)} = I_{\delta_j} \;.$$

Now, for each $j \in \mathbb{Z}$ take a minimal rank decomposition (5.9) of $H^{(j)}_T$ such that (5.13) holds true and consider the system $\tilde{\Sigma} = \big(\tilde{A}(\cdot), \tilde{B}(\cdot), \tilde{C}(\cdot), \tilde{D}(\cdot); \tilde{X}(\cdot)\big)$ defined by (5.10), where the one-sided inverses of Φ_j and Ψ_j are chosen as in (5.15). We claim that $\tilde{\Sigma}$ coincides with the system $\hat{\Sigma} = \big(\hat{A}(\cdot), \hat{B}(\cdot), \hat{C}(\cdot), \hat{D}(\cdot); \hat{X}(\cdot)\big)$ defined by (5.12). Indeed, since $\pi^{(j+1)}_{\mu_{j+1}} \tilde{H}^{(j)}_T \varepsilon^{(j)}_{\nu_j} = \hat{T}^{(j)}_{\mu_{j+1},\nu_j}$ we see that

$$\tilde{A}(j) = \Phi_{j+1}^{(-1)} \tilde{H}^{(j)}_T \Psi_j^{(-1)} = \varphi_{j+1}^{(-1)} \pi^{(j+1)}_{\mu_{j+1}} \tilde{H}^{(j)}_T \varepsilon^{(j)}_{\nu_j} \psi_j^{(-1)} = \hat{A}(j) \;.$$

Also, the equalities

$$\Psi_{j+1} \varepsilon^{(j+1)}_1 = \psi_{j+1,1} = \hat{B}(j), \quad \pi^{(j)}_1 \Phi_j = \varphi_{j1} = \hat{C}(j)$$

are clear. Now the assertion of the theorem follows from Theorem 5.2. □

We conclude this section with two illustrative examples.

EXAMPLE 5.4. Let M_i $(i \in \mathbb{Z})$ be matrices of size $m(i) \times p(i+1)$ and consider the infinite block matrix $T = [t_{ki}]_{k,i=-\infty}^{\infty}$ with $t_{i,i-1} = M_i$ $(i \in \mathbb{Z})$ and $t_{ki} = 0$ if $k \neq i-1$ $(k,i \in \mathbb{Z})$. One immediately sees that $\delta_j := \delta_j(T) = \operatorname{rank} H_T^{(j)} = \operatorname{rank} M_j$ $(j \in \mathbb{Z})$. This shows, in particular, that any sequence of positive integers can be a degree sequence for a double infinite lower triangular block matrix.

We now apply Theorem 5.3 to construct a minimal realization of the matrix T under consideration. We have $\mu_j = \nu_j = 1$ $(j \in \mathbb{Z})$, and hence $T_{\mu_j,\nu_j}^{(j)} = M_j$, $\hat{T}_{\mu_{j+1},\nu_j}^{(j)} = 0$. Write a minimal rank decomposition of M_j:

$$M_j = \varphi_j \psi_j, \quad \varphi_j \in \mathbb{C}^{m(j) \times \delta_j}, \quad \psi_j \in \mathbb{C}^{\delta_j \times p(j-1)} \quad (j \in \mathbb{Z}).$$

It follows from Theorem 5.3 that the system $\Sigma = \left(0, \psi_{j+1}, \varphi_j, 0; \ \mathbb{C}^{\delta_j}\right)$ is a minimal realization of the given matrix T.   □

EXAMPLE 5.5. Let a_j $(j \in \mathbb{Z})$ and α be $m \times m$ matrices and consider the lower triangular block matrix $T = [t_{ki}]_{k,i=-\infty}^{\infty}$ with $t_{ki} = 0$ for $k \leq i$, $t_{0,-1} = \alpha$ and $t_{ki} = a_{k-i}$ for all $(k,i) \neq (0,-1)$ with $k > i$. Thus, the matrix T differs from the lower triangular block–Toeplitz matrix $T^{(a)} = [a_{k-i}]_{k,i=-\infty}^{\infty}$ $(a_{-j} := 0, \ j \geq 0)$ by one entry in position $(0,-1)$. Introduce also the lower triangular block–Toeplitz matrix $T^{(\alpha)} = [\alpha_{k-i}]_{k,i=-\infty}^{\infty}$, where $\alpha_j = a_j$ $(j = 2,3,\ldots)$, $\alpha_1 = \alpha$ and $\alpha_j = 0$ for $j \leq 0$.

We first apply Theorem 5.3 to construct a minimal realization of the matrix T. Note that

$$H_T^{(j)} = [a_{k+\ell-1}]_{k,\ell=1}^{\infty} := H_a \quad (j \in \mathbb{Z}, \ j \neq 0),$$

$$H_T^{(0)} = [\alpha_{k+\ell-i}]_{k,\ell=1}^{\infty} := H_\alpha, \quad \hat{H}_T^{(j)} = [a_{k+\ell}]_{k,\ell=1}^{\infty} := \hat{H} \quad (j \in \mathbb{Z}).$$

Denote $\operatorname{rank} H_a = d$, $\operatorname{rank} H_\alpha = \tilde{d}$, and assume that d and $\tilde{d}$ are finite. Let μ be a positive integer such that the submatrix $T_\mu = [a_{k+\ell-1}]_{k,\ell=1}^{\mu}$ of H_a is of rank d and the submatrix $\tilde{T}_\mu = [\alpha_{k+\ell-1}]_{k,\ell=1}^{\mu}$ of H_α is of rank $\tilde{d}$. Write minimal rank decompositions of T_μ and $\tilde{T}_\mu$:

$$T_\mu = \varphi\psi; \quad \varphi = \operatorname{col}\left(\varphi_i\right)_{i=1}^{\mu}, \ \varphi_i \in \mathbb{C}^{m \times d}; \quad \psi = \operatorname{row}\left(t_j\right)_{j=1}^{\mu}, \ \psi_j \in \mathbb{C}^{d \times m};$$

$$\tilde{T}_\mu = \tilde{\varphi}\tilde{\psi}; \quad \tilde{\varphi} = \operatorname{col}\left(\tilde{\varphi}_i\right)_{i=1}^{\mu}, \ \tilde{\varphi}_i \in \mathbb{C}^{m \times \tilde{d}}; \quad \tilde{\psi} = \operatorname{row}\left(\tilde{\psi}_j\right)_{j=1}^{\mu}, \ \tilde{\psi}_j \in \mathbb{C}^{\tilde{d} \times m}.$$

Denote $\hat{T}_\mu = [a_{k+\ell}]_{k,\ell=1}^{\mu}$. Then according to formulas (5.12) a minimal realization $\Sigma = \left(A(\cdot), B(\cdot), C(\cdot), D(\cdot); X(\cdot)\right)$ of T is given by the following formulas

$$X(j) = \mathbb{C}^d \quad (j \in \mathbb{Z},\ j \neq 0) \quad X(0) = \mathbb{C}^{\tilde{d}};$$

$$A(j) = \varphi^{(-1)} \hat{T}_\mu \psi^{(-1)} \quad (j \in \mathbb{Z},\ j \neq -1,0);$$

$$A(-1) = \tilde{\varphi}^{(-1)} \hat{T}_\mu \psi^{(-1)}, \quad A(0) = \varphi^{(-1)} T \tilde{\psi}^{(-1)};$$

(5.16)
$$B(j) = \psi_1 \quad (j \in \mathbb{Z},\ j \neq -1),\ B(-1) = \tilde{\psi}_1;$$

$$C(j) = \varphi_1 \quad (j \in \mathbb{Z},\ j \neq 0),\ C(0) = \tilde{\varphi}_1;$$

$$D(j) = 0 \quad (j \in \mathbb{Z}).$$

Now we shall express the minimal realization (5.16) in slightly different terms. Let $(A,B,C,0;\mathbb{C}^d)$ and $(\tilde{A},\tilde{B},\tilde{C},0;\mathbb{C}^{\tilde{d}})$ be linear time–invariant systems that are minimal realization of the block–Toeplitz matrices $T^{(a)}$ and $T^{(\alpha)}$, respectively. Then a minimal realization of the given matrix T is given by the system

(5.17)
$$
\begin{cases}
x(k+1) = Ax(k) + Bu(k), \\
y(k) = Cx(k), \quad k \in \mathbb{Z},\ k \neq -1,0; \\
x(0) = A(-1)x(-1) + \tilde{B}u(-1), \\
y(-1) = Cx(-1); \\
x(1) = A(0)x(0) + Bu(0), \\
y(0) = \tilde{C}x(0);
\end{cases}
$$

where
$$A(-1) = \left[\mathrm{col}\left(\tilde{C}\tilde{A}^{i-1}\right)_{i=1}^{\mu}\right]^{(-1)} \mathrm{col}\left(CA^{i-1}\right)_{i=1}^{\mu} A\ ,$$
$$A(0) = A\ \mathrm{row}\left(A^{i-1}B\right)_{k=1}^{\mu}\left[\mathrm{row}\left(A^{i-1}\right)_{i=1}^{\mu}\right]^{(-1)}\ .$$

The realization (5.17) is just another form of writing the realization (5.16). To see this one uses the factorizations

$$T_\mu = \mathrm{col}\left(CA^{i-1}\right)_{i=1}^{\mu}\cdot\mathrm{row}\left(A^{i-1}B\right)_{i=1}^{\mu} \qquad \tilde{T}_\mu = \mathrm{col}\left(\tilde{C}\tilde{A}^{i-1}\right)_{i=1}^{\mu}\cdot\mathrm{row}\left(\tilde{A}^{i-1}\tilde{B}\right)_{i=1}^{\mu}$$

and

$$\hat{T}_\mu = \mathrm{col}\left(CA^{i-1}\right)_{i=1}^{\mu}\cdot A\cdot\mathrm{row}\left(A^{i-1}B\right)_{i=1}^{\mu}\ . \qquad\qquad \square$$

6. THE CLASS OF SYSTEMS WITH CONSTANT STATE SPACE DIMENSION

In what follows $\mathcal{S}_c\big(p(\cdot),m(\cdot)\big)$ denotes the class of all time–varying systems $\Sigma = \big(A(\cdot),B(\cdot),C(\cdot),D(\cdot);X(\cdot)\big)$, for which all the state spaces $X(j)$ have the same dimension: $\dim X(j) = d = \mathrm{const}\ (j \in \mathbb{Z})$. The integer d will be called the state space dimension of Σ. Clearly, by means of a similarity transformation such a

system can be viewed as a time-varying system with a constant state space $\mathbb{C}^d$.
Time-varying systems of this type have been a subject for study by many authors (see
e.g. [13], [6], [5], [3], [2] and references therein).   Important topics in this
study were the structural properties and corresponding decomposition of the systems.
It was generally recognized that a Kalman structural decomposition with reference to
to the pair of properties complete observability/reachability is impossible (even
for periodical systems).   Of course, given a system $\Sigma \in \mathcal{P}_c\big(p(\cdot),m(\cdot)\big)$ Theorems
3.1-3.3 imply that there is a completely observable and completely reachable
reduction Σ_0 of Σ, but generally speaking, the system Σ_0 does not have to be in
$\mathcal{P}_c\big(p(\cdot),m(\cdot)\big)$.   The following example (see [3]) illustrates this.

EXAMPLE.   Consider the system $\Sigma = \big(A(\cdot),B(\cdot),C(\cdot),D(\cdot);X(\cdot)\big)$ with

$$p(j) = m(j) = 1, \quad X(j) = \mathbb{C}^1 \ (j \in \mathbb{Z}),$$

$$A(j) = B(j) = \begin{cases} 0, \ j \text{ even} \\ 1, \ j \text{ odd} \end{cases} ; \quad C(j) = 1, \quad D(j) = 0 \ (j \in \mathbb{Z}).$$

An easy computation shows that for j even the reachable subspace of Σ coincides with
$\mathbb{C}^1$, while for j odd it is the zero space.   It follows that a completely observable
and completely reachable reduction $\Sigma_0 = \big(A^0(\cdot),B^0(\cdot),C^0(\cdot),D^0(\cdot);X^0(\cdot)\big)$ of Σ has
time dependent state space dimensions: dim $X^0(j) = 1$ for j even and dim $X^0(j) = 0$
for j odd.

The theorem below summarizes the results on complete observability and
reachability for systems with constant state space dimensions which follow easily
from Theorems 3.1-3.4.

THEOREM 6.1. *(a) A system* $\Sigma = \big(A(\cdot),B(\cdot),C(\cdot),D(\cdot);X(\cdot)\big) \in \mathcal{P}_c\big(p(\cdot),m(\cdot)\big)$
is completely reachable and completely observable if and only if the degree sequence
$\big\{\delta_j(T_\Sigma)\big\}_{j=-\infty}^{\infty}$ *of its input/output map* T_Σ *is constant, say* $\delta_j(T_\Sigma) = \delta \ (j \in \mathbb{Z})$, *and*
dim $X(j) = \delta \ (j \in \mathbb{Z})$.

(b) If $\Sigma \in \mathcal{P}_c\big(p(\cdot),m(\cdot)\big)$ *and the degree sequence of* T_Σ *is constant, then*
Σ *is a dilation of a completely reachable and completely observable system*
$\Sigma_0 = \big(A^0(\cdot),B^0(\cdot),C^0(\cdot),D(\cdot);X^0(\cdot)\big) \in \mathcal{P}_c\big(p(\cdot),m(\cdot)\big)$ *which is obtained by the*
reduction procedure described in Theorem 3.2.

(c) Two completely reachable and completely observable systems
$\Sigma,\tilde{\Sigma} \in \mathcal{P}_c\big(p(\cdot),m(\cdot)\big)$ *are similar if and only if their input/output maps coincide:*
$T_\Sigma = T_{\tilde{\Sigma}}$.

We pass now to the problem of minimality in the class $\mathcal{G}_c\big(p(\cdot),m(\cdot)\big)$. A system $\Sigma = \big(A(\cdot),B(\cdot),C(\cdot),D(\cdot);X(\cdot)\big) \in \mathcal{G}_c\big(p(\cdot),m(\cdot)\big)$ will be referred to as c-*minimal* if its state space dimension is as small as possible among all systems in $\mathcal{G}_c\big(p(\cdot),m(\cdot)\big)$ with the same input/output map as Σ. First we establish necessary and sufficient conditions for existence of (c-minimal) realizations with constant state space dimensions of a given lower triangular block matrix and present algorithms for constructing such realizations.

THEOREM 6.2. *Let* $T = \big[t_{ki}\big]_{k,i=-\infty}^{\infty}$ *be a lower triangular block matrix with* $m(k)\times p(i)$ *matrix entries* t_{ki}. *Then* T *is an input/output map of a system* $\Sigma \in \mathcal{G}_c\big(p(\cdot),m(\cdot)\big)$ *if and only if*

$$(6.1) \qquad\qquad d(T) := \sup_{j\in\mathbb{Z}} \operatorname{rank} H_T^{(j)} < \infty \; .$$

If this condition is satisfied, then a system $\Sigma = \big(A(\cdot),B(\cdot),C(\cdot),D(\cdot);X(\cdot)\big) \in \mathcal{G}_c\big(p(\cdot),m(\cdot)\big)$ *such that* $T = T_\Sigma$ *can be constructed in the following way. Choose a realization* $\tilde{\Sigma} = \big(\tilde{A}(\cdot),\tilde{B}(\cdot),\tilde{C}(\cdot),\tilde{D}(\cdot);\tilde{X}(\cdot)\big) \in \mathcal{G}\big(p(\cdot),m(\cdot)\big)$ *of the matrix* T *such that*

$$(6.2) \qquad\qquad d := \sup_{j\in\mathbb{Z}} \dim \tilde{X}(j) < \infty \; ,$$

and let X *be a linear space of dimension* d. *For each* $j \in \mathbb{Z}$ *choose an injective linear map* $\tau_j \colon \tilde{X}(j) \to X$ *with left inverse* $\tau_j^{(-1)}$ *and put*

$$(6.3) \qquad
\begin{aligned}
X(j) &= X, & A(j) &= \tau_{j+1}\tilde{A}(j)\tau_j^{(-1)}, & & (j \in \mathbb{Z}) \; . \\
B(j) &= \tau_{j+1}\tilde{B}_j, & C(j) &= \tilde{C}_j\tau_j^{(-1)}, & \tilde{D}(j) &= D(j)
\end{aligned}$$

Then T *is the input/output map of the system* $\Sigma = \big(A(\cdot),B(\cdot),C(\cdot),D(\cdot);X(\cdot)\big)$.

Moreover, if the system $\tilde{\Sigma}$ *is chosen to be a minimal realization of* T *(in which case (6.2) follows from (6.1)), then* Σ *is c-minimal. In particular, the state space dimension of a c-minimal realization of* T *is equal to the number* $d(T)$ *defined by (6.1).*

Proof. If $\Sigma = \big(A(\cdot),B(\cdot),C(\cdot),D(\cdot);X(\cdot)\big)$ is a realization of T and $\Sigma \in \mathcal{G}_c\big(p(\cdot),m(\cdot)\big)$ then Theorem 3.4 implies that $\operatorname{rank} H_T^{(j)} \le \dim X(j) = \text{const}$ $(j \in \mathbb{Z})$, and hence (6.1) holds true.

Conversely, assume that (6.1) is satisfied. Then there is a realization $\tilde{\Sigma} = \big(\tilde{A}(\cdot),\tilde{B}(\cdot),\tilde{C}(\cdot),\tilde{D}(\cdot);\tilde{X}(\cdot)\big)$ such that (6.2) holds true. Indeed, one can choose, for instance, the realization $\tilde{\Sigma}$ to be minimal, in which case, in view of Theorem 3.4, $\operatorname{rank} H_{\tilde{\Sigma}}^{(j)} = \dim \tilde{X}(j)$, and hence (6.1) implies (6.2). Now let $\tilde{\Sigma} = \big(\tilde{A}(\cdot),\tilde{B}(\cdot),\tilde{C}(\cdot),\tilde{D}(\cdot);\tilde{X}(\cdot)\big)$ be a system that satisfies (6.2) and let $\Sigma = \big(A(\cdot),B(\cdot),C(\cdot),D(\cdot);X(\cdot)\big)$ be a system constructed from $\tilde{\Sigma}$ by means of formulas (6.3). We have to show that the matrices

$$\beta_{ki} := C(k)A(k-1)A(k-2)\ldots A(i+1)B(i)$$

coincide with the matrices t_{ki} for all $k > i$. Let $k > i+1$. Since $\tau_i^{(-1)}\tau_i = I_{X(i)}$ ($i \in \mathbb{Z}$) and $\tilde{\Sigma}$ is a realization of T, one deduces from (6.3) that

$$\beta_{ki} = \tilde{C}_k\tau_k^{(-1)}\tau_k\tilde{A}(k-1)\tau_{k-1}^{(-1)}\tau_{k-1}\tilde{A}(k-2)\tau_{k-2}^{(-1)}\ldots\tau_{i+2}\tilde{A}(i+1)\tau_{i+1}^{(-1)}\tau_{i+1}\tilde{B}(i)$$

$$= \tilde{C}_k\tilde{A}(k-1)\tilde{A}(k-2)\ldots\tilde{A}(i+1)\tilde{B}(i) = t_{ki}.$$

For $i = k-1$ we have

$$\beta_{k,k-1} = C(k)B(k-1) = \tilde{C}(k)\tau_k^{(-1)}\tau_k\tilde{B}(k-1) = \tilde{C}(k)\tilde{B}(k-1) = t_{k,k-1}.$$

Thus, comparing with (1.5), we conclude that the given matrix T coincides with T_Σ, the input/output map of the system Σ given by (6.3).

Now assume that the realization $\tilde{\Sigma}$ of T is minimal. Then, in view of Theorem 3.4, dim $\tilde{X}(j)$ = rank $H_T^{(j)}$ ($j \in \mathbb{Z}$), and hence (6.2) holds true. Thus, we can apply the construction described in the theorem to get the system $\Sigma = \left(A(\cdot),B(\cdot),C(\cdot),D(\cdot);X(\cdot)\right)$ by formulas (6.3). To show that this system Σ is c-minimal take any realization $\hat{\Sigma} = \left(\hat{A}(\cdot),\hat{B}(\cdot),\hat{C}(\cdot),\hat{D}(\cdot);\hat{X}(\cdot)\right)$ of T with constant state space dimensions, say dim $\hat{X}(j) = \hat{d}$ ($j \in \mathbb{Z}$), and use (4.5) to write

$$H_{T_{\tilde{\Sigma}}}^{(j)} = N_j(\hat{\Sigma})R_j(\hat{\Sigma}) \quad (j \in \mathbb{Z}).$$

Here the operators $N_j(\hat{\Sigma})$ and $R_j(\hat{\Sigma})$ are defined as in (2.1) and (2.6), respectively, and it is clear that

$$\text{rank } N_j(\hat{\Sigma}) \le \hat{d}, \quad \text{rank } R_j(\hat{\Sigma}) \le \hat{d} \quad (j \in \mathbb{Z}).$$

It follows, therefore, that

$$\text{rank } H_T^{(j)} \le \hat{d} \quad (j \in \mathbb{Z}),$$

and hence $d(T)$, the state space dimension of the realization Σ, is less than or equal to $\hat{d}$. $\square$

Note that to obtain a c-minimal realization Σ of a given lower triangular block matrix T one can take as $\tilde{\Sigma}$ any of the minimal realization of T constructed in Theorems 5.1–5.3, provided the condition (6.1) is satisfied.

Our next result gives a characterization of c-minimality and its connection with observability and reachability.

THEOREM 6.3.   *Let* $\Sigma \in \mathcal{S}_c\left(p(\cdot),m(\cdot)\right)$ *be a system with constant state space dimension* d. *The following assertions are equivalent:*

(i) Σ *is c-minimal;*

(ii) $d = \max\limits_{j \in \mathbb{Z}} \text{rank } H_{T_\Sigma}^{(j)}$;

(iii) Σ *is reachable and observable for some time* j.

Proof. To prove the implication (i)→(ii) assume that $\Sigma = \left(A(\cdot),B(\cdot),C(\cdot),D(\cdot);X(\cdot)\right)$ is a c-minimal system and write the factorizations (4.5) for $H_{T_\Sigma}^{(j)}$:

$$(6.4) \qquad H_{T_\Sigma}^{(j)} = N_j(\Sigma)R_j(\Sigma) \qquad (j \in \mathbb{Z}).$$

It follows that

$$(6.5) \qquad \operatorname{rank} H_{T_\Sigma}^{(j)} \leq \dim X(j) = d \quad (j \in \mathbb{Z}) ,$$

and hence

$$d(T_\Sigma) \leq d ,$$

where $d(T_\Sigma)$ is defined by (6.1). In view of Theorem 6.2 there is a realization of T_Σ with constant state space dimension $d(T_\Sigma)$, and since Σ is c-minimal

$$d \leq d(T_\Sigma) .$$

Thus, $d = d(T_\Sigma)$ and (ii) follows.

To prove the implication (ii)→(iii) assume that for some $j_0 \in \mathbb{Z}$

$$d = \operatorname{rank} H_{T_\Sigma}^{(j_0)} .$$

Then (6.5) implies that

$$(6.6) \qquad \operatorname{rank} H_{T_\Sigma}^{(j_0)} = \dim X(j_0) .$$

Since $\operatorname{rank} R_j(\Sigma) \leq \dim X(j)$ and $\operatorname{rank} N_j(\Sigma) \leq \dim X(j)$ the equalities (6.4) and (6.6) lead to the conclusion that

$$\operatorname{rank} R_{j_0}(\Sigma) = \operatorname{rank} N_{j_0}(\Sigma) = \dim X(j_0) \quad (= d),$$

and in view of Proposition 2.1 and 2.5 we see that the system Σ is observable and reachable at time j_0.

It remains to establish the implication (iii)→(i). To this end assume that the system Σ is observable and reachable at time j_0. Then Propositions 2.1 and 2.5 imply that the operators $R_{j_0}(\Sigma)$ and $N_{j_0}(\Sigma)$ are of full rank $(= \dim X(j_0))$, and in view of (6.4)

$$\operatorname{rank} H_{T_\Sigma}^{(j_0)} = \dim X(j_0) \ (= d).$$

Now let $\tilde{\Sigma} = \left(\tilde{A}(\cdot),\tilde{B}(\cdot),\tilde{C}(\cdot),\tilde{D}(\cdot);\tilde{X}(\cdot)\right) \in \mathscr{S}_c\left(p(\cdot),m(\cdot)\right)$ be any system with constant state space dimension $\tilde{d}$ such that $T_{\tilde{\Sigma}} = T_\Sigma$. Then (6.4) implies that

$$d = \operatorname{rank} H_{T_\Sigma}^{(j_0)} = \operatorname{rank} H_{T_{\tilde{\Sigma}}}^{(j_0)} \leq \dim \tilde{X}(j_0) = \tilde{d} ,$$

and hence Σ is c-minimal. $\qquad\qquad\qquad\qquad\qquad\qquad\qquad\qquad\qquad \square$

In the next theorem we describe how to cut a given system $\Sigma \in \mathscr{S}\left(p(\cdot),m(\cdot)\right)$ to a c-minimal system with the same input/output map as Σ, provided condition (6.1) is satisfied for T_Σ.

THEOREM 6.4.   *Let* $\Sigma = \big(A(\cdot),B(\cdot),C(\cdot),D(\cdot);X(\cdot)\big) \in \mathcal{S}\big(p(\cdot),m(\cdot)\big)$ *be a system such that*

$$(6.7) \qquad\qquad d := \max_{j\in Z} \delta_j(T_\Sigma) < \infty \; ,$$

where $\delta_j(T_\Sigma) = \mathrm{rank}\, H_{T_\Sigma}^{(j)}$ $(j \in Z)$. *Define the decompositions* $X(j) = X^1(j) \oplus X^0(j) \oplus X^2(j)$ $(j \in Z)$ *as in Theorem 3.2 and decompose*

$$(6.8) \qquad\qquad X^1(j) = \tilde{X}^{-1}(j) \oplus \tilde{X}^1(j), \;\; \tilde{X}^2(j) = \tilde{X}^{-2}(j) \oplus \tilde{X}^2(j) \;\; (j \in Z)$$

so that

$$(6.9) \qquad\qquad \dim\big(\tilde{X}^1(j) \oplus \tilde{X}^2(j)\big) = d - \delta_j(T_\Sigma) \;\; (j \in Z) \; .$$

Set $\tilde{X}^0(j) = \tilde{X}^1(j) \oplus X^0(j) \oplus \tilde{X}^2(j)$. *Then relative to the decompositions* $X(j) = \tilde{X}^{-1}(j) \oplus \tilde{X}^0(j) \oplus \tilde{X}^{-2}(j)$ *the following partitionings hold true for each* $j \in Z$:

$$A(j) = \begin{bmatrix} * & * & * \\ * & \tilde{A}^0(j) & * \\ 0 & * & * \end{bmatrix} : \tilde{X}^{-1}(j) \oplus \tilde{X}^0(j) \oplus \tilde{X}^{-2}(j) \to \tilde{X}^{-1}(j{+}1) \oplus \tilde{X}^0(j{+}1) \oplus \tilde{X}^{-2}(j{+}1),$$

$$(6.10) \quad B(j) = \begin{bmatrix} * \\ \tilde{B}^0(j) \\ 0 \end{bmatrix} : \mathbb{C}^{p(j)} \to \tilde{X}^{-1}(j{+}1) \oplus \tilde{X}^0(j{+}1) \oplus \tilde{X}^{-2}(j{+}1),$$

$$C(j) = \begin{bmatrix} 0 & \tilde{C}^0(j) & * \end{bmatrix} : \tilde{X}^{-1}(j) \oplus \tilde{X}^0(j) \oplus \tilde{X}^{-2}(j) \to \mathbb{C}^{m(j)},$$

and the system $\tilde{\Sigma} = \big(\tilde{A}^0(\cdot),\tilde{B}^0(\cdot),\tilde{C}^0(\cdot),D(\cdot);\tilde{X}^0(\cdot)\big)$ *is c–minimal and has the same input/output map as* Σ.

(In (6.10) * stands for unspecified elements.)

Proof. First we write more fine partitionings of the operators $A(j)$, $B(j)$ and $C(j)$ relative to the decomposition $X(j) = \tilde{X}^{-1}(j) \oplus \tilde{X}^1(j) \oplus X^0(j) \oplus \tilde{X}^2(j) \oplus \tilde{X}^{-2}(j)$. In view of Theorem 3.2 and formulas (6.8) the partitioning of the operators $A(j)$ acting from

$$\tilde{X}^{-1}(j) \oplus \tilde{X}^1(j) \oplus X^0(j) \oplus \tilde{X}^2(j) \oplus \tilde{X}^{-2}(j)$$

into

$$\tilde{X}^{-1}(j{+}1) \oplus \tilde{X}^1(j{+}1) \oplus X^0(j{+}1) \oplus \tilde{X}^2(j{+}1) \oplus \tilde{X}^{-2}(j{+}1)$$

is as follows:

$$(6.11) \qquad A(j) = \begin{bmatrix} * & * & * & * & * \\ * & A_{11}(j) & A_{10}(j) & A_{12}(j) & * \\ 0 & 0 & A_{00}(j) & A_{02}(j) & * \\ 0 & 0 & 0 & A_{22}(j) & * \\ 0 & 0 & 0 & 0 & * \end{bmatrix} ,$$

and

$$(6.12) \qquad B(j) = \begin{bmatrix} * \\ B_1(j) \\ B_0(j) \\ 0 \\ 0 \end{bmatrix} : \mathbb{C}^{p(j)} \to \tilde{X}^{-1}(j{+}1) \oplus \tilde{X}^1(j{+}1) \oplus X^0(j{+}1) \oplus \tilde{X}^2(j{+}1) \oplus \tilde{X}^{-2}(j{+}1),$$

$$(6.13) \qquad C(j) = \begin{bmatrix} 0 & 0 & C_0(j) & C_2(j) & * \end{bmatrix} : \tilde{X}^{-1}(j) \oplus \tilde{X}^1(j) \oplus X^0(j) \oplus \tilde{X}^2(j) \oplus \tilde{X}^{-2}(j) \to \mathbb{C}^{m(j)}.$$

In view of Theorem 3.2 the system $\Sigma_0 = \big(A_{00}(j), B_0(j), C_0(j), D(j); X^0(j)\big)$ has the same input/output map as Σ: $T_{\Sigma_0} = T_\Sigma$. Furthermore, a comparison of the partitionings (6.10) with the partitionings (6.11)–(6.13) shows that

$$\tilde{A}^0(j) = \begin{bmatrix} A_{11}(j) & A_{10}(j) & A_{12}(j) \\ 0 & A_{00}(j) & A_{02}(j) \\ 0 & 0 & A_{22}(j) \end{bmatrix} ,$$

$$\tilde{B}^0(j) = \begin{bmatrix} B_1(j) \\ B_0(j) \\ 0 \end{bmatrix} , \qquad \tilde{C}^0(j) = \begin{bmatrix} 0 & C_0(j) & C_2(j) \end{bmatrix} ,$$

i.e., the system $\tilde{\Sigma}$ is a dilation of the system Σ_0. Hence the input/output maps of these two systems coincide: $T_{\tilde{\Sigma}} = T_{\Sigma_0}$, and therefore $T_{\tilde{\Sigma}} = T_\Sigma$.

To complete the proof note that all the state spaces $\tilde{X}^0(j)$ of $\tilde{\Sigma}$ have dimension d, and in view of (6.7) $d = \max_{j \in \mathbb{Z}} \delta_j(T_{\tilde{\Sigma}})$. Theorem 6.3 then implies that $\tilde{\Sigma}$ is c–minimal. $\qquad\qquad\square$

In conclusion of this section we note that two c–minimal systems with the same input/output map need not be similar, as simple examples show (see e.g., [13]). However, if two systems Σ_1 and Σ_2 are completely controllable and completely observable, then in view of Theorem 6.3 they are c–minimal and Theorem 6.1 implies that Σ_1 and Σ_2 are similar, provided $T_{\Sigma_1} = T_{\Sigma_2}$.

7. MINIMALITY AND REALIZATION FOR PERIODICAL SYSTEMS

In this section we specify the general results of Sections 3 and 5 for the case of periodical systems.

First, from Theorem 3.8 one immediately infers that minimality of an N-periodical system Σ is equivalent to minimality of N time-invariant systems. More precisely the following result holds true.

THEOREM 7.1. *An* N-*periodical system* Σ *is minimal if and only if the associated time-invariant systems* $\Sigma^{(j)}$ (j = 0,1,...,N-1) *are minimal.*

Next, for an N-periodical system Σ the reduction procedure of Theorem 3.2 can be actually performed and it consists of a finite number of operations on finite matrices. Moreover, the reduction Σ_0 is an N-periodical system. Indeed, in view of (2.5) and (2.9) the subspaces $N_j(\Sigma)$ and $R_j(\Sigma)$ can be written as follows:

$$
(7.1) \quad
\begin{aligned}
N_j(\Sigma) &= \text{Ker col}\left(C^{(j)}(A^{(j)})^{k-1}\right)_{k=1}^{\dim X(j)} \\
R_j(\Sigma) &= \text{Im row}\left((A^{(j)})^{k-1}B^{(j)}\right)_{k=1}^{\dim X(j)}
\end{aligned}
\qquad (j \in \mathbb{Z})
$$

Since $N_{j+N}(\Sigma) = N_j(\Sigma)$ and $R_{j+N}(\Sigma) = R_j(\Sigma)$, the subspaces $X^1(j)$, $X^0(j)$ and $X^2(j)$ in Theorem 3.2 are N-periodical: $X^i(j+N) = X^i(j)$ (i = 0,1,2; $j \in \mathbb{Z}$). This, together with the N-periodicity of the operators A(j), B(j) and C(j), ensures that the reduction $\Sigma_0 = \left(A^0(\cdot),B^0(\cdot),C^0(\cdot),D(\cdot);X^0(\cdot)\right)$ which is obtained from the partitionings (3.1)–(3.3) is N-periodical. The above considerations lead to the following result.

THEOREM 7.2. *Any* N-*periodical system* $\Sigma = \left(A(\cdot),B(\cdot),C(\cdot),D(\cdot);X(\cdot)\right)$ *is a dilation of a minimal system which is also* N-*periodical. Namely, let* $N_j(\Sigma)$ *and* $R_j(\Sigma)$ *be defined by (7.1). For each* j = 0,1,...,N-1 *put* $X^1(j) = N_j(\Sigma)$, *let* X^0_j *be a direct complement of* $X^1(j) \cap R_j(\Sigma)$ *in* $R_j(\Sigma)$ *and choose a direct complement* $X^2(j)$ *of* $N_j(\Sigma) + R_j(\Sigma)$ *in* X(j). *Define* $X^i(j+kN) = X^i(j)$ (i = 0,1,2; j = 0,1,...,N-1, $k \in \mathbb{Z}$). *Then relative to the decompositions* $X(j) = X^1(j) \oplus X^0(j) \oplus X^2(j)$ *the partitionings* (3.1)–(3.3) *hold true and the system* $\Sigma_0 = \left(A^0(\cdot),B^0(\cdot),C^0(\cdot),D(\cdot);X^0(\cdot)\right)$ *is minimal and* N-*periodical.*

Of course, given an N-periodical system Σ there are non-periodical minimal systems with the same input/output map as Σ. Theorems 3.3 and 7.2 imply that any such system is similar to an N-periodical minimal system with the same input/output map as Σ. Furthermore, it is clear that applying N-periodical similarity to an N-periodical minimal system Σ we obtain again a minimal N-periodical system with the same intput/output map as Σ. The converse statement is also true as the following theorem states.

THEOREM 7.3. *Let* Σ, $\tilde{\Sigma} \in \mathcal{S}_N\bigl(p(\cdot),m(\cdot)\bigr)$ *be minimal* N-*periodical systems. Then the systems* Σ *and* $\tilde{\Sigma}$ *are* N-*periodically similar if and only if they have the same input/output map:* $T_\Sigma = T_{\tilde{\Sigma}}$.

Proof. The "if" part of the theorem is clear. To prove the "only if" part take two minimal systems $\Sigma = \bigl(A(\cdot),B(\cdot),C(\cdot),D(\cdot);X(\cdot)\bigr)$ and $\tilde{\Sigma} = \bigl(\tilde{A}(\cdot),\tilde{B}(\cdot),\tilde{C}(\cdot),\tilde{D}(\cdot);\tilde{X}(\cdot)\bigr)$ with the same input/output map and assume that these systems are N-periodical. Choose an integer ω greater than each of the numbers $\dim X(j)$ (j = 0,1,...,N-1) and $\dim \tilde{X}(j)$ (j = 0,1,...,N-1), and introduce the matrices

$$\Delta_j(\Sigma) = \mathrm{row}\bigl((A^{(j)})^{k-1}B^{(j)}\bigr)_{k=1}^\omega, \quad \Gamma_j(\Sigma) = \mathrm{col}\bigl(C^{(j)}(A^{(j)})^{k-1}\bigr)_{k=1}^\omega, \quad (j = 0,1,...,N-1) ,$$

where $A^{(j)}$, $B^{(j)}$ and $C^{(j)}$ are defined by (1.15). Let $\Delta_j(\tilde{\Sigma})$ and $\Gamma_j(\tilde{\Sigma})$ denote the corresponding matrices for the system $\tilde{\Sigma}$. Since Σ and $\tilde{\Sigma}$ have the same input/output map, we have the relations (4.6). Using the N-periodicity of the systems Σ and $\tilde{\Sigma}$ one easily deduces from (4.6) the following equations

$$\Gamma_j(\Sigma)\Delta_j(\Sigma) = \Gamma_j(\tilde{\Sigma})\Delta_j(\tilde{\Sigma}) \quad (j = 0,1,...,N-1),$$

$$\Gamma_{j+1}(\Sigma)A(j)\Delta_j(\Sigma) = \Gamma_{j+1}(\tilde{\Sigma})\tilde{A}(j)\Delta_j(\tilde{\Sigma}) \quad (j = 0,1,...,N-1).$$

Since Σ and $\tilde{\Sigma}$ are minimal, Theorem 7.1 implies (cf. (2.5),(2.9)) that the operators $\Gamma_j(\Sigma)$ and $\Gamma_j(\tilde{\Sigma})$ (j = 0,1,...,N-1) are left invertible, while the operators $\Delta_j(\Sigma)$ and $\Delta_j(\tilde{\Sigma})$ (j = 0,1,...,N-1) are right invertible. Then setting

$$K(j) := \bigl(\Gamma_j(\tilde{\Sigma})\bigr)^{(-1)}\Gamma_j(\Sigma) = \Delta_j(\tilde{\Sigma})\bigl(\Delta_j(\Sigma)\bigr)^{(-1)} \quad (j = 0,1,...,N-1)$$

and arguing as in Step 4 of the Proof of Theorems 3.1-3.4 we see that (4.8) and (4.9) hold true for j = 0,1,...,N-1. Extending $K(j)$ N-periodically for any $j \in \mathbb{Z}$ we obtain the desired N-periodical similarity of Σ and $\tilde{\Sigma}$. $\qquad\qquad$ □

Concerning the degree sequence $\left\{\delta_j(T_\Sigma)\right\}_{j=-\infty}^\infty$ of the input/output map T_Σ of an N-periodical system Σ, it is clear that this sequence is N-periodical: $\delta_{j+N}(T_\Sigma) = \delta_j(T_\Sigma)$. Furthermore, from the construction of the associated LTI-systems $\Sigma^{(j)}$ (see (1.13)-(1.15)) we know that the state space of $\Sigma^{(j)}$ is $X(j)$. Thus, Theorems 3.4 and 3.1 imply the following result.

THEOREM 7.4. *An* N-*periodical system* $\Sigma = \bigl(A(\cdot),B(\cdot),C(\cdot),D(\cdot);X(\cdot)\bigr)$ *is minimal if and only if*

$$\delta_j(T_\Sigma) = \dim X(j) = \delta(\Sigma^{(j)}) \quad (j = 0,1,...,N-1),$$

where $\delta(\Sigma^{(j)})$ *stands for the McMillan degree of the LTI-system* $\Sigma^{(j)}$.

We consider now the realization problem for N-periodical systems. Assume that $T = \left[t_{ki}\right]_{k,i=-\infty}^{\infty}$ is a lower triangular block matrix, with the following periodicity property

(7.2) $$t_{k+N,i+N} = t_{k,i} \quad (k,i \in \mathbb{Z}).$$

Then, clearly, there are only N possible different matrices $H_T^{(j)} = \left[t_{j+i-1,j-k}\right]_{i,k=1}^{\infty}$, i.e., $H_T^{(j+N)} = H_T^{(j)}$, and therefore one immediately sees that the system Σ constructed in Theorem 5.1 according to (5.3)-(5.6) is N-periodical. Thus, we have the following result.

THEOREM 7.5. *Let* $T = \left[t_{ki}\right]_{k,i=-\infty}^{\infty}$ *be a lower triangular block matrix with* $m(k) \times p(i)$ *matrix entries* t_{ki}. *Then* T *is an input/output map of an N-periodical system* $\Sigma \in \mathscr{S}_N\left(p(\cdot),m(\cdot)\right)$ *if and only if* T *satisfies the periodicity property (7.2) and the numbers* $\delta_j(T) = \text{rank } H_T^{(j)}$ $(j = 0,1,...,N-1)$ *are finite. If these conditions are satisfied, then an N-periodical system* $\Sigma = \left(A(\cdot),B(\cdot),C(\cdot),D(\cdot);X(\cdot)\right) \in \mathscr{S}_N\left(p(\cdot),m(\cdot)\right)$ *such that* $T = T_\Sigma$ *is constructed as in Theorem 5.1 by the formulas (5.3)-(5.6). Moreover, this system* Σ *is minimal.*

In a similar way one easily sees that the periodic counterpart of Theorem 5.2 is the following result.

THEOREM 7.6. *Let* $T = \left[t_{ki}\right]_{k,i=-\infty}^{\infty}$ *be a lower triangular block matrix with* $m(k) \times p(i)$ *entries* t_{ki} *such that condition (7.2) is satisfied and the numbers* $\delta_j(T)$ $(j = 0,1,...,N-1)$ *are finite. For each* $j = 0,1,...,N-1$ *write a minimal rank decomposition of* $H_T^{(j)}$:

(7.3)
$$H_T^{(j)} = \Phi_j \Psi_j$$
$$\Psi_j: \ell_0\left(\left(\mathbb{C}^{p(j-\nu)}\right)_{\nu=1}^{\infty}\right) \to \tilde{X}(j), \quad \Phi_j: \tilde{X}(j) \to \ell\left(\left(\mathbb{C}^{m(j+\nu-1)}\right)_{\nu=1}^{\infty}\right)$$

where $\tilde{X}(j)$ *is a linear space of dimension* $\delta_j(T)$. *Denote* $\tilde{H}_T^{(j)} = \left[t_{j+i,j-k}\right]_{i,k=1}^{\infty}$ *and set*

$$\tilde{A}(j) = \left[\Phi^{(j+1)}\right]^{(-1)}\tilde{H}_T^{(j)}\left[\Psi^{(j)}\right]^{(-1)}, \quad (j = 0,1,...,N-2),$$

$$\tilde{A}(N-1) = \left[\Phi^{(0)}\right]^{(-1)}\tilde{H}_T^{(N-1)}\left[\Psi^{(N-1)}\right]^{(-1)},$$

$$B(j) = \Psi^{(j+1)}\varepsilon_1^{(j+1)} \quad (j = 0,1,...,N-2), \quad \tilde{B}(N-1) = \Psi^{(0)}\varepsilon_1^{(0)},$$

$$\tilde{C}(j) = \pi_1^{(j)}\Phi^{(j)} \quad (j = 0,1,...,N-1), \quad \tilde{D}(j) = t_{jj} \quad (j = 0,1,...,N-1).$$

Define $\tilde{X}(j+kN) = \tilde{X}(j)$, $\tilde{A}(j+kN) = \tilde{A}(j)$, $\tilde{B}(j+kN) = \tilde{B}(j)$, $\tilde{C}(j+kN) = \tilde{C}(j)$, $\tilde{D}(j+kN) = \tilde{D}(j)$ $(k \in \mathbb{Z}, \quad j = 0,1,...,N-1)$. *Then the N-periodical system* $\tilde{\Sigma} = \left(\tilde{A}(\cdot),\tilde{B}(\cdot),\tilde{C}(\cdot),\tilde{D}(\cdot);\tilde{X}(\cdot)\right)$ *is a minimal realization of the matrix T.*

In a similar way one can state the periodical version of Theorem 5.3. Since for a matrix T that satisfies (7.2) there are only N possibly different matrices $H_T^{(j)}$, one can find an integer ν such that for all $j \in \mathbb{Z}$ the finite submatrix $T_\nu^{(j)} := \left[t_{j+i-1,\,j-k}\right]_{i,k=1}^\nu$ of $H_T^{(j)}$ is of rank $\delta_j(T)$, provided the latter number is finite. Thus the following result holds true.

THEOREM 7.7. *Let* $T = \left[t_{ki}\right]_{k,i=-\infty}^{\infty}$ *be a lower triangular block matrix with* $m(k) \times p(i)$ *entries* t_{ki} *such that condition (7.2) holds true and the numbers* $\delta_j :=$ $\delta_j(T)$ *are finite. Let* ν *be an integer such that the matrices* $T_\nu^{(j)} :=$ $\left[t_{j+i-1,\,j-k}\right]_{i,k=1}^\nu$ *are of rank* δ_j $(j = 0,1,\ldots,N-1)$ *and write minimal rank decompositions*

$$T_\nu^{(j)} = \varphi_j \psi_j \quad (j = 0,1,\ldots,N-1) \;,$$

$$\varphi_j = \mathrm{col}\left(\varphi_{ji}\right)_{i=1}^\nu, \quad \varphi_{ji} \in \mathbb{C}^{m(j+i-1)\times\delta_j}; \quad \psi_j = \mathrm{row}\left(\psi_{jk}\right)_{k=1}^\nu, \quad \psi_{jk} \in \mathbb{C}^{\delta_j\times p(j-k)} \;.$$

Denote $\hat{T}_\nu^{(j)} = \left[t_{j+i,\,j-k}\right]_{i,k=1}^\nu,$ *and set*

$$\hat{X}(j) = \mathbb{C}^{\delta_j} \quad (j = 0,1,\ldots,N-1), \quad \hat{X}(\ell+kN) = \hat{X}(\ell) \quad (k \in \mathbb{Z}, \quad \ell = 0,1,\ldots,N-1);$$

$$\hat{A}(j) = \varphi_{j+1}^{(-1)}\hat{T}_\nu^{(j)}\psi_j^{(-1)} \quad (j = 0,1,\ldots,N-2), \qquad \hat{A}(N-1) = \varphi_0^{(-1)}\hat{T}_\nu^{(N-1)}\psi_{N-1}^{(-1)},$$

$$\hat{A}(\ell+kN) = \hat{A}(\ell) \quad (k \in \mathbb{Z}, \quad \ell = 0,1,\ldots,N-1);$$

$$\hat{B}(j) = \psi_{j+1,1} \quad (j = 0,1,\ldots,N-2), \qquad B(N-1) = \psi_{0,1}$$

$$\hat{B}(\ell+kN) = \hat{B}(\ell) \quad (k \in \mathbb{Z}, \quad \ell = 0,1,\ldots,N-1); \qquad \hat{C}(j) = \varphi_{j1} \quad (j = 0,1,\ldots,N-1),$$

$$\hat{C}(\ell+kN) = \hat{C}(\ell) \quad (k \in \mathbb{Z}, \quad \ell = 0,1,\ldots,N-1); \qquad \hat{D}(j) = t_{jj} \quad (j \in \mathbb{Z}).$$

Then the system Σ *is a minimal realization of the matrix* T.

The results of Section 6 about systems with constant state space dimensions have also clear versions for N-periodical systems. We omit the precise statements of these results.

REFERENCES

[1] Ball, J.A., Gohberg, I., Kaashoek, M.A.: Nevanlinna-Pick interpolation for linear time-varying input-output maps: the discrete case, in this volume.

[2] Bittanti, S.: Deterministic and Stochastic Linear Periodic Systems, in *Time Series and Linear Systems*, Brittanti, S. (Ed.), Springer-Verlag, Berlin, 141-182, 1986.

[3]   Bittanti, S., Bolzern, P.: On the structure theory of discrete-time linear systems, *International J. Systems Science* **17** (1986), 33-47.

[4]   Dewilde, P.: A course on the algebraic Schur and Nevanlinna-Pick interpolation problems, in *Algorithms and Parallel VLSI Architectures,* Deprettere, E.F. and van der Ven, A.J., editors, North-Holland, 1991.

[5]   Grasselli, O.M.: A canonical decomposition of linear periodic discrete-time systems, *International J. Control* **40** (1984), 201-204.

[6]   Evans, D.S.: Finite-dimensional realizations of discrete-time weighting patterns, *SIAM J. Appl Math.* **22** (1972), 45-67.

[7]   Kailath, T.: *Linear Systems,* Prentice-Hall, Englewood Cliffs, N.J., 1980.

[8]   Kalman, R.E.: Mathematical description of linear dynamical systems, *SIAM J. Control and Optimization,* 1 (1963), 152-192.

[9]   Kalman, R.E., Falb, P.L., Arbib, M.A.: *Topics in Mathematical Systems Theory,* McGraw-Hill, New York, N.Y., 1969.

[10] Meyer, R.A., Burris, C.S.: A unified analysis of multirate and periodically time-varying digital filters, *IEEE Trans. Circuits Syst.* **22** (1985), 162-168.

[11] Rosenbrock, H.H.: *State Space and Multivariable Theory,* Nelson, London, 1970.

[12] Veen van der, A.J., Dewilde, P.M.: System theory and computational architectures for time-varying networks, to appear.

[13] Weiss, L.: Controllability, realization and stability of discrete-time systems, *SIAM J. Control* **10** (1972), 230-251.

[14] Youla, D.C.: The synthesis of linear dynamical systems from prescribed weighting patterns, *SIAM J. Appl. Math.,* **14** (1966), 527-549.

I. Gohberg
School of Mathematical Sciences
The Raymond and Beverly Sackler
Faculty of Exact Sciences
Tel-Aviv University
Ramat Aviv, Tel Aviv 69989
Israel

M.A. Kaashoek
Dept. of Mathematics and Computer Science
Vrije Universiteit
De Boelelaan 1081a
1081 HV Amsterdam
The Netherlands

L. Lerer
Dept. of Mathematics
Technion-Israel Institute of Technology
Haifa 32000
Israel

Titles previously published in the series
OPERATOR THEORY: ADVANCES AND APPLICATIONS
BIRKHÄUSER VERLAG

1. **H. Bart, I. Gohberg, M.A. Kaashoek:** Minimal Factorization of Matrix and Operator Functions, 1979, (3-7643-1139-8)

2. **C. Apostol, R.G. Douglas, B.Sz.-Nagy, D. Voiculescu, Gr. Arsene** (Eds.): Topics in Modern Operator Theory, 1981, (3-7643-1244-0)

3. **K. Clancey, I. Gohberg:** Factorization of Matrix Functions and Singular Integral Operators, 1981, (3-7643-1297-1)

4. **I. Gohberg** (Ed.): Toeplitz Centennial, 1982, (3-7643-1333-1)

5. **H.G. Kaper, C.G. Lekkerkerker, J. Hejtmanek:** Spectral Methods in Linear Transport Theory, 1982, (3-7643-1372-2)

6. **C. Apostol, R.G. Douglas, B. Sz-Nagy, D. Voiculescu, Gr. Arsene** (Eds.): Invariant Subspaces and Other Topics, 1982, (3-7643-1360-9)

7. **M.G. Krein:** Topics in Differential and Integral Equations and Operator Theory, 1983, (3-7643-1517-2)

8. **I. Gohberg, P. Lancaster, L. Rodman:** Matrices and Indefinite Scalar Products, 1983, (3-7643-1527-X)

9. **H. Baumgärtel, M. Wollenberg:** Mathematical Scattering Theory, 1983, (3-7643-1519-9)

10. **D. Xia:** Spectral Theory of Hyponormal Operators, 1983, (3-7643-1541-5)

11. **C. Apostol, C.M. Pearcy, B. Sz.-Nagy, D. Voiculescu, Gr. Arsene** (Eds.): Dilation Theory, Toeplitz Operators and Other Topics, 1983, (3-7643-1516-4)

12. **H. Dym, I. Gohberg** (Eds.): Topics in Operator Theory Systems and Networks, 1984, (3-7643-1550-4)

13. **G. Heinig, K. Rost:** Algebraic Methods for Toeplitz-like Matrices and Operators, 1984, (3-7643-1643-8)

14. **H. Helson, B. Sz.-Nagy, F.-H. Vasilescu, D.Voiculescu, Gr. Arsene** (Eds.): Spectral Theory of Linear Operators and Related Topics, 1984, (3-7643-1642-X)

15. **H. Baumgärtel:** Analytic Perturbation Theory for Matrices and Operators, 1984 (3-7643-1664-0)

16. **H. König:** Eigenvalue Distribution of Compact Operators, 1986, (3-7643-1755-8)

17. **R.G. Douglas, C.M. Pearcy, B. Sz.-Nagy, F.-H. Vasilescu, D. Voiculescu, Gr. Arsene** (Eds.): Advances in Invariant Subspaces and Other Results of Operator Theory, 1986, (3-7643-1763-9)

18. **I. Gohberg** (Ed.): I. Schur Methods in Operator Theory and Signal Processing, 1986, (3-7643-1776-0)

19. **H. Bart, I. Gohberg, M.A. Kaashoek** (Eds.): Operator Theory and Systems, 1986, (3-7643-1783-3)

20. **D. Amir:** Isometric characterization of Inner Product Spaces, 1986, (3-7643-1774-4)

21. **I. Gohberg, M.A. Kaashoek** (Eds.): Constructive Methods of Wiener-Hopf Factorization, 1986, (3-7643-1826-0)

22. **V.A. Marchenko:** Sturm-Liouville Operators and Applications, 1986, (3-7643-1794-9)

23. **W. Greenberg, C. van der Mee, V. Protopopescu:** Boundary Value Problems in Abstract Kinetic Theory, 1987, (3-7643-1765-5)

24. **H. Helson, B. Sz.-Nagy, F.-H. Vasilescu, D. Voiculescu, Gr. Arsene** (Eds.): Operators in Indefinite Metric Spaces, Scattering Theory and Other Topics, 1987, (3-7643-1843-0)

25. **G.S. Litvinchuk, I.M. Spitkovskii:** Factorization of Measurable Matrix Functions, 1987, (3-7643-1843-X)

26. **N.Y. Krupnik:** Banach Algebras with Symbol and Singular Integral Operators, 1987, (3-7643-1836-8)

27. **A. Bultheel:** Laurent Series and their Pade Approximation, 1987, (3-7643-1940-2)

28. **H. Helson, C.M. Pearcy, F.-H. Vasilescu, D. Voiculescu, Gr. Arsene** (Eds.): Special Classes of Linear Operators and Other Topics, 1988, (3-7643-1970-4)

29. **I. Gohberg** (Ed.): Topics in Operator Theory and Interpolation, 1988, (3-7634-1960-7)

30. **Yu.I. Lyubich:** Introduction to the Theory of Banach Representations of Groups, 1988, (3-7643-2207-1)

31. **E.M. Polishchuk:** Continual Means and Boundary Value Problems in Function Spaces, 1988, (3-7643-2217-9)

32. **I. Gohberg** (Ed.): Topics in Operator Theory. Constantin Apostol Memorial Issue, 1988, (3-7643-2232-2)

33. **I. Gohberg** (Ed.): Topics in Interplation Theory of Rational Matrix-Valued Functions, 1988, (3-7643-2233-0)

34. **I. Gohberg** (Ed.): Orthogonal Matrix-Valued Polynomials and Applications, 1988, (3-7643-2242-X)

35. **I. Gohberg, J.W. Helton, L. Rodman** (Eds.): Contributions to Operator Theory and its Applications, 1988, (3-7643-2221-7)

36. **G.R. Belitskii, Yu.I. Lyubich:** Matrix Norms and their Applications, 1988, (3-7643-2220-9)

37. **K. Schmüdgen:** Unbounded Operator Algebras and Representation Theory, 1990, (3-7643-2321-3)

38. **L. Rodman:** An Introduction to Operator Polynomials, 1989, (3-7643-2324-8)

39. **M. Martin, M. Putinar:** Lectures on Hyponormal Operators, 1989, (3-7643-2329-9)

40. **H. Dym, S. Goldberg, P. Lancaster, M.A. Kaashoek** (Eds.): The Gohberg Anniversary Collection, Volume I, 1989, (3-7643-2307-8)

41. **H. Dym, S. Goldberg, P. Lancaster, M.A. Kaashoek** (Eds.): The Gohberg Anniversary Collection, Volume II, 1989, (3-7643-2308-6)

42. **N.K. Nikolskii** (Ed.): Toeplitz Operators and Spectral Function Theory, 1989, (3-7643-2344-2)

43. **H. Helson, B. Sz.-Nagy, F.-H. Vasilescu, Gr. Arsene** (Eds.): Linear Operators in Function Spaces, 1990, (3-7643-2343-4)

44. **C. Foias, A. Frazho:** The Commutant Lifting Approach to Interpolation Problems, 1990, (3-7643-2461-9)

45. **J.A. Ball, I. Gohberg, L. Rodman:** Interpolation of Rational Matrix Functions, 1990, (3-7643-2476-7)

46. **P. Exner, H. Neidhardt** (Eds.): Order, Disorder and Chaos in Quantum Systems, 1990, (3-7643-2492-9)

47. **I. Gohberg** (Ed.): Extension and Interpolation of Linear Operators and Matrix Functions, 1990, (3-7643-2530-5)

48. **L. de Branges, I. Gohberg, J. Rovnyak** (Eds.): Topics in Operator Theory. Ernst D. Hellinger Memorial Volume, 1990, (3-7643-2532-1)

49. **I. Gohberg, S. Goldberg, M.A. Kaashoek:** Classes of Linear Operators, Volume I, 1990, (3-7643-2531-3)

50. **H. Bart, I. Gohberg, M.A. Kaashoek** (Eds.): Topics in Matrix and Operator Theory, 1991, (3-7643-2570-4)

51. **W. Greenberg, J. Polewczak** (Eds.): Modern Mathematical Methods in Transport Theory, 1991, (3-7643-2571-2)

52. **S. Prössdorf, B. Silbermann:** Numerical Analysis for Integral and Related Operator Equations, 1991, (3-7643-2620-4)

53. **I. Gohberg, N. Krupnik:** One-Dimensional Linear Singular Integral Equations, Volume I, Introduction, 1991, (3-7643-2584-4)

55. **R.R. Akhmerov, M.I. Kamenskii, A.S. Potapov, A.E. Rodkina, B.N. Sadovskii:** Measures of Noncompactness and Condensing Operators, 1992, (3-7643-2716-2)

MIX
Papier aus verantwortungsvollen Quellen
Paper from responsible sources
FSC® C105338

If you have any concerns about our products,
you can contact us on
ProductSafety@springernature.com

In case Publisher is established outside the EU,
the EU authorized representative is:
Springer Nature Customer Service Center GmbH
Europaplatz 3, 69115 Heidelberg, Germany

Printed by Libri Plureos GmbH
in Hamburg, Germany